FORGING A Poison Prevention AND Control System

Committee on Poison Prevention and Control
Board on Health Promotion and Disease Prevention

INSTITUTE OF MEDICINE
OF THE NATIONAL ACADEMIES

THE NATIONAL ACADEMIES PRESS
Washington, D.C.
www.nap.edu

THE NATIONAL ACADEMIES PRESS 500 Fifth Street, NW Washington, DC 20001

NOTICE: The project that is the subject of this report was approved by the Governing Board of the National Research Council, whose members are drawn from the councils of the National Academy of Sciences, the National Academy of Engineering, and the Institute of Medicine. The members of the committee responsible for the report were chosen for their special competences and with regard for appropriate balance.

This study was supported by Contract/Grant No. 240-02-0004 between the National Academy of Sciences and the Health Resources and Services Administration. Any opinions, findings, conclusions, or recommendations expressed in this publication are those of the author(s) and do not necessarily reflect the views of the organizations or agencies that provided support for the project.

Library of Congress Cataloging-in-Publication Data

Forging a poison prevention and control system / Committee on Poison Prevention and Control, Board on Health Promotion and Disease Prevention.
p. ; cm.
Includes bibliographical references and index.
ISBN 0-309-09194-2 (hardcover)
1. Poisoning, Accidental—Prevention—Government policy—United States. 2. Poison control centers—Government policy—United States.
[DNLM: 1. Poison Control Centers—organization & administration—United States. 2. Delivery of Health Care—methods—United States. 3. Health Policy—United States. 4. Poison Control Centers—economics—United States. 5. Poisoning—prevention & control—United States. QV 600 F721 2004] I. Institute of Medicine (U.S.). Committee on Poison Prevention and Control.
RA1224.5.F67 2004
363.17'91—dc22

2004009703

Additional copies of this report are available from the National Academies Press, 500 Fifth Street, NW, Lockbox 285, Washington, DC 20055; (800) 624-6242 or (202) 334-3313 (in the Washington metropolitan area); Internet, http://www.nap.edu.

Printed in the United States of America.

The serpent has been a symbol of long life, healing, and knowledge among almost all cultures and religions since the beginning of recorded history. The serpent adopted as a logotype by the Institute of Medicine is a relief carving from ancient Greece, now held by the Staatliche Museen in Berlin.

"Knowing is not enough; we must apply.
Willing is not enough; we must do."
—Goethe

INSTITUTE OF MEDICINE
OF THE NATIONAL ACADEMIES

Adviser to the Nation to Improve Health

THE NATIONAL ACADEMIES

Advisers to the Nation on Science, Engineering, and Medicine

The **National Academy of Sciences** is a private, nonprofit, self-perpetuating society of distinguished scholars engaged in scientific and engineering research, dedicated to the furtherance of science and technology and to their use for the general welfare. Upon the authority of the charter granted to it by the Congress in 1863, the Academy has a mandate that requires it to advise the federal government on scientific and technical matters. Dr. Bruce M. Alberts is president of the National Academy of Sciences.

The **National Academy of Engineering** was established in 1964, under the charter of the National Academy of Sciences, as a parallel organization of outstanding engineers. It is autonomous in its administration and in the selection of its members, sharing with the National Academy of Sciences the responsibility for advising the federal government. The National Academy of Engineering also sponsors engineering programs aimed at meeting national needs, encourages education and research, and recognizes the superior achievements of engineers. Dr. Wm. A. Wulf is president of the National Academy of Engineering.

The **Institute of Medicine** was established in 1970 by the National Academy of Sciences to secure the services of eminent members of appropriate professions in the examination of policy matters pertaining to the health of the public. The Institute acts under the responsibility given to the National Academy of Sciences by its congressional charter to be an adviser to the federal government and, upon its own initiative, to identify issues of medical care, research, and education. Dr. Harvey V. Fineberg is president of the Institute of Medicine.

The **National Research Council** was organized by the National Academy of Sciences in 1916 to associate the broad community of science and technology with the Academy's purposes of furthering knowledge and advising the federal government. Functioning in accordance with general policies determined by the Academy, the Council has become the principal operating agency of both the National Academy of Sciences and the National Academy of Engineering in providing services to the government, the public, and the scientific and engineering communities. The Council is administered jointly by both Academies and the Institute of Medicine. Dr. Bruce M. Alberts and Dr. Wm. A. Wulf are chair and vice chair, respectively, of the National Research Council.

www.national-academies.org

COMMITTEE ON POISON PREVENTION AND CONTROL

Bernard Guyer *(Chair)*, Zanvyl Krieger Professor of Children's Health, Johns Hopkins University, Baltimore

Jeffrey A. Alexander, Professor, Health Management and Policy and Organizational Behavior and Human Resources, School of Business; Faculty Associate, Survey Research Center; Senior Associate Dean, School of Public Health, University of Michigan, Ann Arbor

Paul Blanc, Professor of Medicine; Endowed Chair in Occupational and Environmental Medicine; University of California, San Francisco

Dennis Emerson, Emergency Nurse, St. Luke's Regional Medical Center, Boise, ID

Jerris R. Hedges, Professor and Chair, Department of Emergency Medicine, Oregon Health and Science University, Portland

Mark Scott Kamlet, H. John Heinz III Professor of Economics and Public Policy and Provost, Carnegie Mellon University, Pittsburgh

Angela Mickalide, Program Director, National SAFE KIDS Campaign, Washington, DC

Paul Pentel, Professor of Medicine and Pharmacology; Chief, Division of Clinical Pharmacology, Hennepin County Medical Center, University of Minnesota, Minneapolis

Barry H. Rumack, Clinical Professor of Pediatrics, University of Colorado School of Medicine, and Director Emeritus, Rocky Mountain Poison and Drug Center, Denver Health Authority

David P. Schor, Chief, Division of Family and Community Health Services, Ohio Department of Health, Columbus

Daniel A. Spyker, Director of Clinical Pharmacology, Genentech, Inc., San Francisco

Andy Stergachis, Professor of Epidemiology and Affiliate Professor of Pharmacy, University of Washington, Seattle

David J. Tollerud, Professor and Associate Director, Institute for Public Health Research, University of Louisville, KY

Deborah Klein Walker, Associate Commissioner for Programs and Prevention, Massachusetts Department of Public Health, Boston

Liaison from the Board on Children, Youth, and Families

Mary Jane England, President, Regis College, Weston, MA

Study Staff

Anne Mavor, Study Director
Susan McCutchen, Research Associate
Elizabeth Townsend, Senior Project Assistant

IOM Board on Health Promotion and Disease Prevention

Rose Marie Martinez, Board Director
David Butler, Senior Program Officer
Rita Gaskins, Board Administrative Assistant
Jim Banihashemi, Financial Associate

Consultant

Delon Brennen, Johns Hopkins University

BOARD ON HEALTH PROMOTION AND DISEASE PREVENTION

Kathleen E. Toomey, Director, Division of Public Health, Georgia Department of Human Resources, Atlanta

William Vega, Director, Behavioral Research and Training Institute, University Behavioral HealthCare, New Brunswick, NJ

Patricia Wahl, Dean and Professor of Biostatistics, School of Public Health and Community Medicine, University of Washington, Seattle

Lauren Zeise, Chief, Reproductive and Cancer Hazard Assessment, Office of Environmental Health Hazard Assessment, Oakland, CA

Liaison from the Institute of Medicine

Jeffrey P. Koplan, Vice President for Academic Health Affairs, Emory University, Atlanta

Reviewers

This report has been reviewed in draft form by individuals chosen for their diverse perspectives and technical expertise, in accordance with procedures approved by the NRC's Report Review Committee. The purpose of this independent review is to provide candid and critical comments that will assist the institution in making its published report as sound as possible and to ensure that the report meets institutional standards for objectivity, evidence, and responsiveness to the study charge. The review comments and draft manuscript remain confidential to protect the integrity of the deliberative process. We wish to thank the following individuals for their review of this report:

Kelly J. Devers, Center for Studying Health System Change
Susan S. Gallagher, Education Development Center, Inc.
Lewis R. Goldfrank, Bellevue Hospital Center
Maxine Hayes, State of Washington Department of Health
Henri Manasse, Jr., American Society of Health System Pharmacists
L. Joseph Melton III, Mayo Clinic College of Medicine
Harold Pollack, University of Chicago
Henry W. Riecken, University of Pennsylvania, Professor Emeritus
Robert B. Wallace, University of Iowa

Although the reviewers listed above have provided many constructive comments and suggestions, they were not asked to endorse the conclusions or recommendations nor did they see the final draft of the report before its release. The review of this report was overseen by Enriqueta C. Bond, Burroughs Wellcome Fund, and George F. Sheldon, The University of North Carolina. Appointed by the National Research Council and Institute of Medicine, they were responsible for making certain that an independent examination of this report was carried out in accordance with institutional procedures and that all review comments were carefully considered. Responsibility for the final content of this report rests entirely with the authoring committee and the institution.

Acknowledgments

The Committee is grateful to the many individuals who contributed information contained in this report and who were helpful to us throughout the study process (see Appendix A for a complete list of contributors). We would particularly like to thank the directors and staffs of the poison control centers for their willingness to describe current operations and programs and share insights on the challenges ahead. Without their cooperation, this report would not have been possible. We also extend our special thanks to the American Association of Poison Control Centers, whose staff provided the Committee with data from their 2000 and 2001 surveys and gave informative briefings on the current status of the Toxic Exposure Surveillance System and the national efforts in public education for poison prevention and control.

The Committee also wishes to particularly acknowledge those individuals who prepared written material and contributed to the collection and preliminary analysis of data. Miriam Cisternas, MCG Data Services, prepared the material for the morbidity analysis used in Chapter 3. Lois Fingerhut, National Center for Health Statistics, prepared the mortality analysis material also found in Chapter 3. Holly Hackman and Jessica Cates from the Massachusetts Department of Public Health, contributed to sections describing federal agencies and their participation in poison prevention and control found in Chapter 9. Laura Copeland, Tracy Finlayson, Maureen Metzger, and Soheil Soliman from the University of Michigan assisted in collecting and summarizing data used by the Committee to characterize current poison control centers.

The Committee is most grateful to sponsors Carol Delany and Byron Bailey of the Maternal and Child Health Bureau of the Health Resources and Services Administration for their continued interest in and support of our work.

Staff of the Institute of Medicine (IOM) and the National Research Council (NRC) made important contributions to our work in many ways. The Committee wishes to acknowledge Rose Martinez of the IOM for her good counsel throughout the study. We extend particular thanks to NRC staff member Susan McCutchen for her efforts as our research associate and the part she played in the preparation of this report. We are also grateful to NRC staff member Elizabeth Townsend, the Committee's senior project assistant, who was indispensable in organizing meetings, arranging travel, compiling agenda materials, and in managing the production aspects of this report. We would like to thank Delon Brennen, consultant, who assisted us in our efforts throughout the study. Finally, we wish to thank Laura Penny, our editor.

Contents

EXECUTIVE SUMMARY 1

PART I: OVERVIEW

1 Introduction 23
2 Toward a Poison Prevention and Control System 34

PART II: CURRENT STATUS AND OPPORTUNITIES

3 Magnitude of the Problem 43
4 Historical Context of Poison Control 80
5 Poison Control Center Activities, Personnel, and Quality Assurance 106
6 Current Costs, Funding, and Organizational Structures 136
7 Data and Surveillance 176
8 Prevention and Public Education 201
9 A Public Health System for Poison Prevention and Control 269

PART III: CONCLUSIONS AND RECOMMENDATIONS

10 Conclusions and Recommendations 305

REFERENCES 318

APPENDIXES

A Contributors 329

B Committee and Staff Biographies 332

INDEX 339

Executive Summary

The Institute of Medicine (IOM) was asked by the Maternal and Child Health Bureau (MCHB) of the Health Resources and Services Administration (HRSA) to assist in developing a more systematic approach to understanding, stabilizing, and providing long-term support for poison prevention and control services. Within this context the Committee was asked to examine the future of poison prevention and control services in the United States. The specific tasks included in the charge are provided in Box ES-1. In order to respond fully and specifically to the charge, the Committee adopted the very language used by HRSA: to consider the "future of poison prevention and control services" and to develop a "systematic" approach. Therefore, we examined the role of poison control services within the context of the larger public health system, the injury prevention and control field, and the fields of general medical care and medical and clinical toxicology.[1] Furthermore, we examined how poison control centers function relative to the functions performed by other health

[1]The term *toxicologist* is a general description of an individual dealing with any aspect of acute or chronic poisonings, and it does not have a specific definition or implication with regard to training or job description. For example, this term may be used to describe individuals whose activities range from molecular biology to epidemiology, as long as they deal in some way with the toxic effects of chemicals. The term *clinical toxicologist* implies a more clinical orientation, but likewise has no specific definition or implications. *Medical toxicologists* are physicians with specific training and board certification in the subspecialty of medical toxicology, which focuses on the care of poisoned patients.

BOX ES-1
Committee Charge

The Institute of Medicine was asked by the Maternal and Child Health Bureau of the Health Resources and Services Administration to assist in developing a more systematic approach to understanding, stabilizing, and providing long-term support for poison prevention and control services. Within this context the Committee was asked to examine the future of poison prevention and control services in the United States. The specific tasks included in the charge are to review:

1. The scope of services provided, including consumer telephone consultation, technical assistance, and/or hospital consultation for the care of patients with life-threatening poisonings, and education of the public and professionals;
2. The coordination of poison control centers with other public health, emergency medical, and other emergency services;
3. The strengths and weaknesses of various organizational structures for poison control centers and services, including a consideration of personnel needs;
4. Approaches to providing the financial resources for poison prevention and control services;
5. Methods for assuring consistent, high-quality services, including the certification of centers and methods of evaluation; and
6. Current and future data systems and surveillance needs.

The Committee was asked to consider these questions in light of future demographic and population trends, and in the context of the threats of biological and chemical terrorism.

care agencies and government organizations at the federal, state, and local levels.

Poisoning is a much larger public health problem than has generally been recognized, and no comprehensive system is in place for its prevention and control. To address its charge of creating such a system, the IOM Committee faced two major, overarching issues. The first of these was a definitional problem—there is simply no universally agreed upon definition of poisoning from either a clinical or epidemiological perspective. Thus, in order to assess the magnitude, scope, and boundaries of the area under study, the Committee adopted an operational definition of poisoning without attempting to resolve all the classification disputes about specific elements of the definition. The second major issue concerned the historical development of the poison control centers and their position in the broader fields of public health and emergency medical services. In order to make recommendations about stabilizing and providing long-term support to the network of centers, the Committee developed a vision for the future organization, structure, and funding of a poison prevention and control system.

THE DEFINITION, SCOPE, AND MAGNITUDE OF POISONING

The Committee's operational definition of poisoning subsumes "damaging physiological effects of ingestion, inhalation, or other exposure to a range of pharmaceuticals, illicit drugs, and chemicals, including pesticides, heavy metals, gases/vapors, and common household substances, such as bleach and ammonia" (Centers for Disease Control and Prevention, 2004, p. 233). The Committee's approach to defining poisoning is addressed in Box ES-2. A broad clinical definition of human poisoning, as noted above, captures any toxin-related injury. However, each agency that collects data or provides services in this arena has evolved its own particular definitional boundaries of the poisoning problem. Furthermore, definitions of a poisoning and its place among other medical diagnoses vary from the 9th to the 10th revisions of the *International Classification of Diseases*, the system that drives health data categorization at both the federal and state levels. Finally, the network of poison control centers has evolved its own operational definition of what constitutes an "exposure" to a poisonous substance. As a result, the Committee adopted an operational definition of poisoning that could be used to analyze the available datasets to better understand the magnitude of the poison problem (see Chapter 3 for expanded discussion of this question).

The Committee estimates that more than 4 million poisoning episodes (actual or suspected exposures) occur in the United States annually, with approximately 300,000 cases leading to hospitalization. The poisoning death rate increased by 56 percent between 1990 and 2001 (Centers for Disease Control and Prevention, 2004). In 2001, poisoning was the second leading cause of injury-related mortality, accounting for an estimated 30,800 deaths annually. A conservative estimate of the economic burden of poisoning not including costs related to alcohol deaths is $12.6 billion per year (2002 dollars), based on the societal lifetime cost of injury.

Poisoning is a public health problem across the entire lifespan. It is well recognized that unintentional exposure to hazardous household substances (including medications found in the home) occurs mainly among preschool-aged children; the majority of these exposures can be treated in the home and the associated mortality rate is low. It is less well appreciated that the burden of unintentional drug overdose and suicide deaths is more likely to occur among adolescents and young adults, and that the elderly are at high risk for poisoning due to scenarios such as mixing medications or taking the wrong dosage. Finally, new concerns about biological and chemical terrorist acts have elevated poisoning to a national

BOX ES-2
Defining Poisoning

"All things are poison and not without poison; only the dose makes a thing not a poison"

Paracelsus (1493–1541)

There is no standard definition of poisoning that is universally accepted and applied in clinical practice, in data collection, and in public health policy settings.

Clinical Definition

Human poisoning subsumes any toxin-related injury. The injury can be systemic or organ-specific (e.g., neurological injury or hepatotoxicity). The source of the toxin can be a synthetic chemical or a naturally occurring plant, animal, or mineral substance. Thus poisoning can include the toxic effects of a classic toxin (e.g., cyanide), an overdose of a prescription medication (e.g., an antidepressant), an overdose of an over-the-counter preparation (e.g., headache tablets), or a complementary treatment (such as an herbal medicine or dietary supplement).

Classification Complexities

Disagreement over the classification of certain poisoning events leads to discrepancies in the estimates of poison-related mortality and morbidity; prominent among these disagreements are:

- Exposures that fall in and out of various classification schemes (e.g., envenomation from a rattlesnake or black widow spider might be grouped with non-toxic bites).
- Medical misadventure/adverse effects at therapeutic levels; medication responses that are not dose related but idiosyncratic, with or without allergic component.
- Delayed versus acute toxic effects.
- Illness from naturally occurring toxins derived from microorganisms (e.g., seafood-related toxins).
- Toxic effects from ethanol (e.g., rapid ingestion, withdrawal, chronic).
- Exposure to a potential toxin without a defined clinical effect (as when parents telephone a poison control center about a possible ingestion by their child).

The Committee's Operational Definitions

To arrive at reasonable estimates of the magnitude of poisoning, the Committee adopted the definitions used by key federal health agencies and organizations that monitor poisoning in the population (see Chapter 3 for details).

- Morbidity estimates used definitions from the National Interview Health Survey, National Ambulatory Medical Care Survey, National Hospital Ambulatory Care Survey, National Hospital Discharge Survey, and National Electronic Injury Surveillance Survey.
- Exposure estimates were derived from the Toxic Exposure Surveillance System.
- Mortality estimates used the classification of the National Center for Health Statistics.

security issue of public health importance. Poison control centers respond to calls from the public in all of these areas; although approximately 50 percent of calls concern possible exposures to children 5 years of age and under, approximately 7.6 percent are suspected suicides, and another 3.5 percent are cases of substance misuse or abuse. Furthermore, 3 percent of the calls are categorized as alcohol related.

The national goals for reducing poisoning mortality and morbidity, established by *Healthy People 2010,* did not fully recognize this broader picture of the importance of poisoning in the United States. The specific objectives as cited are to reduce nonfatal poisonings to 292 per 100,000 population (based on emergency department visit incidence) and deaths caused by poisoning to 1.5 per 100,000 population. According to the Committee's estimates of the current level of poisoning (2001 data)—530 poisonings per 100,000 population and 8.5 deaths per 100,000 population—these goals are unlikely to be reached by 2010.[2] The Committee concludes that the national efforts to reduce poisoning must be linked to a national agenda for public health promotion and disease prevention. We envision a future Poison Prevention and Control System that is integrated with the medical care system and public health and that includes a network of poison control centers as a vital, but not exclusive, element.

BACKGROUND

In approaching its work, the Committee recognized that the public-access peer-reviewed literature on poison control centers did not provide an adequate evidentiary base to answer the charge. As a result, the Committee conducted a series of analyses using existing databases and engaged in primary data collection to develop a more in-depth understanding of current poison control center services and organizational structures. The review and analysis focuses on, but is not limited to, the current characteristics of poison control centers and the challenges for the future regarding prevention, service delivery, and surveillance.

The current network of poison control centers in the United States has developed to meet local needs and is supported for the most part by local resources. There is no coordinated national system. The evolution from the earliest center in 1953 has been individualized and chaotic; at one point, in 1978, there were as many as 661 poison control centers, many of them serving relatively small populations. Now there are 63 poison control centers covering various regions that collectively serve nearly the

[2]The Committee's higher estimates are based on multiple sources that include but are not limited to those used in *Healthy People 2010*. In addition, the Committee's analyses drew on dates between 1997 and 2001, whereas *Healthy People 2010* estimates are for 1997.

entire U.S. population. These centers offer a critical set of services to the public and health care professionals by providing timely, professional treatment advice in response to telephone queries concerning poisoning exposures. According to the American Association of Poison Control Centers (AAPPC), in 2002 more than 2.3 million human exposure calls were received by all centers combined. As noted earlier, calls to poison control centers are classified as human exposure (to poison) if a member of the public or health care community is reporting an actual or suspected poisoning exposure. Thus, not all human exposure calls are poisonings. For each such call, both the suspected exposure reported by the caller and the treatment response by poison control center staff are recorded. Thus a wealth of data on reported poisoning exposures is generated. Finally, poison control centers provide an important training ground for medical toxicologists, nurses, nurse managers, pharmacists, and other health care professionals.

Unfortunately, the current "network" of poison control centers suffers a number of shortcomings. First, it is financially unstable, with each center drawing its support from numerous federal, state, and local sources that are frequently undergoing fiscal challenges and budget adjustments. The Poison Control Center Enhancement and Awareness Act of 2000, amended in 2003, was enacted to stabilize center operations. Although these funds are intended to provide an emergency safety net, their magnitude and focus on supporting new activities rather than existing staff and infrastructure do not ensure consistent, effective, and efficient delivery of poison prevention and control services to the U.S. population. In the past year alone, two poison control centers lost their funding and were forced to close; other centers expend considerable time and effort obtaining needed support. Second, the current network of poison control centers operates, in key aspects, in a manner that could be characterized as a collection of independent organizations rather than as a "system." As a result, there is insufficient sharing of strategies and resources. Third, there is no effective link to the nation's public health system that provides a seamless net of services in prevention, injury control, and all-hazards emergency preparedness. Fourth, the current poison control center data collection and reporting system, known as the Toxic Exposure Surveillance System (TESS), functions as a proprietary system that is not fully available to the work of federal and state agencies engaged in protecting the population from consumer product or intentional hazards.

CONCLUSIONS AND RECOMMENDATIONS

The Committee concluded, based on its research and discussions, that the current network of poison control centers does not constitute the complete "system" of poison prevention and control services needed by the nation in the 21st century. Such a system must provide the best prevention and patient care services for the diverse population of Americans who are exposed to hazardous substances and protect the nation from the threats associated with biological and chemical terrorist events and other emerging public health emergencies. Therefore, the Committee based its report on a proposed Poison Prevention and Control System, including within it a network of poison control centers as a vital, but not exclusive, element. The Committee also concluded that in order to fulfill their pivotal role in the overall system, poison control centers must be more stable financially and better integrated and coordinated for performance of their public health roles.

The Committee considered the strengths and weaknesses of a variety of options for the number and distribution of poison control centers in a Poison Prevention and Control System. Although modern telecommunications technology makes it feasible to consider one single, highly efficient, large center serving the entire country, the Committee found a number of weaknesses with that model. A single national center would have difficulty appreciating local variations in poisonous substances such as plants and insects. In addition, a single center would concentrate all the expertise in one location, thereby eliminating important and timely local medical consultations. Finally, a single center is vulnerable to practical problems of power failures, limited surge capacity, and potential transmission lags during times of high volume. The Committee also considered a national model that would have a single poison control center in each state. This model was also rejected as inconsistent with the current realities. A number of states with relatively small and dispersed populations have chosen to contract with larger centers to meet their needs. Also, in large states like California, there is a statewide system with multiple centers because one single center alone cannot meet the entire need. Thus, the Committee concluded that a system of regional centers would provide an appropriate balance of size and responsiveness.

The rationale for a regionalized system includes the following elements. Poison control centers must be large enough to sustain an adequate-sized staff to meet usual demands and the surge capacity required to respond to situations of mass poisoning or suspected terrorism events.[3] A regional distribution of such centers would satisfy the need to distribute medical toxicological leadership across the United States to address

[3]In 2002, the Presidential Task Force on Citizen Preparedness in the War on Terrorism recommended that poison control centers provide emergency information in the event of a terrorist event involving biological, chemical, or nuclear toxins.

the diversity of poison exposures and to provide firsthand consultation to hospitals and physicians. The interaction among regionally based centers would promote innovation and the sharing of best practices. Finally, a regionalized system should provide enough redundancy in skills and resources to meet surge needs and potential equipment failures.

The Committee concluded that decisions about the number of centers should be based on considerations of population coverage, telecommunication capabilities, and types of funding. While the currently available data are not adequate to prescribe a specific size or geographical coverage for centers, the Committee believes there may be economies of scale and scope that can be achieved through a regionalized system. Defining a set of core services will support the development of a federal funding formula for regionalized poison control centers. Ultimately, the needs assessment data must be developed to define the financial and services base for developing contractual agreements for poison control services. The Committee believes that the concept of regionalized national poison control centers is critical to the development of the Poison Prevention and Control System.

The Committee's recommendations form the basis for the Poison Prevention and Control System. They are grouped according to the areas listed in the Committee's charge:

- Scope of core poison prevention and control activities
- Coordination of poison control centers with other public health entities
- Strengths and weaknesses of poison control center organizational structures
- Financial support for the Poison Prevention and Control System
- Assurance of high-quality poison control center services
- National data system and surveillance needs

Scope of Core Poison Prevention and Control Activities

The Committee identified a core set of activities that constitutes the essential functions of the network of poison control centers within the larger system envisioned by the Committee. Although these activities are already being carried out, it is essential to identify them as a set of core activities so that they become the basis for consistent funding under the aegis of the proposed expanded federal legislation. These activities are considered by the Committee to be core because (1) they represent critical components of current and future poison control efforts; (2) the structure of poison control centers and expertise of their staffs make them uniquely capable of performing these activities (i.e., there are no other organizations in the public health and health care arena that can perform these activities at the same level of excellence and cost); and (3) they provide an infrastructure to which other related activities can readily be added as required. The notion of core activities does not imply that poison control centers should confine their activities solely to these areas. The addition of other activities should be based on local capabilities and opportunities for funding. Examples include undertaking clinical toxicology research or providing training for health care students who are not specifically focused on careers in medical or clinical toxicology.

Recommendations

1. All poison control centers should perform a defined set of core activities supported by federal funding that is tied to the provision of these activities. The core activities include: (1) manage telephone-based poison exposure and information calls; (2) prepare and respond to all-hazards emergency needs (especially biological or chemical terrorism or other mass exposure events); (3) capture, analyze, and report exposure data; (4) train poison control center staff, including specialists in poison information and poison information providers; (5) carry out continuous quality improvement; and (6) integrate their services into the public health system. In addition, a subset of poison control centers should train medical toxicologists; this is considered a core activity for only a subset of poison control centers because their involvement is necessary for the certification of this specialty. A subset of poison control centers should also assist in the training of pharmacists through clinical toxicology fellowships that prepare them for poison control center management positions.

2. Poison control centers should collaborate with state and local health departments to develop, disseminate, and evaluate public and professional education activities. Poison control centers alone cannot fulfill the need for public and professional education related to poisoning prevention and treatment and all-hazards response. Public health agencies already have the authorities, networks, and administrative mechanisms to carry out broad educational efforts, as they do for the prevention of other injuries and for other public health campaigns.

Coordination of Poison Control Centers with Other Public Health Entities

The mission of public health is to assure conditions in which people can be healthy. As noted earlier, meeting the ambitious national objectives for poisoning prevention set by the U.S. Department of Health and Human Services in *Healthy People 2010*, particularly with the potential burden of biological and chemical attacks, requires the combined efforts of public health agencies and the proposed regional system of poison control centers. The public health system, through its *Essential Services of Public Health* and core functions of assessment, policy development, and assurance, offers a useful framework for providing and coordinating poison prevention services (see Table ES-1).

To achieve the ultimate goal of preventing poisonings, as well as to improve the outcomes for those who are poisoned, the Committee envisions the need for a clear, single point of accountability at each level of government. The responsible agencies would assure the accomplishment of all public health core functions or essential services as they relate to

TABLE ES-1 Core Functions and Essential Services of Public Health as Applied to Poison Prevention and Control Services

Core Functions	10 Essential Services
Assessment Collection, assembly, analysis, and distribution of information on the community's health	1. Monitor health status to identify community problems. 2. Diagnose and investigate health problems and the health hazards in the community. 3. Evaluate the effectiveness, accessibility, and quality of personal and population-based health services.
Policy development Development of comprehensive policies based on scientific knowledge and decision making	4. Inform, educate, and empower people about health issues. 5. Mobilize community partnerships to identify and solve health problems. 6. Develop policies and plans that support individual and community health efforts.
Assurance Determination of needed personal and communitywide health services, and provision of these services by encouraging action by others, by requiring action by others, or by direct provision	7. Assure a competent public health and personal health care workforce. 8. Enforce laws and regulations that protect health and ensure safety. 9. Link people to needed personal health services and assure the provision of health care when otherwise unavailable.
Assessment, policy development, and assurance	10. Research for new insights and innovative solutions to health problems.

SOURCE: Adapted from the IOM report, *The Future of Public Health* (1988).

poison prevention and control. This does not mean that the responsible agencies would perform all the functions within their respective agencies. However, they would (1) take responsibility for developing the plan to accomplish the activities needed to ensure that the system is in place, with a set of uniform standards across the country; (2) convene and work with the other agencies, including the existing poison control center network, to implement the plan; and (3) work in partnership to develop a set of performance standards for all components of the system. One possible model for the development of performance measures for a state-federal partnership is the Title V Maternal and Child Health (MCH) Block Grant, which is administered by states, and the federal grants for MCH activi-

Examples as Applied to Poison Prevention and Control Services

1. Monitor population frequency of poisonings across the lifespan. Assess outcomes.
2. Assess factors contributing to poisonings. Develop policies and services for primary and secondary prevention.
3. Evaluate public education activities related to poisonings. Continuously review and evaluate poison control center functions and their efficiency and effectiveness. Ensure the availability and accessibility of poison control information to the entire public.

4. Assess and enhance the public's knowledge about poison impact, prevention, and control.
5. Establish effective communication with community members regarding poisonings.
6. Apply population-based data to policy development for poison prevention and control.

7. Create and maintain a workforce that is competent in poison prevention and control. Educate health professionals on subjects related to poisonings.
8. Develop laws, statutes, and regulations that provide for optimal use of poison control centers and protect individuals in the workplace.
9. Create provisions for high-quality, culturally competent poison control center services. Ensure linkages among all parts of the public health and medical systems with poison control centers.

10. Identify best practices for poison control centers. Contribute to the evidence base for poison prevention and control through the funding and generation of new knowledge.

ties, which are administered by the MCHB in HRSA. This partnership has been in place for 5 years and has successfully developed and implemented performance criteria and data reporting mechanisms.

Recommendations

3. The U.S. Department of Health and Human Services (DHHS) and the states should establish a Poison Prevention and Control System that integrates poison control centers with public health agencies, establishes performance measures, and holds all parties accountable for protecting the public. At the federal level, the Secretary of Health and Human Services should designate the lead agency for this purpose; at the state level, the governor of each state should formally designate the appropriate lead (e.g., injury prevention directors from the public health entity).

a. The Secretary of DHHS should assure integration of the existing regional network of poison control centers with the public health system.

b. The Secretary of DHHS should create a single national repository of legislation, model prevention and education programs, website designs, and best practices material. Technical assistance should be provided for website design, content, navigation, and maintenance, maximizing the individual centers' identity and contributions. Materials should be evaluated for quality and impact on intended audiences. For maximum effectiveness, their content should reflect the range of cultures and languages in the United States.

c. The governor should assure that relevant all-hazards emergency preparedness and response activities are integrated with the Poison Prevention and Control System.

4. The Centers for Disease Control and Prevention (CDC), working with HRSA and the states, should continue to build an effective infrastructure for all-hazards emergency preparedness, including bioterrorism and chemical terrorism. A specific activity of this effort is to evaluate, through an objective structured review, the use of TESS as a source of case detection to all-hazards surveillance.

Strengths and Weaknesses of Poison Control Center Organizational Structures

Early in its information gathering, the Committee decided that the existing data should be adequate to address the questions raised by HRSA about the organization and financing of the centers. Unfortunately, as the analysis progressed, we found that no data on service quality and outcomes had been systematically collected by the centers and that data on local variations in salaries and rent were not readily available. As a result, the Committee's analysis provided only preliminary findings. The Committee found a wide range of service-delivery models, organizational structures, and financing arrangements among poison control centers that successfully deliver core services. Although an earlier study conducted on six poison control centers suggested possible economies of scale for service areas of 2 million people or more, the Committee found little conclusive evidence, in its own analysis, that economies of scale operate with respect to size of population served and poison control center costs. Costs were best predicted by variables related to staffing patterns and wage rates rather than hardware expenses, population served, or funding source. More complete data are needed to further explore this important concern.

The Committee's qualitative analysis of 10 poison control centers indicated that the more efficient centers had lower staff turnover rates with fewer concerns about salaries and were more likely to (1) participate in partnerships or joint ventures in the community, (2) have written strategic plans specific to the poison control center, and (3) be organizationally affiliated with a private institution. Furthermore, the more efficient centers were less likely to cite problems related to complex reporting and accountability and problems of balancing core poison control functions with other activities such as research and bioterrorism response and preparedness. It is important to note that the analyses were based solely on population served, cost per human exposure call, and penetrance.

The existing data are insufficient for the development of either contractual specifications or performance measures for a new Poison Prevention and Control System. The Committee suggests new data-gathering efforts to obtain original financial and performance data from existing poison control centers. These data are needed to guide future public funding of core activities.

Recommendation

5. HRSA should commission a systematic management review focusing on organizational determinants of cost, quality, and staffing of poison control centers as the foundation for the future funding of this program. This analysis should include the following elements:

a. The development of new indicators of quality and impact of poison control center services.

b. The implications of different organizational structures and funding accountabilities on service quality and impact.

c. The role of center size and governance in poison control center service quality and impact.

d. The impact of regional differences on poison control center operational cost.

e. How staffing patterns, recruitment, and retention of poison control center staff affect cost, quality, and impact of poison control centers.

f. An economic evaluation of poison control centers to determine whether economies of scale exist among them.

Financial Support for the Poison Prevention and Control System: Poison Control Centers and State and Local Infrastructures

Poison control centers are currently funded by a patchwork of sources (including federal, state, institutional, and private) that are subject to budget cuts and changing priorities every year. Across the states there are 29 separate funding sources. Some examples include federal and state Medicaid programs, federal block grants, federal grants, state line-item appropriation, state-funded universities, telephone surcharges, private hospitals, and private donations. As financial pressures on state governments and health systems have risen, the willingness of traditional funders to continue to provide revenues has diminished, leaving many centers facing great uncertainty, budget pressures, and cutbacks. In 2001, AAPCC reported $104 million in total funding for poison control centers. In a separate analysis, the Committee estimated a similar amount by multiplying the cost per human exposure call[4] by call volume. The Committee concludes that the most effective approach to stabilization is through federal funding of approximately $100 million to support the core activities. This funding could reduce or replace the support for core activities provided by many of the current funding sources; however, it would not reduce the need for state and local funding to support non-core services.

Recommendation

6. Congress should amend the current Poison Control Center Enhancement and Awareness Act to provide sufficient funding to support the proposed Poison Prevention and Control System with its national network of regional poison control centers. Support for the core activities at the *current* level of service is estimated to require more than $100 million annually. Extension of services to include the growing all-hazards emergency needs (especially biological or chemical terrorism) and enhancements

---

[4]Cost per human exposure call represents all poison control center expenses divided by the number of human exposure calls.

to current surveillance and data collection activities will require additional support and should be supplemented as appropriate to such mandates. The funding could be channeled either through a direct federal grant or a federal-state matching process. Performance measures for poison control center services must be specified and monitored by the funding agencies involved. Separate funding will be required to support activities performed at the federal and state levels.

In addition to the funds required by each poison control center to implement the core activities, the Committee estimates an amount roughly on the magnitude of $30 million to assure that all the essential services of public health related to poisoning are accomplished. This estimate includes approximately $10 million in the form of $200,000 grants to each state to support a poison prevention coordinator's office whose responsibilities would include coordination of public education efforts and a plan for their evaluation and $20 million for federal-level activities, including (1) development and maintenance of quality assurance and improvement mechanisms for every component of the Poison Prevention and Control System; (2) training activities for health providers outside the poison control centers who require training in toxicology, such as emergency department workers and emergency medical technicians; (3) a clearinghouse for primary prevention materials and resources; and (4) research and the translation of research and evaluation studies into best practices and regulatory changes. Federal estimates are based on similar public health programs funded by the CDC and HRSA.

Recommendation

7. Congress should amend existing public health legislation to fund a state and local infrastructure to support an integrated Poison Prevention and Control System. The Committee at this time is not able to provide a precise estimate of the required level of support for such a federal and state program. The Committee recommends that the Secretary of Health and Human Services should develop a budget proposal to support the costs of training, research, data archiving and reporting, quality assurance, and public education (including state-level coordination of prevention education and the creation of a central repository of best model programs). This amount is in addition to the $100 million needed to support poison control core services.

Assure High-Quality Poison Control Center Services

Certification of poison control centers is currently the responsibility of AAPCC, and the centers are required to join this organization to become certified. A more accepted model for certification of health care professionals or programs is for it to be the responsibility of an independent agency, rather than an organization in which the applicants are paying members. (For example, medical toxicologists are certified by a board that is a member of the American Board of Medical Specialties rather than by a toxicology organization.) With the continued development of poison control centers and their increased integration into the public health system, alternative certification processes will offer advantages over the current system, including greater independence of the process from the participants, wider input from the health care community, and wider recognition of the skills and contributions of poison control centers and their personnel.

Recommendation

8. A fully external, independent body should be responsible for certification of poison control centers and specialists in poison information. This body should be separate from the professional organizations representing them.

National Data System and Surveillance Needs

A Uniform Definition of Poisoning

Among the most important functions of the Poison Prevention and Control System will be the collection and provision of poison exposure and surveillance data to the nation's health authorities. The Committee found many barriers to the effective operation of a comprehensive data and surveillance system and to the provision and utilization of the information by agencies at the federal, state, and local levels. The steps to ameliorate this situation are complex, but there is a pressing need for change. The Committee recommends that these be addressed at the same time that the legislative, financing, and organizational reforms are being implemented.

Recommendation

9. The Secretary of Health and Human Services should instruct key agencies to convene an expert panel to develop a definition of poisoning that can be used in surveillance activities (including the Toxic Exposure Surveillance System) and ongoing data collection studies. Furthermore:

a. The Secretary should ask the World Health Organization to review and reform the *International Classification of Diseases* codes for poisoning, thereby addressing the discrepancies and complexities identified in the current classification.

b. The Secretary should require agencies that sponsor existing surveillance and data collection instruments to use a common definition of poisoning that allows comparability across data collection efforts.

c. The National Center for Health Statistics (NCHS) should review the methodology of its existing surveys to maximize the value of their survey data for poison prevention and control.

d. Other agencies collecting health-related data at the federal level outside NCHS, and at the state level, should enhance their surveys or surveillance data systems to better gather and interpret data related to poisoning injury and risk factors.

Privacy Barriers to Data Collection

New patient protections provided by the Health Insurance Portability and Accountability Act (HIPAA) and state privacy regulations have placed substantial limitations on sharing health care data. This situation is exacerbated by the fact that there are many misconceptions among health care professionals regarding the conditions under which such data are available.

Recommendation

10. DHHS should undertake a targeted education effort to improve health provider awareness of poisoning data collection as it relates to the Health Insurance Portability and Accountability Act (HIPAA) and state privacy regulations to mitigate their unintended chilling effect on poison control center consultation, including follow-up. DHHS should review and resolve the negative impact of HIPAA and state privacy regulations on poison control center functions, including toxicology consultations and outcomes evaluation.

Availability of TESS Data

The Toxic Exposure Surveillance System is a proprietary data and surveillance system owned by AAPCC. Using funding from CDC, AAPCC has recently developed a capability to provide real-time surveillance through TESS based on input from the poison control centers. The Committee recognizes that this system was established and has been significantly strengthened through the initiative of AAPCC. However, there is now enough evidence to suggest that a private system cannot meet the national need for timely data in this area. Despite federal funding, the computer code for TESS is owned by a private company, further complicating its use and distribution.

Recommendation

11. The Director of the Centers for Disease Control and Prevention should ensure that exposure surveillance data generated by the poison control centers and currently reported in the Toxic Exposure Surveillance System are available to all appropriate local, state, and federal public health units and to the poison control centers on a "real-time" basis at no additional cost to these users. These data should also be publicly accessible with oversight mechanisms and privacy guarantees and at a cost consistent with other major public use systems such as those currently managed by the National Center for Health Statistics.

Research Needs

The Committee made an attempt, within the constraints of the available literature and data systems, to document the magnitude of the poisoning problem and its cost, in terms of health care outcomes, to the nation. We concluded that despite limitations in the data, poisoning is a far greater problem than has been generally recognized and deserves a higher level of scrutiny and support. The Committee recommends a baseline assessment of the magnitude and cost of poisoning. Furthermore, the Committee found a dearth of research on poisoning and poison control center operations and encourages funding of research in this area.

Recommendation

12. Federally funded research should be provided for (1) studies on the epidemiology of poisoning, (2) the prevention and treatment of poisoning and drug overdose, (3) health services access and delivery, (4) strategies to improve regulations and facilitate researchers' input into regulatory procedures, and (5) the cost efficiency of the new Poison Prevention and Control System on population-based outcomes for general and specific poisonings.

a. CDC should take the lead in marshalling the relevant data pertaining to the epidemiology of poisoning. It should produce a comprehensive report estimating the national incidence of poisoning morbidity and mortality, exploiting its existing data sources. Within the centers, the National Center for Injury Prevention and Control (NCIPC) could lead this effort, coordinating data needs with NCHS. Data sources should include TESS, the National Health Interview Survey, the National Electronic Injury Surveillance System, the Drug Abuse Warning Network, MedWatch, and others.

b. The Agency for Healthcare Research and Quality (AHRQ) and CDC should be directed to undertake a rigorous economic analysis of the overall direct and indirect health care costs of poisoning and drug overdose.

c. The Secretary of Health and Human Services should encourage funding by appropriate agencies, such as CDC and the Consumer Product Safety Commission, to ensure the needed flow of information from toxicology researchers in poison control centers on prevention problems and strategies to regulators from toxicology researchers in poison control centers and to encourage the study and development of new regulatory strategies and initiatives to reduce poisonings.

d. Researchers should be funded through grants from appropriate institutes such as the National Institutes of Health, the National Library of Medicine, AHRQ, and CDC/NCIPC, to study prevention and treatment of poisonings and drug overdose, health service access and delivery, and the cost efficiency and clinical impact of the Poison Prevention and Control System.

Part I

Overview

1

Introduction

"Alle Ding sind Gift und nichts ohne Gift; alein die Dosis macht das ein Ding kein Gift ist"

"All things are poison and not without poison; only the dose makes a thing not a poison"

Paracelsus (1493–1541)

BACKGROUND

The field of poison prevention provides some of the most celebrated examples of successful public health interventions, yet paradoxically, the poison control "system" today is little more than a network of poison control centers that is poorly integrated into the larger spheres of public health or injury prevention. Reviews of the history of effective injury prevention strategies frequently highlight the introduction of the "baby aspirin" poisoning legislation in 1966 and the Poison Prevention Packaging Act in 1970. These legislative and regulatory successes were among the first achievements of the modern consumer movement and were consolidated under the jurisdiction of the new Consumer Product Safety Commission in 1973. The introduction of packages containing less than a toxic dose and childproof safety closures on hazardous substances heralded the so-called "passive" methods to prevent injuries, measures requiring little direct behavioral input from the potential victim.

The establishment of the first poison control center, which preceded the legislative actions by some years, represented an innovation in pediatric health care delivery, one that was immediately hailed and replicated across the nation. Poison control center telephone numbers were promoted by pediatricians and adorned family refrigerator doors. A new cadre of highly trained poison information specialists and clinical toxicologists evolved to help staff these centers. The number of phone calls—currently more than 2 million exposure calls annually—attests to the need for and popularity of these services.

Despite these early successes in both the implementation of effective poison control legislation and the development of this new model of poison treatment service, the evolution of the poison control network has been chaotic and uneven. The early growth of poison control services was encouraged and supported by the Emergency Medical Services Systems Act of 1973 (Pub. L. No. 93–154), but little federal funding was available to plan or promote this growth in the 1970s. Ultimately, more than 600 sites in the United States identified themselves as poison control centers. They varied from little more than a designated telephone in small community hospitals with no dedicated staff to centers in academic medical institutions with 24-hour dedicated staff and nationally recognized medical and clinical toxicologist[1] backup. No federal public health agency took responsibility for the oversight of this patchwork poison control network, and no systematic sources of governmental funding emerged to support these heavily utilized health care services.

Poison control centers remained generally peripheral to the expansion of the injury control system during the 1980s and 1990s. Little epidemiological research emerged from such centers to inform the policy and public health practice communities of the magnitude of the poisoning problem or of its place in the greater domain of injury prevention. Some individual poison control centers played a role in injury control, but most struggled financially to sustain the staff and infrastructure to answer phone calls and provide appropriate follow-up. Successful poison control centers supported their operations through ad hoc funding arrangements, in some cases receiving funding from state maternal and child health and emergency medical services agencies, by providing service in occupa-

[1]The term *toxicologist* is a general description of an individual dealing with any aspect of acute or chronic poisonings, and it does not have a specific definition or implication with regard to training or job description. For example, this term may be used to describe individuals whose activities range from molecular biology to epidemiology, as long as they deal in some way with the toxic effects of chemicals. The term *clinical toxicologist* implies a more clinical orientation, but likewise has no specific definition or implications. *Medical toxicologists* are physicians with specific training and board certification in the subspecialty of medical toxicology, which focuses on the care of poisoned patients.

tional medicine settings, and through contractual agreements with the pharmaceutical and other chemical companies. Ultimately, the vast majority of centers closed, leaving the current 63 to cover the U.S. population. Furthermore, the poison control centers were not involved with or incorporated into the development of the emergency medical services for children system. The 1999 Institute of Medicine (IOM) report, *Reducing the Burden of Injury: Advancing Prevention and Treatment*, barely mentions poison control; nor does this issue feature prominently in *Healthy People 2000* or *2010* (http://www.healthypeople.gov).

Onto this background of a small, innovative field struggling to survive, four important developments emerged in the past 10 years:

- First, the American Association of Poison Control Centers (AAPCC), although founded in 1958, emerged as the developer of a critical poison exposure data collection system, the certifier of poison control centers and their key personnel, and the principal advocate for federal legislation and funding.
- Second, two federal agencies of the U.S. Department of Health and Human Services—the Health Resources and Services Administration/Maternal and Child Health Bureau (HRSA/MCHB) and the Centers for Disease Control and Prevention (CDC)—entered the poison control center picture. For the first time, federal legislation (Poison Control Center Enhancement and Awareness Act of 2000 [Pub. L. No. 106–174]) authorized substantial funding for a variety of poison control services, including education, medical toxicology, enhanced data collection, and a national toll-free number. These federal agencies began examining the functioning of the poison control network and the place of poison control centers in public health.
- Third, advances in telecommunication made it possible to answer telephone calls from anywhere in the country and to triage calls to appropriate centers. A national "800" telephone number created a single point of contact for consumers. In addition, a real-time, electronic submission of poison exposure data enabled the rapid assessment of toxic exposures handled by poison control centers across the United States.
- Fourth, the national program of homeland security and the imperative of preparing the public health system to address the risks of biological and chemical terrorism provided new opportunities for poison prevention and surveillance. The availability of the Toxic Exposure Surveillance System (TESS) data propelled the poison control centers into a potentially crucial position in all-hazards/public health preparedness. The anthrax attack demonstrated how a concerned public looked to poison control centers for information and advice (see Appendix 5-A for description).

CHARGE TO THE IOM COMMITTEE

HRSA/MCHB asked the IOM to convene the Committee on Poison Prevention and Control to assist it in developing a more systematic approach to understanding, stabilizing, and providing long-term support for poison prevention and control services. Specifically, HRSA/MCHB charged the Committee to consider the future of poison prevention and control services in the United States by reviewing the past and current approaches to the provision of these services in terms of:

1. The scope of services provided, including consumer telephone consultation, technical assistance and/or hospital consultation for the care of patients with life-threatening poisonings, and education of the public and professionals;
2. The coordination of poison control centers with other public health, emergency medical, and other emergency services;
3. The strengths and weaknesses of various organizational structures for poison control centers and services, including a consideration of personnel needs;
4. Approaches to providing the financial resources for poison prevention and control services;
5. Methods for assuring consistent, high-quality services, including the certification of centers and methods of evaluation; and
6. Current and future data systems and surveillance needs.

Furthermore, the Committee was asked to consider these questions in light of future demographic and population trends, and in the context of the threats of biological and chemical terrorism.

In order to respond fully and specifically to the charge, the Committee adopted the language used by HRSA, that is, to consider the "future of poison prevention and control services" and to develop a "systematic" approach. We believe that HRSA chose this language carefully, asking the Committee to do more than review the current poison control centers in isolation. Therefore, the Committee examined the role of poison control services within the context of the larger public health system, the injury prevention and control field, and the fields of general medical care and medical and clinical toxicology. As part of this approach, the Committee further examined how poison control centers function (e.g., respond to the public and health care professions regarding poisoning exposures, provide toxicosurveillance, potentially detect bioterrorism, train medical and clinical toxicologists) in light of the functions performed by other health care agencies and governmental organizations at federal, state, and local levels.

POISONING: A MATTER OF DEFINITION

This chapter begins with a quote from Paracelsus: "All things are poison . . . only the dose makes a thing not a poison." This statement goes to the heart of a definitional dilemma that faced the Committee throughout its work. As discussed more completely in Chapter 3, there is no single agreed-upon definition of a "poisoning." Each agency that collects data or provides services in this arena has evolved its own definitional boundaries of the poisoning problem. The definition of a poisoning and its place among other medical diagnoses vary from the 9th to the 10th revisions of the *International Classification of Diseases*, which drives data collection at several levels of federal and state government. The poison control centers have their own operational definition of what constitutes an "exposure" to a poisonous substance. Various authorities and authors may decide to include or exclude from the operational definition such important components as intentionally self-inflicted poisoning (as in the act of suicide), overdoses and intoxications from alcohol and illicit drugs, envenomation by insects, illness caused by toxic infectious agents, and ingestions of the right prescription medicine taken at the wrong dose, among others. The implication of these inconsistent definitions is profound for the measurement of the magnitude of poisonings and for the development of public policy and practice in this area. The Committee adopted an operational definition of poisoning that could be used to analyze the available datasets in order to better understand the problem. This definition subsumes "damaging physiological effects of ingestion, inhalation, or other exposure to a range of pharmaceuticals, illicit drugs, and chemicals, including pesticides, heavy metals, gases/vapors, and common household substances, such as bleach and ammonia" (Centers for Disease Control and Prevention, 2004, p. 233). Definitional issues are discussed further in the following sections and in more detail in Chapter 3.

MAGNITUDE OF POISONINGS: A PUBLIC HEALTH PROBLEM

The Committee discovered that estimating the magnitude and cost of poisoning as a public health problem is more complex than generally appreciated, requiring special analyses of available mortality, morbidity, and cost data from separate sources. While more detailed analyses of both the epidemiology of poisoning and the costs and benefits of poison control will be presented in later chapters, we focus initially on three important points: first, that poisoning is a larger and more important public health problem than has generally been recognized; second, that poisonings generate a high cost to the United States; and third, that the population at risk of poisoning is broader than that of young children.

Poisoning: The Second Leading Cause of Injury-Related Death in the United States

The Committee estimates that in 2001 (the most recent year for which data from all sources were available), there were 30,800 poisoning-related deaths in the United States (based on published figures and specially provided estimates from Lois Fingerhut at the National Center for Health Statistics, 2003). This estimate makes poisoning the second leading cause of injury-related death in the United States, behind motor vehicle deaths (N = 42,443) and ahead of gun-related deaths (N = 29,573). Our estimate is higher than that usually reported because it combines (1) the number of deaths in which poisoning is the reported underlying cause, along with (2) deaths in which alcohol or illicit drugs of abuse are the reported underlying cause. The Committee believes that including the alcohol and illicit drug deaths in its estimate is justified because these poisonings come to the attention of the poison control centers and because the revisions of the coding systems seem to be moving toward the inclusion of these cases.

Poisoning Morbidity

As measures of morbidity, the Committee examined poison-related hospitalizations and overall exposures. Again, each of these estimates has its own unique definitional limitations, but the numbers presented in Table 1-1 are a way of showing the order of magnitude of poisoning as a cause of morbidity and health care system use. In 2001, there were 282,012 hospitalizations, as reported through the National Hospital Discharge Data Set, and more than 2.3 million human poison exposures (includes both actual and suspected exposures), as reported to TESS (Watson et al., 2003) by the poison control centers. It should be pointed out that TESS human exposure reports include both actual and suspected exposures to poisonous substances of all types. If anything, these are likely to be underestimates of true death and hospitalization numbers.

TABLE 1-1 Summary of Poisoning Mortality and Morbidity (2001)

Level of Poisoning Severity	Number of Deaths
Poison-related deaths	30,800
Underlying cause	(24,173)
Alcohol and drugs	(6,627)
Number of poison-related discharges from short-stay hospitals	282,012
Human exposure calls to poison control centers	2,267,979

Cost of Poisoning

The annual cost of poisoning, not including costs related to alcohol deaths, to the United States (based on lifetime cost of injury) was estimated in 1989 to be $8.5 billion (Rice et al., 1989). A major definitional limitation of this study was that it excluded costs related to poisoning from alcohol and other illicit drugs. Unfortunately, the Committee found no more recently published data. Adjusting the cost estimates from the Rice study to 2003 dollars using the Consumer Price Index provides an estimate of $12.6 billion for the current cost of poisoning.

Population at Risk

Not only have the magnitude and cost of the poisoning problem been underappreciated, but the diverse nature of poisonings and the populations at risk have changed over time. While poisoning was initially viewed as a problem of young children, it now emerges as a concern across the entire lifespan. Half of all *poison exposures* reported to TESS occur among children 5 years of age; however, only 8 percent of the moderate to major effects from poisonings occur among those in the 5 years and under age group. Approximately 71 percent of moderate and major exposures occur in those over 19 years of age.

Unintentional death from exposure to hazardous household substances occurs primarily among children and youth, the group that also has the highest level of exposure to poisonous substances. However, suicide by poison and alcohol and illicit drug-related poison deaths occur in older adolescent and young adult populations (approximately 7.6 percent of the poison exposures reported to TESS are suspected suicides; another 3.5 percent are from intentional substance misuse or abuse). Death in the workplace from exposure to hazardous substances occurs primarily among working adults. Pesticide deaths are likely to be concentrated in rural farm populations, including immigrant and illegal laborers. Finally, the elderly are at risk of taking the wrong medications or the right medication at the wrong dose. Approximately 8 percent of poisoning exposures reported to TESS are from individuals over 59 years of age. In addition, elderly persons may be the source of medications that inadvertently poison young children.

This changing pattern of poisoning in the U.S. population has important implications for the provision of prevention and control services. Poison control centers were developed to respond primarily to parental concerns about the exposure of their young children to potential poisons and rely on telephone communication. Increasingly, these centers have become involved with the additional situations described above—suicide attempts, alcohol intoxications, medication errors, hazards evaluations—that arise from requests from emergency medical services and emergency department personnel, police and fire officials, and homeland security staff. A future system for poison prevention and control may need to be more appropriately designed and organized to respond to this variety of demands.

THE COMMITTEE'S APPROACH TO THE PROBLEM

Early in its deliberations, the Committee realized that in order to address its charge, it must step back from a focus on the poison control centers alone and reexamine the overall context for preventing and treating poisoning. Reviewing the history of poison prevention in the United States reminded us that, in the past, a broad array of societal strategies, including safe product packaging and consumer legislation, had been used to reduce the risks posed by potentially hazardous substances. Adopting such a broad analysis led us to view poison control centers as part of a public health system intended to improve the health of communities and populations.

The argument for a broad public health approach to poisoning begins with the recognition that the United States has set specific year 2010 objectives (*Healthy People 2010*) for reduction of nonfatal poisonings to no more than 292 per 100,000 population, from the baseline of 349 in 1997, and deaths caused by poisoning to 1.5 per 100,000, from 6.8 in 1997. These ambitious objectives cannot be achieved by the poison control center network alone. There is no evidence that these centers, despite their critical role in poison control management, have reduced the incidence of poisoning in the population. There is good evidence, however, that hazardous substance packaging and regulation have had a primary preventive impact on poisoning (Rodgers, 1996). Furthermore, the body of evidence from the broader field of injury control indicates that reducing the burden of injury in the population (Bonnie et al., 1999) requires an integrated strategy of active behavioral, passive regulatory and engineering, and medical management strategies. Drawing on this broad perspective leads us to propose in Chapter 2 the creation of a Poison Prevention and Control System.

Currently there is no comprehensive system of poison prevention and control. Although poison control centers operate on a common set of certification standards, they form at best a loosely organized network. Each center has grown up in its own culture, has created its own organization and procedures, and cultivates its own mixture of funding sources. The absence of a "system" has led, in part, to the uneven and unstable development of the field. The current poison control center network needs to be integrated into a larger system, just as burn centers operate in a broader context.

Based on its own record of hearings, the Committee came to recognize that its very formation and charge created a high level of anxiety in the poison control center community. It is not surprising that a field that has struggled for survival, and that only recently has had national policy successes in obtaining federal legislation and funding, would view the Committee's charge with apprehension. Therefore, the Committee committed itself to an analysis that would provide the strongest basis for a Poison Prevention and Control System that could be sustained and well integrated into the health care system.

This report will begin with a summary of the future Poison Prevention and Control System as envisioned by the Committee (Chapter 2). That system encompasses, but is not restricted to, the role and position of the poison control centers. Rather, our proposal is based on, first, an analysis of the broad public health functions that underlie all aspects of poison control (e.g., primary prevention through consumer product regulation and public education and secondary prevention through telephone-based poison consultation); second, an analysis of the core functions of a poison prevention and control system and, within these, the core functions of poison control centers; and third, a proposal for a national approach to the organization, funding, and accountability for such a system.

The Committee recognizes that this is an ambitious task.

THE COMMITTEE'S METHODS

The Committee met six times between February 2003 and January 2004. These meetings were used to plan our work and gather firsthand information on the poison control system from key informants and from a site visit to the Rocky Mountain Poison and Drug Center, Denver, Colorado (see Appendix A for the list of contributors). At its first three meetings, the Committee received both solicited and unsolicited opinions and information from representatives of professional organizations and poison control centers. At our second meeting, we heard testimony from the directors of four poison control centers (Rocky Mountain Poison and Drug Center, California Poison Control System, DeVos Children's Hospital Regional Poison Center, and Middle Tennessee Poison Center) representing the range of large and small centers located in a variety of organizational settings. In addition, visits by one or more Committee members and staff were made to the National Capital Poison Control Center and the Maryland Poison Control Center.

Early in the process, the Committee recognized that the literature on poison control centers and the poison control system could not provide the evidentiary base sufficient to fully answer its charge. As a result, the methods used by the Committee were different than those generally used by IOM committees. Rather than relying solely on limited peer-reviewed publications, we focused our efforts on analyses of existing datasets and, where necessary, on primary data collection.

To investigate the organizational and financing aspects of poison control centers, the Committee made use of historical reports, including the work by Zuvekas et al. (1997) that provided an excellent analytical framework. Unfortunately, these data are of limited value because they are based on only six centers. Secondary analyses of the administrative data collected annually by AAPCC were used to examine the range of programmatic indicators and costs from all member poison control centers. We wish to thank AAPCC for its cooperation in providing data from the 2000 and 2001 surveys. Unfortunately, data from the 2002 survey were not available until after the Committee had completed its deliberations.

Based on the analysis of these data, the Committee carried out a qualitative survey of 10 poison control centers to better understand the relationship among various organizational arrangements and effective and efficient service provision. Survey interviews were conducted by telephone with poison control center directors and their staff. These centers were a stratified, nonprobability sample based on cost per human exposure call handled in 2001, population served, and penetrance.

To investigate the epidemiology of poisoning, the Committee recruited the help of Lois Fingerhut at the National Center for Health Statistics. Ms. Fingerhut provided special analyses of poisoning data from the national death statistics for the use of the Committee. In addition, we commissioned Miriam Cisternas to prepare a paper on the epidemiology of poisoning, contrasting and comparing the data from multiple public use data sources. This paper forms the basis for much of the analysis presented in Chapter 3. Another source of data on the epidemiology of poisoning was provided by Monique Sheppard at the third Committee meeting. She reported on the integration of poison data from multiple sources based on an analysis of data from eight northeastern states.

To understand the goals and related programs of the federal agencies, the Committee interviewed representatives from the Consumer Product Safety Commission, U.S. Food and Drug Administration, U.S. Department of Agriculture, Health Resources and Services Administration/Maternal and Child Health Bureau, and Centers for Disease Control and Prevention. A subcommittee spent a day in Atlanta being briefed by CDC staff from the National Center for Injury Prevention and Control, National Center for Environmental Health, Agency for Toxic Substances and Disease Registry, National Institute for Occupational Safety and Health, and Office of Terrorism Preparedness and Emergency Response. Federal contracts and program guidance materials were reviewed. The Committee also drew upon two membership organizations of state agency directors, the Association of Maternal and Child Health Programs and the State and Territorial Injury Prevention Directors Association, for information about the relationship of state agencies to poison control. These organizations conducted voluntary surveys of their member state organizations and provided us the information. Finally, the state plans for the National Bioterrorism Hospital Preparedness Program were reviewed for information on poison control center involvement.

OVERVIEW OF REPORT CHAPTERS

This report is presented in three parts. Part I begins with this introductory chapter followed by Chapter 2, which provides an overview of the Committee's proposal for a future Poison Prevention and Control System; a system does not exist at the moment and will need to be created.

In Part II, we review the historical development of the poison control network, the current status of poisoning as a public health problem, and the principal functional elements of the system. Chapters 3 through 9 describe the evidence and the analyses we used in reaching our conclusions and recommendations. Chapter 3 presents data estimating the magnitude of poisoning in the United States. Chapter 4 provides a historical context for the development and growth of poison control services through 2001. Chapters 5 through 9 examine the current status of poison control centers in terms of functions (including core services), personnel, quality assurance, organization, cost, funding, data and surveillance, prevention and public education, and linkages to federal, state, and local agencies.

Part III summarizes the argument for a new Poison Prevention and Control System by focusing on the Committee's conclusions and recommendations. In Chapter 10, the concluding chapter, we link our analysis to our conception for the future system.

2

Toward a Poison Prevention and Control System

In this early chapter of the report, the Committee presents its concept for a Poison Prevention and Control System in order to guide the reader through the analyses presented in subsequent chapters. This approach is necessary because the recommendations for such a system depart considerably from the picture of the current network of poison control centers.

The Committee's argument for a Poison Prevention and Control System follows directly from the charge given to it by the Health Resources and Services Administration/Maternal and Child Health Bureau to develop a "systematic approach to understanding, stabilizing, and providing long-term support for . . . poison prevention and control services in the United States," emphasizing "the coordination of poison control centers with other public health, emergency medical and other services." The Committee views its recommendation to create a Poison Prevention and Control System as central to addressing its charge.

The Committee also believed it essential to define at the outset the complex term "poisoning." Recognizing that there is no single definition agreed upon by all of the relevant professional bodies, we adopted the operational definitions of poisoning used by the agencies that sponsor the various systems that capture and report on poisoning data. For federal datasets capturing morbidity, we used the International Classification of Diseases (ICD)-9 definitions; in the specific case of poisoning mortality, however, the more up-to-date ICD-10 definition of underlying causes of death was used. In the case of exposure calls to poison control centers, we adopted the idiosyncratic Toxic Exposure Surveillance System (TESS) definitions, as established by the American Association of Poison Control Centers (AAPCC).

THE "SYSTEMS" CHARACTERISTICS OF A POISON PREVENTION AND CONTROL SYSTEM

The Committee recognized that comprehensively addressing the issue of poisoning required a "systems approach," with a broad array of government and nongovernment health agencies, including the poison control centers, working together to achieve the common goal of reducing the incidence and severity of poisonings in the U.S. population. Essential to the system is this shared goal and coordinated activities. As described later (in Chapter 9), the other agencies central to such a system include state and local health departments, emergency medical services units, and federal agencies including the Centers for Disease Control and Prevention, Health Resources and Services Administration, Consumer Product Safety Commission, Food and Drug Administration, and others. Furthermore, the components of the Poison Prevention and Control System must share information freely so that each can assess its contributions and achievements. Shared information, including data from TESS, and feedback and evaluation are at the heart of an effectively functioning system. Later in the report, we make the case that the system needs to be integrated with the broader U.S. public health system.

THE ROLE OF POISON CONTROL CENTERS IN A POISON PREVENTION AND CONTROL SYSTEM

The Committee concluded that poison control centers are essential components and building blocks of a Poison Prevention and Control System. These centers are on the front line of meeting the needs of the public and the health care community for information and guidance concerning poison exposures. Public satisfaction with this free service appears to be high. The centers have played an important historical role in providing consumer services and have provided data supporting new packaging and labeling regulations. Thus, poison control centers serve a vital public health function, accentuated by the public concerns about exposure to biological and chemical agents of terrorism.

In the future, however, poison control centers must be more stable financially, population based, and better integrated and coordinated with other stakeholders involved in the protection of the public from hazardous substances. This conclusion reflects the Committee's assessment that the current network emerged historically with little planning and no consistent state or federal agency support. Each center evolved its own unique set of funding sources, depending on the largesse of sponsoring hospitals in some cases, and on idiosyncratic state and local funding sources in other cases. The current size, number, and location of poison control centers reflect this historical evolution that took place without any overall planning or underlying principles of organization.

THE CASE FOR REGIONAL POISON CONTROL CENTERS

The Committee envisions a system of regional poison control centers across the country, each serving a defined population. Current experience shows that centers can effectively serve regions as small as large metropolitan areas or as large as multiple states. Furthermore, a poison control center may be located within its region or, as is currently the case, may contract to serve regions at some distance from the center. Effective and efficient examples of each type of arrangement currently exist. In cases in which a center contracts from a distance, there must be strong links to state and local public health agencies concerned with poisoning and poisoning prevention. Modern telecommunications technology supports a variety of cooperative arrangements. Such technology also provides the opportunity for one or more centers to assist another center with high call-volume surges or periods of personnel absence or equipment failure.

The Committee considered the strengths and weaknesses of a variety of options for the number and distribution of poison control centers in a Poison Prevention and Control System. Although modern telecommunications technology makes it feasible to consider one single, highly efficient, large center serving the entire country, the Committee found a number of weaknesses with that model. A single national center would have difficulty appreciating local variations in poisonous substances such as plants and insects. In addition, a single center would concentrate all of the expertise in one location, thus eliminating important and timely local medical consultations. Finally, a single center is vulnerable to practical problems of power failures, limited surge capacity, and potential transmission lags during times of high volume.

The Committee also considered a national model that would have a single poison control center in each state. This model was also rejected as inconsistent with the current realities. A number of states with relatively small and dispersed populations have chosen to contract with larger centers to meet their needs. Also, in large states like California, there is a statewide system with multiple centers because one center alone cannot meet the entire need. Thus, we concluded that a system of regional centers provided an appropriate balance of size and responsiveness.

The specific rationale for a regionalized system includes the following elements. Poison control centers must be large enough to sustain an adequately sized staff to meet usual demands and the surge capacity required to respond to situations of mass poisoning or suspected terrorism events. A regional distribution of such centers will satisfy the need to distribute medical toxicological leadership across the United States to address the diversity of poison exposures and to provide firsthand consultation to hospitals and physicians. The interaction among regionally based centers will promote innovation and the sharing of best practices. Finally, a regionalized system should provide enough redundancy in skills and resources to meet surge needs and potential equipment failures.

The Committee concluded that decisions about the number of centers should be based on considerations of population coverage, telecommunication capabilities, and types of funding. Although the currently available data are not adequate to prescribe a specific size or geographical coverage for centers, the Committee believes there may be economies of scale and scope that can be achieved through a regionalized system. Defining a set of core services will support the development of a federal funding formula for regionalized poison control centers. Ultimately, the needs assessment data must be developed to define the financial and services base for developing contractual agreements for poison control services. We believe that the concept of regionalized national poison control centers provides the structural basis for development of a Poison Prevention and Control System.

THE CORE FUNCTIONS OF POISON CONTROL CENTERS

A fundamental component of this proposal is the specification of the core functions of a poison control center functioning within the Poison Prevention and Control System. Chapter 5 provides more detailed definitions of center core functions. Briefly, the core functions of a regional poison control center will include:

- Manage telephone-based poison exposure and information calls;
- Prepare for and respond to all-hazards emergency needs (including biological or chemical terrorism or other mass exposure events) in cooperation with other organizations at local and state levels;
- Capture, analyze, and report exposure data;
- Train poison control center staff, including specialists in poison information and poison information providers;
- Carry out continuous quality improvement;

• Integrate services into the public health system; and

• Train medical and clinical toxicologists in a subset of poison control centers.

CONNECTIONS BETWEEN THE POISON CONTROL CENTERS AND THE BROADER PUBLIC HEALTH SYSTEM

The Committee envisions a far stronger set of connections between poison control centers and public health agencies than is currently the case. The rationale for these connections is discussed in detail in Chapter 9 and is based heavily on the concept of a public health system developed by the Institute of Medicine in its landmark report, *The Future of Public Health* (1988). Furthermore, we believe the Poison Prevention and Control System must be well connected to emergency medical services (EMS) so that emergency medical technicians can be dispatched rapidly where needed, information on exposures and hazards can be shared, and treatment guidelines can be put into place. Finally, the Poison Prevention and Control System will become central to the states' public health preparedness for bioterrorism or other emergency all-hazards events.

Data on poison exposure cannot be kept privately, but rather must be publicly available in real time from the system. The Committee's conclusions about the collection, ownership, and dissemination of data on poisonings and poison exposure are among the most important aspects of this report. We recognize that TESS was established and strengthened through the initiative of AAPCC, but we believe there is enough evidence now to suggest that a private system cannot meet the national need for timely data in this area. Federal agencies must oversee the collection and management of this system and make the data available to state and local agencies as needed for policy decisions and public health practice.

To accomplish these broader goals, the Committee believes that some current poison control center functions will be better carried out by federal, state, or local public health agencies. In this set of activities we include:

• Primary prevention efforts through public education;

• Consumer protection through continuous monitoring of poison exposures and translation into regulation of hazardous products;

• Rapid analysis of exposure data (toxicosurveillance) to detect "outbreaks" and effective use of such data by public health agencies to assure public safety;

• Program and national policy development and implementation; and

• Links to EMS and emergency preparedness organizations.

FEDERAL FUNDING FOR THE CORE SERVICES OF A POISON PREVENTION AND CONTROL SYSTEM

The Committee believes that a stronger, better organized, and accountable system for poison prevention and control is in the national interest and must be supported by federal funding and oversight. This will probably require the passage of federal legislation (or the amendment of the current poison control center stabilization legislation) to define specific roles for federal agencies, a funding formula, the definition of contractual obligations, and mechanisms for accountability of the system. There also must be mechanisms in place for effective collaboration and cooperation among the federal agencies with responsibilities in this area and effective linkages to the counterpart state and local agencies. Stakeholders at the federal level include representatives from the Health Resources and Services Administration, Centers for Disease Control and Prevention, Agency for Toxic Substances and Disease Registry, National Center for Health Statistics, U.S. Food and Drug Administration, U.S. Consumer Product Safety Commission, and emergency medical services.

THE RESEARCH NEEDS OF A POISON PREVENTION AND CONTROL SYSTEM

Finally, the Committee sees a need for a much stronger knowledge base related to poison prevention and control. We have already identified the need to improve the research capacity of the poison control centers themselves. Realizing this goal depends on a federal commitment to fund toxicological, epidemiological, and health services research in this field. Such a commitment must be built into the research mandates of the National Institutes of Health, Centers for Disease Control and Prevention, and Agency for Healthcare Research and Quality.

SUMMARY

In summary, the proposed Poison Prevention and Control System will be different from what currently exists in several key areas:

- The component agencies, including the poison control centers, will work cooperatively to reduce the burden of poisoning;
- The federal legislative base will provide a national mandate and federal core funding;
- Poison exposure data will be publicly available in real time for agency decision making and for merging with other data sources;

- Poison prevention and control will be integrated into the broader injury prevention and public health systems;
- The performance of the system will be held accountable; and
- Primary prevention through public education and hazardous substance regulation will be strengthened.

Part II

Current Status and Opportunities

3

Magnitude of the Problem

The purpose of this chapter is to provide an overview of the occurrence of poisonings in the United States and to describe the distribution of poisoning reports in terms of a variety of demographic characteristics such as age, gender, and race. To provide such an overview, the chapter will also present a working definition of poisoning and drug overdose, highlighting the epidemiological implications of inclusion and exclusion of various categories of events from this classification. Even if definitions vary (as will be discussed in the following section), poisoning is an important problem of national scope. As noted in Chapter 1, more than 2 million people contact poison control centers annually for advice on poisoning exposures (Watson et al., 2003). In addition, poisoning is a leading cause of injury-related morbidity and mortality in the United States. The total health care costs associated with poisoning (see Chapter 6) are also substantial.

Temporal trends may affect the societal impact of poisoning and drug overdose in a variety of ways, given that the U.S. population is growing larger, older, and more ethnically diverse. Changing ethnic distributions, marked by an increasing proportion of Hispanics and Asian Americans, and an increasing proportion of the elderly population (http://www.census.gov) are important considerations for the future of poison prevention and control, particularly in light of research indicating that these groups have been relatively underserved by the existing poison control system (Kelly et al., 1997, 2003). Providing effective access to care for ethnically diverse groups will require overcoming both cultural and language barriers. In addition, culturally related health practices—including patterns of self-treatment with potentially hazardous herbals or other complementary (alternative) medications—may be directly relevant to poisoning prevention and treatment.

For all these reasons, accurate data are key to delineating the magnitude of the poisoning problem, yet there is no single source of incidence data that fully illuminates this picture. Figure 3-1 provides a schematic representation of the universe of poisoning and drug overdose and the relationships among mortality, poisoning resulting in hospitalization (that may or not result in death), and cases that come to the attention of poison control centers, emergency departments, and private physicians (that may or may not lead to hospitalizations). This diagram illustrates that within the universe of poisonings, there is likely to be varying overlap between poisonings captured in different service delivery and data surveillance systems (see Chapter 7). It is largely for this reason that no single data system captures the totality of these data. Thus, the following sections will describe data on poisoning and drug overdose incidence derived from key primary sources. Furthermore, we will attempt to integrate the estimates they yield to provide an overall picture of the magnitude of the problem.

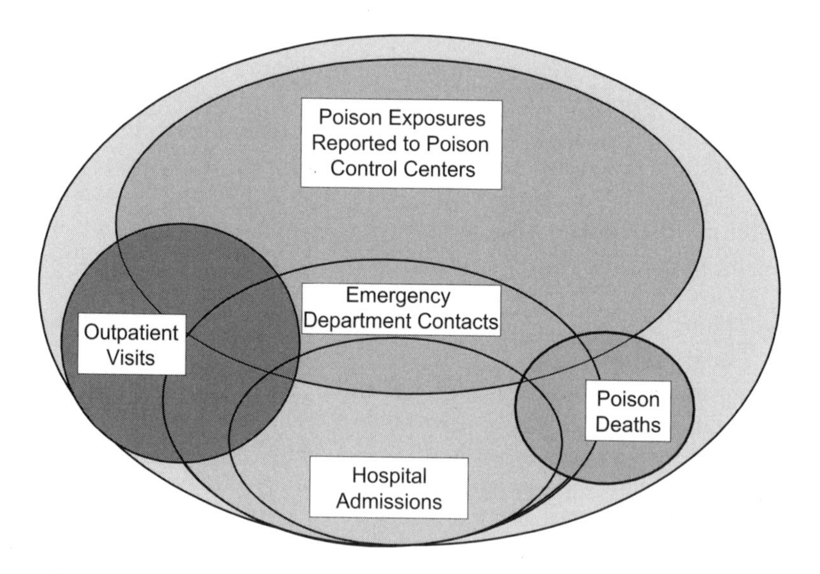

FIGURE 3-1 Poison exposures in the United States.
NOTE: Not drawn to scale.

Defining Poisoning

A fundamental challenge in estimating the magnitude of poisoning and drug overdose is delineating the types of conditions that should be included under this rubric. It is important to acknowledge that there is no standard definition of poisoning that is universally accepted and applied in clinical practice, in data collection, and in public health policy setting. Even within data collection systems, different definitions of eligibility for the purposes of case reporting may apply in various surveillance schemes (see Chapter 7).

In clinical terms, human poisoning subsumes any toxin-related injury. Such injury can be systemic or organ specific (e.g., neurological injury or hepatotoxicity). As important, the source of the toxin can be a synthetic chemical or a naturally occurring plant, animal, or mineral substance. Thus poisoning can include the toxic effects of a classic toxin (e.g., cyanide), an overdose of a prescription medication (e.g., an antidepressant), or an overdose of an over-the-counter preparation (e.g., headache tablets) or a complementary treatment (such as an herbal medicine or dietary supplement).

Although defining the foregoing events as poisoning is fairly straightforward, other classes of exposure may fall in or out of different classification schemes. "Envenomation" from a rattlesnake or a black widow spider clearly falls within the clinical context of poisoning and, therefore, is covered in depth in standard toxicological texts (Goldfrank et al., 2002; Olson et al., 2003). Envenomation may also overlap in some categorizations, however, with insect stings or "bites" that might not be considered toxic, but may be complicated by allergic responses, including fatal anaphylaxis.

A parallel set of issues is associated with medication responses that may not be dose related, but instead are idiosyncratic, with or without an allergic component. Clinical definitions of poisoning generally take into account unusual toxic responses that may involve susceptible subpopulations (e.g., toxic responses related to alternative metabolic pathways clinically relevant in only a subset of the population). Although this may overlap with the mechanisms of other types of poisoning, many definitional schemes separately tally or exclude altogether illnesses defined as adverse therapeutic events, such as drug toxicity that results from multidrug interactions, increased susceptibility or true allergic sensitivity, or dosing error, all of which can be classified as "adverse drug effects."

The toxic effects of ethanol present a specific set of definitional challenges. Acute ethanol toxicity in the context of frank overdose (e.g., rapid ingestion of a large amount of alcohol in a naïve drinker) can be lethal.

Ethanol withdrawal is also associated with severe morbidity and mortality (see Osborn, 2004). Nonetheless, the frequency of acute and chronic ethanol intoxication and the myriad complications that may result from or be associated with ethanol ingestion complicate the use and interpretation of the designation "ethanol poisoning" as it may pertain to the overall incidence of poisoning and drug overdose.

Illness from naturally occurring toxins derived from microorganisms can also lead to definitional confusion. Seafood-related toxins whose ultimate source was from microorganisms, such as those causing paralytic shellfish poisoning, are typically categorized as poisons. In contrast, bacterially derived toxins may or may not be categorized in this manner. In practice, the diagnosis and management of botulism, tetanus, and, more recently, anthrax, has been considered to be a form of "poisoning" relevant to the discipline of clinical toxicology, although these illnesses are not included in most epidemiological definitions of poisoning.

Lay definitions of poisoning are also relevant because they can drive health-care-seeking behavior and self-reporting of conditions, both of which can impact incidence estimates. Lay terms such as "food poisoning" (which could reflect an infectious gastroenteritis or a toxin-related condition), "poison oak" (a form of allergic contact dermatitis), and even "sun poisoning" (which could refer to sunburn or heat stroke) do not conform to biomedical concepts of poisoning, but may still be unavoidably captured in some incidence estimates.

Factors of intent, that is, whether an exposure occurred with the purpose of causing a toxic response, do not define poisoning per se, but these factors may impact how such events are reported. Defining adverse events associated with drugs of abuse is a particularly salient issue in this regard. For example, some events may or may not be categorized as a poisoning or drug overdose by health care providers, depending on whether the presenting medical complaint is viewed as an intended end-point effect.

Toxin exposure without an attributable and defined or discrete clinical effect presents yet another source of heterogeneous definitions. The absence of a documented clinical effect may reflect the true absence of a substantive exposure (e.g., a person seeking health care because of a potential for exposure to a toxin or because of exposure to a substance perceived to be dangerous by the lay public that has little or no actual toxicity); a subtle effect that may not be manifest by acute symptoms but may have serious long-term potential effects (e.g., a body burden of lead elevated above the population norm); or circumstances that do not allow determination of a causal relationship (e.g., postmortem carbon monoxide determination in a burn victim with both fire and smoke exposure). Although the standard definition of clinical poisoning does not include exposure without disease, the importance of these scenarios in terms of primary, secondary, or tertiary prevention (see Chapter 8) is clearly relevant to the overall magnitude of the poisoning problem.

Definitions Used in This Report

The following analyses attempt to be consistent in the coding that has been used to categorize the poisoning estimates derived and to highlight areas in which there are substantive differences in coding or case definition that might be likely to affect the estimates provided. Further methodological details and a discussion of the coding of poisoning and drug overdoses are also provided in Appendix 3-A. All ICD-9 (*International Classification of Diseases*—9th edition) defined morbidity estimates have used a definition of poisoning and drug overdose that includes envenomations of all kinds (including insect stings). All ICD-9 estimates exclude the specific category of "ethanol toxicity," but include other alcohol types, such as methanol.

In contrast, the ICD-10 mortality analysis includes ethanol deaths, but breaks out this subtotal in key tabular presentations. Another key difference is that the ICD-10-based mortality analysis excludes envenomation-caused mortality of all types (snakebite mortality is rare in the United States; bee sting anaphylaxis is also excluded). This analysis also yields one other specific estimate of fire and smoke deaths in which carbon monoxide toxicity was listed as a contributing cause. Both the ICD-9 and ICD-10 derived estimates excluded therapeutic misadventure and adverse drug reactions.

There is no defined grouping of ICD codes that establishes a single category subsuming all poisoning events. In theory, the multiple coding options allow choice in defining poisoning based on the specific codes selected. In practice, the level of resolution provided by certain codes may not allow for discrimination within certain subcategories of toxins.

A limited number of analyses also allow for the side-by-side examination of two coding schemes, one based on the ICD-9 system and the other based on a narrative descriptor of the patient's chief complaint related to the event in question. For consistency, these analyses relied on the ICD-9 codes for case definition and neither included nor excluded cases based solely on these supplemental narratives. They are presented, however, in part to demonstrate how definitions and terminology may cloud interpretation of "poisoning" incidence. Detailed study of the patterns of overlap between the narrative "chief complaint" for patient visit and its categorization as coded by an ICD-9 code was beyond the scope of this analysis. Follow-up study of the sensitivity and specificity of the "chief complaint" nosology may be relevant to a larger review of potential approaches that might be used by the National Center for Health Statistics (NCHS) in revised survey methods to estimate poisoning and drug overdose incidence.

EPIDEMIOLOGY OF POISONING

Estimating the incidence of poisoning is a complex and difficult exercise. First, in order to gain a general understanding of the magnitude of the poisoning problem, the Committee commissioned a paper on the epidemiology of poisoning. Cisternas (2003) provides annual estimates of poisoning incidence through an analysis of data from multiple sources available for public use through NCHS. These data were used to generate annual estimates of overall incidence as well as annual incidence stratified by age, gender, race, and geographic region. In addition, level of medical care received and outcome status (where available) were used as an indirect severity measure. Second, summary data for total incidence from two additional data sources were also included to supplement a final tabulation of morbidity and mortality. These supplemental summary totals were derived from the American Association of Poison Control Centers' (AAPCC's) annual Toxic Exposure Surveillance System (TESS) data report and the Centers for Disease Control and Prevention (CDC)-Consumer Product Safety Commission (CPSC) National Electronic Injury Surveillance System (NEISS). Finally, in order to characterize poisoning and drug overdose deaths, a separate analysis of U.S. mortality data was carried out by Lois Fingerhut of CDC's NCHS (Personal communication, L. Fingerhut, December 2003). These data were analyzed by demographic and geographic strata, as well as type and intent of poisoning.

Data Sources

Four core data sources were used in the first part of the analysis. Wherever possible, multiple years of data were combined in order to increase the stability of the estimates (see Table 3-1 for a summary of the number of poison observations extracted from each data source). Appendix 3-A contains a detailed description of each of these four data sources. The sources are:

- *National Health Interview Survey (NHIS):* This annual population-based survey collects health status and demographic information from a sample of households and their family members selected from and meant to estimate for the entire civilian, noninstitutionalized U.S. population (approximately 275 million persons over the period analyzed).
- *National Ambulatory Medical Care Survey (NAMCS):* NAMCS is a

TABLE 3-1 Public Use Data Sources Analyzed for Morbidity Analyses

Data Source	Years Used	Number of Poisonings
National Health Interview Survey	2000–2001	269
National Ambulatory Medical Care Survey	1997–2001	188
National Hospital Ambulatory Medical Care Survey Outpatient File	1997–2001	315
National Hospital Ambulatory Medical Care Survey Emergency Department File	1997–2001	1,810
National Hospital Discharge Survey	1997–2001	11,533

SOURCE: Cisternas (2003) analysis carried out for this Committee. These data form the basis of Tables 3-2 through 3-9 and Table 3-11.

national probability sample survey of patient visits made in the United States to the offices of nonfederally employed physicians classified by the American Medical Association and the American Osteopathic Association as working in settings that are "office-based patient care."

• *National Hospital Ambulatory Medical Care Survey (NHAMCS):* NHAMCS is the hospital ambulatory complement to NAMCS; it is a national sample of ambulatory visits to hospital outpatient centers and emergency departments (EDs). The outpatient center and emergency department records are disseminated in separate files, as the survey questions differ for these two sites of care.

• *National Hospital Discharge Survey (NHDS):* NHDS covers discharges from a sample of short-stay hospital visits that are noninstitutional and are not federal. In order to be included in the survey, hospitals must have six or more beds staffed for patient use.

Summary data were used from two data sources in which data were not available for reanalysis (direct analysis of raw data beyond published summaries). They include:

• *National Electronic Injury Surveillance System—All Injury Program (NEISS-AIP):* CPSC operates a surveillance system known as the National Electronic Injury Surveillance System. In 2000, CPSC expanded the system to collect data on all injuries, not just product-related incidents. NEISS-AIP data are gathered from a sample of 100 hospital emergency departments.

• *Toxic Exposure Surveillance System:* AAPCC compiles TESS data on poison exposure phone calls received at U.S. poison control centers. Summary data reports are provided free of charge through the AAPCC website and published annually (Litovitz et al., 2002). Additional details of this system and its data access are discussed in Chapter 7.

An analysis of mortality data was carried out using the following source:

- *Mortality Vital Statistics:* Electronic mortality vital statistics data are derived from a national file of death certificate-derived data maintained by NCHS. This data file is designed to capture all deaths on a yearly basis. The deaths analyzed were from the single year 2001. This is the "universe" of observations, not a selected sample from which estimates of true incidence are derived. The denominator population (unlike the four surveys described earlier) is the entire U.S. population (not limited to the noninstitutionalized).

Multiple other data sources are potentially relevant to the incidence of poisoning and drug overdose, particularly to certain subsets of events, beyond the seven sources included in this analysis (see Chapter 7). Because these sources do not include a range of events comparable to the sources used (e.g., the Food and Drug Administration's MedWatch program captures voluntary reports of medication-related adverse events, while the Drug Abuse Warning Network (DAWN) system is designed to best capture events associated with medications of abuse potential and illicit drugs), these are not part of this analysis. Nonetheless, they are clearly relevant to more targeted epidemiological questions that could not be addressed here.

It is important to note that none of the sample-based sources of data on poisoning and drug overdose has sufficient observations to provide adequate estimates by specific causes. Thus these data sources do not, in themselves, form a basis for evaluating responses for highly targeted intervention strategies such as the reduction of antidepressant medications for overdose incidence or the prevention of spider envenomation.

Data Coding

A general discussion of the definition of poisoning-related coding issues is presented in Appendix 3-A. Specific to this analysis, all data sources used in the primary analysis of morbidity contained E-codes and ICD-9-CM diagnosis code fields. The definition of poisoning for the analysis includes ICD-9-CM diagnostic and external cause of injury ("E") codes: 960.0–964.5, 964.9–979.0, 980.1–989.9, E850.0–E858.9, E860.1–E869.9, E950.0–E952.9, E961.0–E962.9, and E980.0–E982.9. The diagnostic or E-codes for ethanol intoxication, ethanolism, or its sequelae were excluded, as were adverse drug reactions or related diagnoses and diagnoses related to bacterial food poisoning.

NAMCS and NHAMCS files also included up to three "reason for visit" fields based on the patient's chief complaint. Any relating to poisoning (5900.2—Unintentional poisoning: Ingestion, inhalation, or exposure to potentially poisonous products, 5820.1—Overdose, intentional, and 5910.0—Adverse effect of drug abuse) were examined, but were only included if confirmed by a consistent concomitant ICD-9-CM diagnosis or E-codes as listed previously.

The two datasets from which summary data only are reported use their own poisoning codes that are not based on the ICD scheme. Thus the data presented rely on these systems' inclusion and inclusion criteria whose potential selection effects are discussed briefly below as well as in Chapter 7 in relation to surveillance.

Poisoning mortality for 2001 was defined by ICD-10 using the codes for underlying cause of death. The codes included X40–X49, X60–X69, X85–X90, Y10–Y19, and Y35.2. In addition, ICD-10 codes for deaths due to mental and behavioral disorders attributed to psychoactive substance use, F10–F16 and F18–F19, are also included because these can be driven by poisoning mortality according to current coding procedures. "T" series codes were not relevant to this analysis because they should be superceded by "X," "Y," or "F" series codes for the underlying cause of death in fatal poisoning. No deaths occurred in 2001 that were coded as U01.6 or U01.7, terrorism-related poisoning designations.

Findings

National Health Interview Survey

A total of 269 injury episode observations were identified by ICD-9-CM and E-codes from 2000–2001 NHIS injury/episode files. Table 3-2 includes estimates of annual poisoning episodes overall and stratified by various demographic characteristics and whether direct treatment was given. Based on sampling weights, which allow mathematical calculation of the population frequency based on the observations (see Appendix 3-A for details), the number of annual poisoning episodes (as contrasted with exposures) in the United States is estimated to be 1,575,000 for the 275.25 million persons in the noninstitutionalized population, yielding a poisoning-related episode rate of 570 per 100,000 per year.

Females were more likely to be poisoned than males (690 versus 450 per 100,000, respectively), and were more likely to have direct contact with a health provider for their episode than males (530 versus 420 per 100,000, respectively). Children (under 18 years of age) were more likely

TABLE 3-2 Annual Number and Rates of Poisoning Episodes by Respondent Self-Report from National Health Interview Survey (NHIS), 2000–2001

	Sample Number	Number of Poisoning Episodes (thousands)	Total Population (millions)	Episodes[a] per 100,000 Persons per Annum
All episodes reported	269	1,575	275.25	570
Gender				
Male	107	601	134.26	450
Female	162	974	140.99	690
Age				
Under 18	86	528	72.49	730
18–64	153	847	169.98	500
65 and over	30	200	32.78	610
Race				
White	220	1,355	216.99	620
Black	24	95[b]	33.61	280[b]
Other	25	124[b]	24.64	500[b]
Region				
Northeast	37	226	52.66	430
Midwest	71	441	66.50	660
South	94	584	98.08	600
West	67	324	58.01	560
Treated by direct clinician contact[c]	204	1,171	275.25	430
Gender				
Male	77	428	134.26	320
Female	127	743	140.99	530
Age				
Under 18	49	272	72.49	380
18–64	133	746	169.98	440
65 and over	22	153	32.78	470
Race				
White	165	996	216.99	460
Black	21	81[b]	33.61	240[b]
Other	18	94[b]	24.64	380[b]
Region				
Northeast	30	184	52.66	350
Midwest	54	340	66.50	510
South	74	429	98.08	440
West	46	218	58.01	380
Treated by phone only	59	375	275.25	140
Gender				
Male	29	169	134.26	130
Female	30	206	140.99	150
Age				
Under 18	36	250	72.49	340
18 and over	23	125	202.76	60

[a]The term "episode" is used to refer to an event reported by the interviewee.
[b]Estimate has low statistical reliability (relative standard error >30 percent or sample N <30).
[c]Respondent indicated that poisoned individual received ambulatory or inpatient care.
NOTE: NHIS is an annual population-based survey of approximately 101,000 individuals ascertaining poisoning via respondent self-report.
SOURCE: Cisternas (2003).

to be poisoned than either working age adults or those 65 years of age and over, but were less likely to receive direct treatment. Whites were more likely to report a poisoning episode than all others. Differences in episode numbers and rates by race, however, should be viewed cautiously because the nonwhite estimates exhibit low statistical reliability due to small sample sizes. For this reason the data as shown are collapsed to show rates for whites, blacks, and all others (combining multiple other groups). Respondents from the Northeast were the least likely to report poisoning episodes (430 per 100,000), while those from the Midwest were the most likely (660 per 100,000). The most common ICD-9 codes included in the poisoning subset were "toxic effect of other substances (venom; bites of venomous snakes, lizards, and spiders; tick paralysis)" (989.5), "accidental poisoning from poisonous foodstuffs and plants" (E865.9), and "toxic effect of noxious substances eaten as food (unspecified)" (988.9).

As shown in Table 3-2, important differences emerge when the data are analyzed separately based on whether a direct medical contact took place, as opposed to telephone consultation only. Among those with direct medical contact, the general patterns of race and regional differences remain. The age distribution changes dramatically. The higher overall incidence rate for persons under 18 years of age is explained by cases treated by telephone call only; the rate in this age category is actually lower among clinician-treated cases. This group has a nearly equal gender mix, indicating that the gender gap in total cases estimated by the NHIS is explained by medically treated episodes of poisoning. Although estimation by narrower age strata (e.g., children 6 years of age and under) might further highlight these trends, this was beyond the scope of the analysis presented here.

All but 6 of the 269 episodes in the NHIS 2000–2001 poisoning subset had answers provided to questions about poisoning treatments (Table 3-3). About one-quarter were treated only by telephone calls to a doctor's office and/or a poison control center, but had no ambulatory or inpatient visits reported (16 percent reported a telephone call to a poison control center, regardless of whether a subsequent ambulatory or inpatient visit was made; data not shown in table). Approximately three-quarters of respondents had one or more visits to a doctor's office, clinic, or hospital outpatient or emergency department. The number reporting a visit to the hospital was quite small and not statistically reliable.

The NHIS-based estimates of the proportion of cases that received medical care and the overlap among the various levels of medical management for these poisoning episodes provide a benchmark for utilization estimates from other data sources that were also examined. For example, based on the NHIS, at least 568,000 annual visits at outpatient non-ED sites (as might be reflected in the NAMCS outpatient and NHAMCS sur-

TABLE 3-3 Treatments Mentioned for Annual Poisoning Self-Report Episodes, National Health Interview Survey (NHIS), 2000–2001

	Sample Number	Number of Episodes per Annum (thousands)	Percentage of Total
Total	263[a]	1,546	100
Phone call only (doctor's office or poison control center)	59	375	24
Ambulatory or inpatient visit	204	1,171	76
Visit to doctor's office/clinic/outpatient department	97	568	37
Visit to doctor's office/clinic/outpatient department without hospitalization or emergency department treatment	90	521	34
Visit to doctor's office	70	419	27
Visit to clinic/outpatient department	28	154[b]	10
Visit to emergency department	107	611	40
Visit to emergency department without hospitalization	97	558	36
Visit to hospital	17	92[b]	6

[a]Six respondents refused to answer this question.
[b]Estimate has low statistical reliability (relative standard error >30 percent or sample N <30).
NOTE: NHIS is an annual population-based survey covering approximately 101,000 individuals that ascertains poisoning based on respondent self-report. Thus the level of care reported (doctor's office visit, emergency department visit, or hospitalization) is defined by interview report, not by medical record extraction.
SOURCE: Cisternas (2003).

veys if their estimates were combined) and another 611,000 annual emergency department visits (as might be found in the NHAMCS emergency department visit file) should be expected. Of interest, the NHIS survey data indicate that there was relatively little overlap between the ED and non-ED doctor and clinic/outpatient categories. As would be anticipated, the majority of episodes in which the survey respondent reported that inpatient treatment (hospital admission) had occurred also indicated either that an emergency department or doctor/outpatient visit had occurred for that episode. Thus, as a crude approximation of prevalence of poisoning associated with ambulatory visits, the NHIS data suggest that the visit estimates from NAMCS and both NHAMCS files can be added together presuming little overlap (multiple counting of the same event). It can further be extrapolated that these visits represent about three-quarters of all poisoning episodes, based on the 24 percent of estimated NHIS episodes that were treated solely by a telephone call.

National Ambulatory Medical Care Survey

A subset of 188 records (out of a total of 120,464), including an ICD-9-CM diagnosis or E-code for poisoning, was extracted from the National Ambulatory Care Medical Survey 1997–2001 data files, resulting in an annual estimate of approximately 1,582,000 visits (Table 3-4). This represents approximately 0.2 percent of all doctor's office visits annually estimated through this survey. This estimate is nearly four times higher than what might be expected given the estimates from the NHIS (as shown in Table 3-3). Patterns of rates among the various demographic groups demonstrated similarities and differences compared with the NHIS data. Male and female patients in NAMCS had similar rates of poisoning-related visits, as opposed to the lower rates for males in the NHIS.

Rates for the various ethnic groups are presented in Table 3-4, but are too sparse in the nonwhite categories to be estimated with precision. In addition, 18 percent of the patient visits were associated with unknown race/ethnicity. Visit rates were highest in the Midwest and West and were lowest in the South. Although the Midwest was also highest in the NHIS (see Table 3-2), the other regions appear to differ in their rank order based on the NAMCS data.

Two-thirds of these visits were associated with an ICD-9 external cause of injury E-code of poisoning; a slightly smaller proportion (56 percent) was associated with an ICD-9 diagnosis code of poisoning (multiple codes possible for the same event). The most common ICD-9 codes were "toxic effects of other substances (venom; bites of venomous snakes, lizards, and spiders; tick paralysis)" (989.5), followed by "accidental poisoning by unspecified substance" (E866.8) and "accidental poisoning by unspecified drug" (E858.9). Observations are too sparse to generate reliable incidence estimates by category of specific ICD-9 code.

The NAMCS survey is one in which "patient reason for visit" data could be present, coded (not by the ICD-9 scheme) from an open-ended "chief complaint" or main symptom from the patient's perspective. Despite this option, in practice concomitant "patient reason" poisoning codes were relatively infrequent. It should also be noted again that cases were not selected for inclusion or exclusion in the principal analysis based on patient reason codes (see Methods in Appendix 3-A). Had this been a basis for inclusion (e.g., not confirmed by a concomitant ICD-9 diagnosis or E-code for poisoning), only 12 observations would have been added, an increase of 6 percent (total estimate of 1,689,000 visits, rather than 1,575,000). The relative rank of "patient reason for visit" responses will be

TABLE 3-4 Annual Number and Rates of Poisoning Doctor Visits as Confirmed by ICD-9-CM Codes, National Ambulatory Medical Care Survey (NAMCS), 1997–2001

	Sample Number	Number of Poisoning-Related Visits (thousands)		Total Population (millions)[a]	Visits per 100,000 Persons per Annum
Total	188	1,582		271.56	580
Gender					
Male	88	771		132.52	580
Female	100	811		139.05	580
Age					
Under 18	28	301		72.02	420
18–64	123	1,001		167.09	600
65 and over	37	281		32.45	870
Race					
White	164	1,385		215.97	640
Black	17	151[b]		33.11	460
Other	7	46[b]		21.91	210
Region					
Northeast	29	267		52.44	510
Midwest	46	461[b]		66.46	690
South	66	478		96.68	490
West	47	376		55.99	670
Reason/symptoms included poisoning (percentage)	15	160	(10)	271.56	60
Poison E-code included (percentage)	131	1,055	(67)	271.56	390
Poison diagnosis code included (percentage)	89	885	(56)	271.56	330

[a]Estimated from NHIS 1997–2001 person files.
[b]Estimate has low statistical reliability (relative standard error >30 percent or sample N <30).
NOTE: NAMCS is an annual survey of approximately 1,200 office-based physicians who record visits over a 2-week period.
SOURCE: Cisternas (2003).

presented in a later section, pooled with comparable data from the other outpatient surveys using the same supplementary coding scheme.

Survey questions concerning follow-up planned for patient visits were included in the 1999–2001 NAMCS questionnaires. Reports on more than half the poisoning visits indicated that a follow-up visit was planned. In 2001, a question was added to NAMCS concerning whether the visit was an initial or follow-up visit for the problem in question. Although the sample size for 2001 (a total of 32 cases) is too small to make population-based estimates, the 2001 data indicate that at least half may have occurred as a follow-up visit. These data suggest that true incidence estimates for *new* events should discount visits by about 50 percent. Only 2 of the 106 cases (1.9 percent) for the 1999–2000 period indicated that the patient would be admitted to the hospital.

National Hospital Ambulatory Medical Care Survey

Outpatient subset The National Hospital Ambulatory Care Survey 1997–2001 outpatient files (non-ED visits) included 315 observations with a poisoning ICD-9-CM diagnosis or E-code of poisoning, representing an annual estimate of 163,000 or 0.2 percent of all outpatient visits (Table 3-5). This estimate is in line with that expected based on the NHIS data previously presented.

As was the case with the NAMCS data, poisoning-related visit rates for males and females were virtually identical (820 versus 810 per 100,000, respectively). Unlike NAMCS, however, the visit rates for blacks were higher than for whites (110 versus 60 per 100,000, respectively). This could be consistent with a pattern of ambulatory care in which minority populations may be more likely to be served by hospital-based outpatient clinics than private physicians' offices. Distribution by region also differs between the outpatient NHAMCS file and both the NAMCS and NHIS files, with the visits from the outpatient file having the highest rates in the Northeast.

As was the case with NAMCS, NHAMCS data are too sparse to estimate annual visits associated with specific ICD-9 classifications. The most common ICD-9 code was "toxic effects of other substances (venom; bites of venomous snakes, lizards, and spiders; tick paralysis)" (989.5), followed by "accidental poisoning by unspecified drug" (E858.9) and "accidental poisoning by lead and its compounds and fumes" (E866.0).

Similar to the NAMCS, only a small proportion of visits had "patient reason" for visit information associated with poisoning and no case was included based on this information alone, absent a consistent ICD-9 code. Including these in the incidence estimates, even if not associated with a concomitant ICD-9 code, would have added only 24 observations to the analysis, resulting in an increase of 6 percent (yielding an estimate of 172,000 visits rather than 163,000). Rank order of "patient visit reason" will be presented in a later section.

The majority of respondents to the 2001 episode of care question (parallel to the NAMCS item, as discussed previously) indicated that this was an initial visit, while only an estimated 14 percent of the visits in that year were categorized as a follow-up visit. By extrapolation, this would sug-

TABLE 3-5 Annual Number and Rates of Poisoning Outpatient/Clinic Visits Confirmed by ICD-9-CM Codes, National Hospital Ambulatory Medical Care Survey (NHAMCS) Outpatient File, 1997–2001

	Sample Number	Number of Poisoning Visits (thousands)		Total Population (millions)[a]	Visits per 100,000 Persons per Annum
Total	315	163		271.56	60
Gender					
Male	152	81		132.52	60
Female	163	82		139.05	60
Age					
Under 18	93	43		72.02	60
18–64	190	98		167.09	60
65 and over	32	22		32.45	70
Race					
White	237	123		215.97	60
Black	66	36		33.11	110
Other	12	4[b]		21.91	20[b]
Region					
Northeast	77	40		52.44	80
Midwest	68	44		66.46	70
South	100	49		96.68	50
West	70	30		55.99	50
Reason/symptoms included poisoning (percentage)	37	16	(10)	271.56	10
Poison E-code included (percentage)	204	95	(59)	271.56	40
Poison diagnosis code included (percentage)	174	106	(65)	271.56	40

[a]Estimated from NHIS 1997–2001 person files.
[b]Estimate has low statistical reliability (relative standard error >30 percent or sample N <30).
NOTE: NHAMCS is an annual survey of hospitals with outpatient and emergency departments that record visits over a 4-week period.
SOURCE: Cisternas (2003).

gest a lower discounting rate to convert these poisoning visits into incident episodes compared with NAMCS, where follow-up visits formed a larger proportion of the sample. The disposition data for the 1999–2001 period indicated that an estimated 15 percent of visits in that period had a follow-up visit planned, and 25 of the 197 visits in the sample (population estimate 9,300 out of 150,000 or 6 percent) for that period were to be admitted to a hospital.

Emergency department subset A total of 1,810 records with a poisoning ICD-9-CM diagnosis or E-code of poisoning were extracted from the 1997–2001 NHAMCS emergency department data subset, yielding an estimate of 1,428,000 visits (Table 3-6). This represents 1.5 percent of all

TABLE 3-6 Annual Number and Rates of Emergency Department Visits as Confirmed by ICD-9-CM Codes, National Hospital Ambulatory Medical Care Survey (NHAMCS) Emergency Department File, 1997–2001

	Sample Number	Number of Visits (thousands)		Total Population (millions)[a]	Poisoning-Related Visits per 100,000 Persons
Total	1,810	1,428		271.56	530
Gender					
Male	853	665		132.52	500
Female	957	763		139.05	550
Age					
Under 18	432	352		72.02	490
18–64	1,240	975		167.09	580
65 and over	138	100		32.45	310
Race					
White	1,383	1,121		215.97	520
Black	378	274		33.11	830
Other	49	33		21.91	150
Region					
Northeast	485	286		52.44	540
Midwest	368	321		66.46	480
South	564	510		96.68	530
West	393	311		55.99	560
Discharge status (percentage of total)					
Referred to another physician or clinic	722	557	(39)	271.56	200
Admitted to hospital	355	256	(18)	271.56	90
Transferred to other facility	156	115	(8)	271.56	40
Reason/symptoms included poisoning	730	570	(40)	271.56	210
Poison E-code included (percentage of total)	1,410	1,089	(76)	271.56	400
Poison diagnosis code included	1,223	1,009	(71)	271.56	370

[a]Estimated from NHIS 1997–2001 person files.
SOURCE: Cisternas (2003).

emergency department visits estimated for the noninstitutionalized population for that period. This estimate was twice as high as would be expected from the NHIS-derived estimate presented earlier.

Unlike NAMCS and the NHAMCS outpatient file, females (550 per 100,000) had a slightly higher rate of visits than males (520 per 100,000), although this difference was far narrower than in the NHIS estimates. The age distribution differed from the NHIS, NAMCS, and NHAMCS outpatient subset, with those 65 years of age and over having the lowest rate of all three age groups. Once again, examination of narrower age strata, especially for those younger than 18 years of age, was beyond the scope of this analysis. As was the case with the outpatient NHAMCS file, the visit rate for whites (520 per 100,000) is lower than that for blacks (830 per 100,000). Rates of poisoning-related visits ranged from a low of 480 per 100,000 persons in the Midwest to a high of 560 per 100,000 persons in the West, a regional pattern that, once again, varied in comparison to each of the other datasets. Of these emergency department visits, 18 percent resulted in a subsequent admission to a hospital, and an additional 8 percent were transferred to another facility, while 39 percent were referred to another physician or clinic.

In 2001, questions were added to the survey concerning whether the patient was seen in the emergency department in the past 72 hours and whether the visit was initial or follow-up. Of those with nonmissing data, 97 percent had *not* been seen in the emergency department in the past 72 hours, and 95 percent were an initial visit (data not shown in Table 3-6).

The most common ICD-9 codes were "toxic effects of other substances (venom; bites of venomous snakes, lizards, and spiders; tick paralysis)" (989.5), "poisoning by unspecified drug or medicinal substance" (977.9), and "suicide and self-inflicted poisoning by tranquilizers" (E950.3). A much higher percentage of emergency department poisoning visits were associated with a reason or symptom of poisoning (40 percent) than from either the NAMCS or NHAMCS outpatient files. Nonetheless, as was the case with NAMCS and the outpatient NHAMCS file, the increase in observations and estimated visits that would be obtained by adding in records with a patient reason or symptom of poisoning that lacked a concomitant ICD-9 code remains negligible: an additional 6 percent to the estimate (1,514,000 as opposed to 1,428,000 cases).

The percentage of total emergency department visits estimated to be associated with poisoning in this analysis is slightly higher here than one published by McCaig in 1996 using 1993–1996 NHAMCS emergency department files (1.1 percent of all visits in that study compared with 1.5 percent here) (McCaig and Burt, 1999). This difference is likely due to differences in the definition of poisoning used in the earlier study, which was limited to visits with a poisoning-related E-code rather than utilizing

ICD-9 disease codes as well. Based only on E-codes, the 1997–2001 data yield a similar proportional estimate to that of McCaig and Burt (1999).

Pooling of NAMCS and NHAMCS Data

When observations from the NAMCS and NHAMCS surveys are pooled, a picture of poisonings treated in the outpatient setting as a whole can be obtained, including emergency departments (Table 3-7). Based on ICD-9 E-codes, poisoning by venomous animals or plants is the primary cause of one-fifth of all 3.1 million outpatient visits, followed by accidental poisoning by unspecified substances (13 percent) and other drugs (10 percent). The fourth leading cause of poisoning is self-inflicted (9 percent), followed by unintentional versus purposely inflicted, and not determined (5 percent). These five primary ICD-9 E-code classified causes subsumed 58 percent of all poisoning visits.

The top primary diagnosis by ICD-9-CM coding was primary effects of other substances (nonmedicinal), which accounted for nearly one-fifth of all primary diagnoses associated with outpatient visits. Poisoning by other and unspecified drugs was a distant second at 6 percent. The third, fourth, and fifth top primary ICD-9-CM codes were adverse effects not elsewhere classified (5 percent); poisoning by analgesics, antipyretics, and antirheumatics (which would include opiates) (4 percent); and poisoning by psychotropic agents (which would include amphetamines and hallucinogens) (3 percent), respectively.

Overall, the limited specific toxin-related information that emerges from these ICD-9-coded data is notable, based on either external cause of injury ("E") or general diagnostic codes. For example, identifiable categories of drugs of abuse appear to be minimal, but may be subsumed in nonspecified categories. This may reflect the limitations of this coding schema and its application in practice, as well as the need for targeted surveillance or supplemental sampling to generate reliable toxin-specific poisoning incidence estimates.

Examination of the primary "patient reason" for visit can provide an understanding of how poisoned individuals describe these episodes in their own terms. Sixteen percent describe their main reason for visit as an insect bite, while 10 percent indicate their primary reason as unintentional poisoning. The third primary reason/symptom was skin rash (7 percent), followed by intentional overdose (5 percent) and adverse effect of drug abuse (4 percent). These five reasons accounted for 43 percent of the cases estimated by the pooled sample, taking into account sampling weights. This analysis is limited to primary reason code; additional patient visit reasons could be listed but were not analyzed here. It is also important to note again that for none of these surveys was a primary

TABLE 3-7 Annual Number and Rates of Ambulatory Poisoning Visits Confirmed by ICD-9 Codes, National Ambulatory Medical Care Survey (NAMCS) and National Hospital Ambulatory Medical Care Survey (NHAMCS) Outpatient and ED Files, 1997–2001

	Sample Number	Number of Poisoning Visits (thousands)	Percentage of Total
Total	2,313	3,173	100
Top 5 primary ICD-9 3-digit E-codes			
E905—Poisoning caused by venomous animals and plants	392	652	21
E866—Accidental poisoning by other and unspecified solid and liquid substances	168	409	13
E858—Accidental poisoning by other drugs	291	302	10
E950—Suicide and self-inflicted poisoning by solids or liquids	384	301	9
E980—Poisoning by solids or liquids, accidental versus purposely inflicted not determined	146	163	5
Subtotal	1,381	1,826	58
Top 5 primary ICD-9 3-digit diagnosis codes			
989—Toxic effects of other substances, chiefly nonmedicinal	422	603	19
977—Poisoning by other and unspecified drugs and medicinal substances	198	178	6
995—Certain adverse effects not elsewhere classified	83	154	5
965—Poisoning by analgesics, antipyretics, and antirheumatics	130	120	4
969—Poisoning by psychotropic agents	100	98	3
Subtotal	933	1,154	36
Top 5 primary patient visit reason/symptom codes			
5755.0—Insect bites	328	519	16
5900.2—Unintentional poisoning: Ingestion, inhalation, or exposure to potentially poisonous products	300	333	10
1860.0—Skin rash	77	218	7
5820.1—Overdose, intentional	186	144	5
5910.0— Adverse effect of drug abuse	193	140	4
Subtotal	1,084	1,354	43

NOTE: NAMCS and NHAMCS are annual surveys of office-based physicians and hospital outpatient and emergency departments, respectively.
SOURCE: Cisternas (2003).

patient reason present in a majority of the poisoning sample, that is, in each case where an unrelated ICD-9 code was also present.

National Hospital Discharge Survey

A total of 11,533 records with an ICD-9-CM diagnosis or E-code of poisoning was extracted from NHDS 1997–2001 files, representing an estimated total of 291,000 annual hospitalizations, or 0.8 percent of all estimated discharges (Table 3-8). Poisoning-related hospitalizations were more likely for females than males (120 per 100,000 compared with 90 per 100,000). Rates of hospitalization increased with age.

Because 25 percent of all discharges were associated with "unknown" race/ethnicity, even though the available data for patient race imply that whites are less likely to be hospitalized for poisoning than blacks, this observation must be viewed with caution. Hospitalization visit rates did not demonstrate substantive variation by region. Discharge status information was available for most of the visits, indicating that the majority of poisoning cases are discharged home. An estimated 9 and 6 percent of hospital visits were discharged to other short- and long-term care facilities, respectively. Because the short-term care facilities included in the discharge status variable could include some (but not all) health care facilities not actually incorporated into the NHDS sampling frame (long-term care facilities as a category are excluded), 9 percent is too high an estimate of multiple hospitalizations per episode to be used as a discounting rate to convert these poisoning hospitalizations into episodes.

A source of admission variable was added to NHDS in 2001. Examination of all hospitalizations by the source of admission (Table 3-9) indicates that 65 percent of all cases (186,000 visits) were admitted from an emergency department, followed by 11 percent (31,000) from physician referral and 2 percent (5,000) from another hospital. Because 19 percent of all hospitalizations in the file were missing the source of admission information, the emergency department, physician, and hospital transfer sources of admission may actually be higher. Thus a figure of 3 percent for hospital transfers is used for later incidence estimates, which would represent the difference between episodes and true incidence. This presumes, however, that readmission of an individual for the same poisoning episode is extremely infrequent.

Two-thirds of the poisoning inpatient visits had a poisoning ICD-9 code as the principal diagnosis on the discharge abstract, with the remainder as a secondary listing (e.g., a primary diagnosis of aspiration pneumonia in a concomitant drug overdose). The six most common ICD-9-CM codes listed were "suicide and self-inflicted poisoning by tranquilizers" (E950.3), "poisoning by benzodiazepine-based tranquilizers" (969.4), "poi-

TABLE 3-8 Annual Number and Rates of Hospitalizations Defined by ICD-9-CM Codes, National Hospital Discharge Survey (NHDS), 1997–2001

	Sample Number	Number of Poisoning Hospitalizations (thousands)		Total Population (millions)[a]	Hospitalizations per 100,000 Persons per Annum
Total	11,533	291		271.56	110
Gender					
Male	4,896	124		132.52	90
Female	6,637	167		139.05	120
Age					
Under 18	2,020	43		72.02	60
18–64	8,048	199		167.09	120
65 and over	1,465	49		32.45	150
Race					
White	6,344	184		215.97	90
Black	1,678	35		33.11	110
Other	593	11		21.91	50
Unknown	2,918	61			
Region					
Northeast	2,411	57		52.44	110
Midwest	3,655	72		66.46	110
South	3,883	105		96.68	110
West	1,584	57		55.99	100
Discharge status (percentage of total)					
Routine/ discharged home	7,750	200	(69)	271.56	7.4
Left against medical advice	387	11	(4)	271.56	0.4
Discharged/ transferred to short-term facility	793	27	(9)	271.56	1.0
Discharged/ transferred to long-term care institution	545	17	(6)	271.56	0.6
Alive, disposition not stated	1,702	28	(10)	271.56	1.0
Dead	138	3.8	(1)	271.56	0.1
Not stated or not reported	218	3.7	(1)	271.56	0.1

[a]Estimated from NHIS 1997–2001 person files.

NOTE: NHDS is an annual survey of a sample of short-stay hospitals that provide data for a sample of their discharge records.

SOURCE: Cisternas (2003).

TABLE 3-9 Source of Admission for Poisoning-Related Hospitalizations, National Hospital Discharge Survey (NHDS), 2001

	Sample Number	Number of Visits (thousands)	Percentage of Total
Total	2,469	286	
Source of admission			
Physician referral	165	31	11
Clinical referral	9	2[a]	1
Health maintenance organization referral	4	1[a]	0
Transfer from a hospital	49	5[a]	2
Transfer from skilled nursing facility	3	0[a]	0
Transfer from other health facility	28	2[a]	1
Emergency department	1,438	186	65
Court/law enforcement	10	2[a]	1
Other	15	3[a]	1
Not available	748	53	19

[a]Estimate has low statistical reliability (sample N < 60).
NOTE: The NHDS is an annual survey of a sample of short-stay hospitals that provide data for a sample of their discharge records.
SOURCE: Cisternas (2003).

soning by antidepressants" (969.0), "suicide and self-inflicted poisoning by tranquilizers" (E950.0), "poisoning by aromatic analgesics not elsewhere classified" (965.4), and "suicide and self-inflicted poisoning by other specified drugs" (E950.4). Nonetheless, these six specific codes only represented 38 percent of all poisoning ICD-9-CM codes listed.

National Electronic Injury Surveillance System—All Injury Program

NEISS-AIP provides a summary estimate of 742,606 poisoning episodes presenting at emergency departments in 2002. However, this figure excludes poisoning from insect stings or other venomous animal bites. Such exposures are incorporated in a category entitled "other bite/sting nonfatal injuries," which includes all bites and stings from all insects and animals excluding dogs, estimated at 910,481. Although it is not possible to estimate with precision the percentage of that category that should be included under the poisonous exposure rubric, for the purposes of this estimate this subset has been conservatively discounted to 80 percent, presuming that one out of five would be considered a sting or envenomation (consistent with the proportion of such cases estimated elsewhere) and then added back to the subtotal that had excluded bites and stings. Of the NEISS-AIP 742,606 poisoning episodes, 172,931 (23 percent) resulted in hospitalization; for bites and stings, a far smaller proportion was hospitalized (1.4 percent).

Toxic Exposure Surveillance System

The 2001 TESS annual report included analysis of data from poison control centers in 48 states and the District of Columbia. Less than 2 percent of the 2,267,979 human exposures reported were caused by food poisoning; the remaining appear to be within the definition of poisoning used for the public data sources, although some primary ethanol-related events are also likely to be included. TESS follow-up on patients with more severe exposures indicates that of the 2.3 million human exposures reported to poison control centers in 2001, 1,736,010 (77 percent) were managed onsite in a non-health care facility,[1] while 498,524 (22 percent) received care from a health care provider. Of those who received care from a health provider, 272,286 (12 percent of all episodes) were treated in an outpatient setting and then released (including hospital-based and freestanding outpatient service settings), and 147,891 (7 percent of all episodes) required subsequent hospitalizations.

Analysis of follow-up call TESS data indicates that 1,074 exposures resulted in known fatalities. Because follow-up calls are routinely carried out only in more severe cases, there is likely some underreporting, even within the population referred for consultation. Repeat entry of the same case into the dataset through referral to more than one poison control center is possible, but is likely to be negligible. There is probably also a small amount of duplication of records for a single poisoning episode due to calls to a single center by both the patient and a health care provider without subsequent linking to a single event. Nonetheless, TESS summary figures were not discounted to generate an incidence frequency lower than the episode totals.

Vital Statistics Mortality Data

An unpublished analysis of 2001 death certificate data prepared by NCHS (Personal communication, L. Fingerhut, 2003) provides an estimate of 24,173 poisoning-related fatalities per annum. As noted earlier, the definition and classification of poisoning based on ICD-10 (introduced for mortality data in 1999) differs from the ICD-9 used for morbidity in the other estimates. Critically, this means that information is available on

---

[1]Of the patient episodes, 74.5 percent were managed onsite, and another 2 percent of patients refused a referral to a health care facility.

manner or intent of the poisoning, but not on the specific substance or agent. In addition, deaths for which the underlying cause was a mental or behavioral disorder due to psychoactive drug use are included (Table 3-10).

Based on this coding, poisoning deaths were classified as unintentional (63 percent), suicides (23 percent), and undetermined intent (13 percent), with the remainder as homicides or legal interventions. External cause codes for poisoning also describe the type of substance involved (e.g., drugs, alcohol, other solids and liquids, or gases and vapors). For example, of the external cause poisoning deaths that were classified as unintentional or of undetermined intent, 93 and 95 percent, respectively, were drug related; of the suicides, 69 percent were drug related and 28 percent were due to exposure to gases and vapors.

Approximately three-quarters (77 percent) of all deaths involving poisoning or toxic effects had at least one mention of drugs, medicaments, or biological substances (Table 3-10). The type of drug mentioned as contributing to the death varied with intent of the death. For example, narcotics and psychodysleptics were mentioned in 50 percent of the unintentional and 64 percent of the undetermined intent deaths involving poisonings and toxic effects. Narcotics and psychodysleptics accounted for only 20 percent of suicides involving poisoning and toxic effects. Cocaine was more commonly listed than other narcotic drugs. In contrast, antiepileptic, sedative-hypnotic, and anti-parkinsonism drugs and antidepressants were more likely to be associated with suicides than with unintentional or undetermined intent deaths.

TABLE 3-10 Poisoning and Drug Overdose Mortality, 2001

	Category of Substance Involved		
Underlying Cause of Death by Intent	Drug	Other	All Medicament
Unintentional	13,024	1,054	14,078
Suicide	3,559	1,632	5,191
Homicide	42	22	64
Undetermined intent	2,769	140	2,909
Drug-related mental or behavioral disorder	1,931	NA	1,931
Subtotal	21,325	2,848	24,173
Alcohol-related behavioral disorder	NA	6,627	6,627
Total	21,235	9,475	30,800[a]

[a]Excludes fire and smoke deaths with carbon monoxide poisoning listed as a contributing cause.

SOURCE: Fingerhut (2003).

Nearly 30 percent of deaths involving poisoning or toxic effects had at least one mention of the toxic effects of substances that were chiefly nonmedicinal. The toxic effects of alcohol and of carbon monoxide were more likely to be listed on death certificates than other toxic, nonmedicinal substances. About one-fourth of suicides involving poisoning and toxic effects had mention of carbon monoxide poisoning, and 9 percent of unintentional deaths involving poisoning and toxic effects included mention of alcohol.

Poisoning death rates increased with age from less than 1 per 100,000 for persons under 15 years of age to 19/100,000 at 35 to 44 years of age, and then declined again with age (Figure 3-2). For persons 15 to 19 and 20 to 24 years of age, death rates for males were about three times the rates for females; for those age 25 and older, the ratio was closer to 2:1 (data not shown).

The age-adjusted death rate for poisoning was 8.5/100,000, with rates ranging from a low of 1.8 for the Asian and Pacific Islander population to 9.8/100,000 for blacks. For each racial and ethnic group and both sexes, age-specific rates were higher for persons ages 35 to 44 and 45 to 54 than for those younger or older, with the highest rate for black males ages 45 to 54 (Figure 3-3).

Poisoning death rates are highest in the Mountain states (with New Mexico's rate being the highest [16.4/100,000] in the United States) and lowest in the West North Central states (particularly North Dakota, South Dakota, Iowa, and Minnesota) and Pacific states (dominated by the low rate in California, 4.1/100,000). In each geographic division except for New England, the highest death rates were associated with unintentional poisoning, followed by poisoning by suicide. In New England, rates for poisoning of undetermined intent were higher than other poisoning death rates (Figure 3-4). The East North Central states had relatively high rates associated with psychoactive drug use mortality.

SYNTHESIS

By combining the estimates of poisoning from the various data sources described, it is possible to develop a more complete estimate of poisoning incidence than can be obtained from any single source. It is interesting to note that while the distribution of estimated poisonings varies across regions for several of the data sources, these differences are blunted when all sources are combined. A comparison of annual episode estimates from the various data sources by level of care is presented in Table 3-11.

Both the NHIS and TESS provide estimates of poisoning events in which no direct (face-to-face) clinical evaluation or treatment occurred.

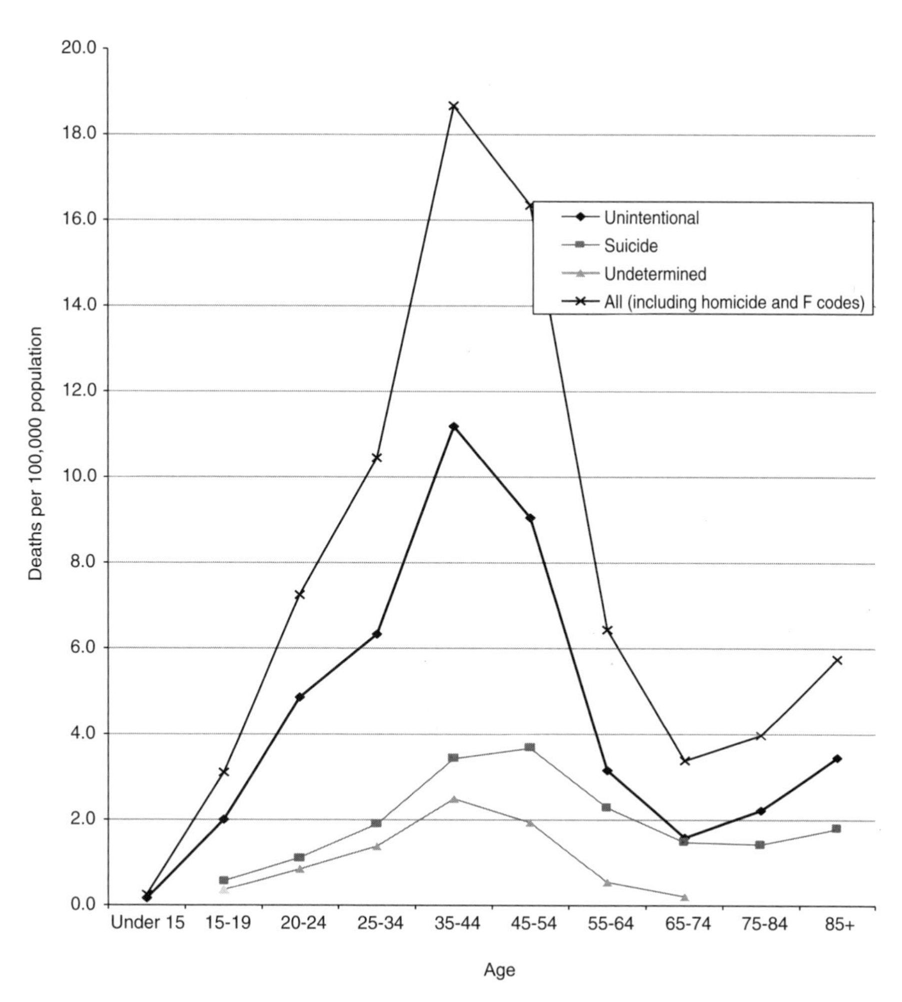

FIGURE 3-2 Death rates due to poisoning by intent and age, 2001.
SOURCE: Fingerhut (2003).

TESS data, by definition, only include poisoning cases for which a call was made to a poison control center. Although a few geographic areas are wholly excluded from TESS, the TESS experience of 1.7 million cases annually managed by telephone consultation alone (no subsequent clinical care) outstrips the NHIS estimate of less than 400,000 such events, even though the NHIS estimate also includes survey respondents who contacted a physician's office by telephone, but did not call a poison control center.

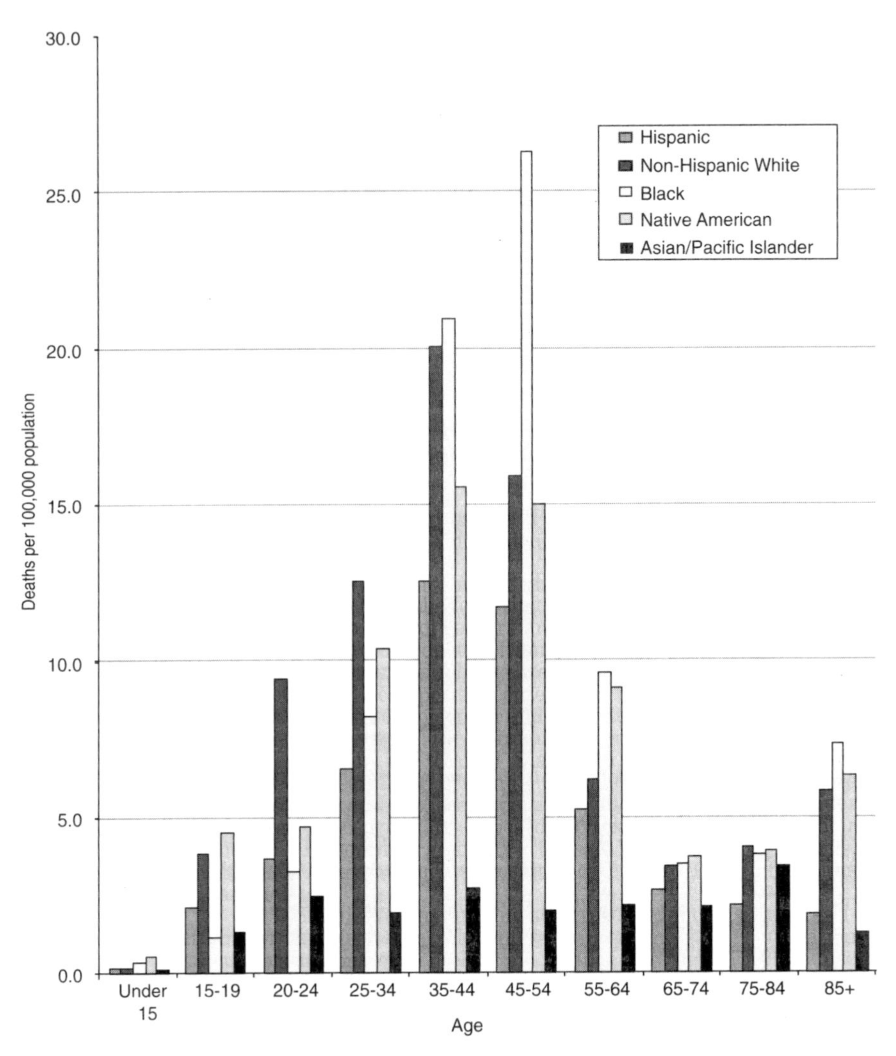

FIGURE 3-3 Poisoning death rates, 2001 (rates based on external cause codes as F codes).
SOURCE: Fingerhut (2003).

The NHIS also appears to underestimate the total incidence of poisonings that are directly treated by health care providers. The NHIS-derived estimate of approximately 1.2 million differs substantially from the upper-end estimate of 2.3 million cases annually based on combined data from NAMCS/NHAMCS. The ratio of the NHIS to TESS "telephone contact only" cases is 0.21; the ratio of the NHIS to NAMCS/NHAMCS health care provider-treated cases is 0.51.

A number of factors may drive underestimates from the NHIS that could differentially impact telephone consultation as compared with directly treated cases. Of interest are visits in which a patient's primary "reason for visit" formed the sole basis for defining a poisoning event (data not shown in Table 3-11). This might be comparable to a question-

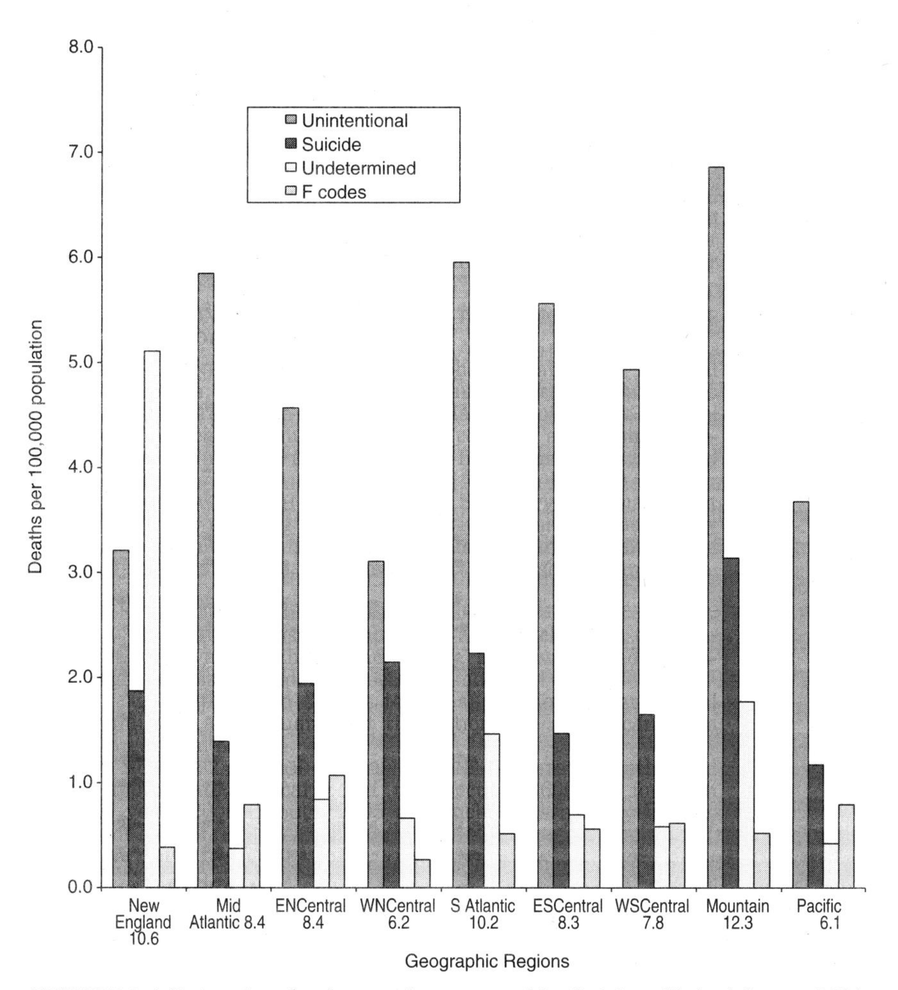

FIGURE 3-4 Poisoning death rates by geographic division, United States, 2001. (Undetermined includes the 64 deaths classified as homicides.)
SOURCE: Fingerhut (2003).

TABLE 3-11 Estimates of Annual Poisoning Episodes by Level of Care from All Sources

Level of Care	Data Source	Annual Incidence
Total episodes	NHIS	1,575,062
	TESS[a]	2,267,979
Telephone contact only	NHIS	374,794
	TESS[a]	1,736,010
Total treated by direct health care provider contact	NHIS	1,170,970
	TESS[a]	498,524
	NAMCS/NHAMCS	2,287,771[b]
	NEISS-AIP[c]	924,702
Seen at doctor's office/clinic or outpatient facility without subsequent emergency department contact	NHIS	520,782
	NAMCS/NHAMCS	922,877[b]
Seen in emergency department without subsequent hospitalization	NHIS	557,914
	NHAMCS	1,112,320[b]
	NEISS-AIP[c]	749,245
	TESS[a]	244,513
Hospitalized	NHIS	92,274[d]
	NHAMCS	265,714
	NHDS	282,012[b]
	TESS[a]	147,891
	NEISS-AIP[c]	175,457
Died	NHDS	3,770
	TESS[a]	1,074
	NCHS	24,173[e]

[a]Toxic Exposure Surveillance System: 2 percent of TESS exposures are associated with food poisoning. TESS hospitalizations include psychiatric admissions.
[b]Visit estimate discounted to account for possible multiple visits per episode, as follows: NAMCS—50 percent, NHAMCS outpatient—14 percent, NHAMCS ED—5 percent, NHDS—3 percent.
[c]National Electronic Injury Surveillance System—All Injury Program. This source does not include envenomation in its poisoning category; 20 percent of "other bite/sting" episodes are therefore included in this table.
[d]Estimate has low statistical reliability (relative standard error > 30 percent or sample N < 30).
[e]Excludes 6,627 cases coded with alcohol-related behavioral disorder as the underlying cause of death.
SOURCE: Cisternas (2003).

naire assessment of poisoning by subject self-report. This definition would reduce the NAMCS/NHAMCS estimated incidence to 746,000 rather than 2.3 million ambulatory visits for poisoning annually and may explain, in part, the lower rate of poisoning by self-report generated by the NHIS. To the extent that the public may have a different definition of poisoning than clinicians, respondents might differentially report such events when queried in standard items used in the current NHIS. It would not explain, however, the NHIS underreporting relative to TESS data, which could be attributable to other factors such as the NHIS respondents who may not mention attempted suicide or drugs of abuse misadventures when answering an injury/poisoning screener question (i.e., due to perceived stigma or even fear of legal exposure). Recall effects, in which events leading to medical care may be more likely to be reported relative to an event leading to a telephone call to a poison control center, may also result in varying proportional underestimation.

There is an even wider gap between the TESS experience of nearly 500,000 poisoning cases per annum that are treated by providers and the NAMCS/NHAMCS estimate of approximately 2.3 million episodes (ratio = 0.22 based on the data in Table 3-11). This ratio, however, is well within the range of that observed in selected studies that have attempted to determine the proportion of ED cases of poisoning or drug overdose that are reported to poison control centers (see Chapter 7). Nonetheless, summing the number of visits from NAMCS and NHAMCS could potentially overestimate the incidence of treated poisoning due to having more than one visit per episode within a source. For example, some ED-treated cases are referred to outpatient follow-up. To the extent that these were not coded as "follow-up," but rather as new visits, non-ED incidence may have been overestimated. If the 39 percent ED-referral rate resulted in 10 percent follow-up recorded as a new outpatient visit, the 922,877 outpatient estimate should be discounted to 700,413, and the total combined ED/outpatient incidence to 2,605,307. There is also the risk of overestimation of episodes due to individuals being seen at more than one of the ambulatory settings; however, the NAMCS, NHAMCS, and NHIS suggest that this is not widespread. Based on the disposition and episode of care information available from the most recent years of NAMCS and NHAMCS, discounting the number of visits by 50, 14, and 3 percent for NAMCS, NHAMCS outpatient, and NHAMCS emergency department subsets, respectively, provides a reasonable estimate of poisoning episodes treated in ambulatory settings. The incidence figures presented in Table 3-11 take into account these discounted rates.

If the TESS "telephone consultation only" figure and the NAMCS/NHAMCS health care provider-treated estimates are combined, an alternative total annual U.S. poisoning estimate of approximately 4 million cases is obtained. This estimate of annual incidence of poisoning is nearly twice as high as that estimated by the NHIS data and more than 60 percent higher than that based on TESS data alone.

Patients who receive inpatient care for poisoning are almost always admitted to a hospital through the emergency department, physician/clinic referral, or via transfer from another institution. Thus, NHDS data should not add to the overall estimate of poisoning episodes. This source yields an estimate of 282,012 annual episodes, which is consistent with the discharge status of hospital admission for 265,714 visits provided by the disposition information from the NHAMCS outpatient data and emergency department files combined, and with the assumption already discussed. This estimate excludes any contribution to the hospitalization total from NAMCS, for which hospitalization was noted for two sample observations only (1.9 percent, unweighted). Even if an additional 30,000 hospitalizations were added from this source, the combined NAMCS/NHAMCS estimate would remain similar (and would be even closer) to the NHDS figure.

Estimates of fatal poisonings range from 1,074 for TESS data to 24,173 for the NCHS analysis of death certificate data for 2001 (climbing to 30,800 when alcohol behavioral disorder coded deaths are included). Death certificate data are generally considered the most reliable source for such data as they also include out-of-hospital deaths (see Chapter 7 for a detailed discussion of the strengths and limitations of death certificate data). It is noteworthy that only one in four in-hospital deaths (based on NHDS) appear to be reported through TESS, compared with a 1:5 ratio of TESS to NAMCS/NHAMCS for poisoning cases receiving direct health care. This suggests that case severity alone does not drive poison control center case consultation as reflected in TESS reporting (also discussed in Chapter 7).

It is important to acknowledge that varying approaches to case definition and coding inclusion may impact the estimates cited above. For example, the inclusion of envenomations of various kinds may have led to inflated survey-based estimates, particularly for nonhospitalized poisoning events. The category of bites/sting is also included in TESS estimates, accounting for 85,713 cases (3.8 percent of the total) in that system in 2001. TESS totals also include adverse drug reactions (35,634; 1.6 percent) and "food poisoning" (41,319; 1.8 percent), categories that were excluded from the other analyses. The inclusion of 6,627 alcohol behavioral abuse coded deaths in the NCHS analysis should also be viewed in the context of TESS reporting, which in the same year reported only 15 ethanol deaths, only 5 of which were not combined with another co-ingestion.

It is also important to acknowledge that these estimates are based on selected major national surveys and databases. We did not attempt to derive estimates from a wider range of possible surveillance data sources, an undertaking that would have been beyond the scope of this chapter. Another potential limitation of this analysis is that it does not include incidence data that might be inferred from events coded solely as abnormal laboratory findings but not coded as overt illness. To a limited extent, such events could be estimated from some databases, for example, an event coded as an elevated medication level, but not coded on the basis of a concomitant symptom complex leading to a diagnosis of drug overdose. Although such events are generally not considered poisonings per se, tracking such data can be useful from a public health perspective. See Chapter 7 for a detailed description of multiple surveillance resources relevant to various types of poisoning and drug overdose events.

In summary, these analyses suggest that a conservative estimate of the annual incidence of poisoning episodes in the United States is 4 million cases per annum. One in four cases do not appear to lead to any direct ambulatory or inpatient treatments. Approximately 300,000 cases may be hospitalized, 7.5 percent of all events and approximately 13 percent of all those seen by a health care provider at any site. An estimate of fatal poisonings is at least 24,000, which represents 0.8 percent of all poisoning incidents; including ethanol-coded deaths increases this proportion to approximately 1 percent.

These estimates also suggest that the United States has a longer way to go in reaching its 2010 objectives than had been originally anticipated. Our estimate of 8.5 fatal poisonings per 100,000 population is far above the national 2010 objective of 1.5, and even higher than the 1997 estimate of 6.8 used as a baseline. This discrepancy may reflect differences in definitions used. Furthermore, our estimate of nonfatal poisonings associated with emergency department visits (identified through NHAMCS) of 530 per 100,000 population in 2001 is nearly twice the national 2010 objective of 292 per 100,000, and again even higher than the 1997 baseline estimate of 349 nonfatal poisonings per 100,000.

Appendix 3-A

Additional Detail on Survey Sources and Frequency Estimations

NATIONAL HEALTH INTERVIEW SURVEY

This survey has several core data files that include questions asked in every year. For this study, the core files of interest are the person file, which contains demographic information for every individual included in the survey, and the injury/poisoning file, which includes information on injury and poisoning episodes that can be merged back to the person-level information. The injury/poisoning episode section of the questionnaire asks one family member to respond on behalf of the family.

The wording of the question has changed since this section was first administered in 1997. The wording of the screener question was identical in 2000 and 2001 and asked whether anyone in the family was injured or poisoned seriously enough to get medical treatment or advice in the previous 3 months. Weighted estimates must be multiplied by four to obtain annual figures. In addition, specific questions about how the poisoning occurred and the type of poisoning episode (i.e., ICD-9-CM E-codes and diagnosis codes) were only ascertained and coded starting in 2000. For this reason, we only use the NHIS for poisoning data from 2000–2001.

The key NHIS questionnaire items used for this analysis queried survey participants regarding any injury or poisoning to themselves or a household family member over the previous 3 months (specifically "injured or poisoned seriously enough that [you/they] got medical advice or treatment?"). A follow-up item ascertained the nature of the treatment with the following close-ended, mutually exclusive selections: (1) did not receive medical treatment or advice; (2) phone call to doctor or health care professional; (3) phone call to poison control center; (4) visit to doctor's office; (5) visit to clinic or outpatient department; (6) visit to emergency department; (7) visit to hospital (stayed at least one night); (8) refused; and (9) don't know. In the NHIS, the kind of injury or poisoning was coded by ICD-9 diagnosis and external cause (E) codes.

The strength of this source as it pertains to understanding the epidemiology of poisoning is that it is population based. In addition, the NHIS can be used to create the denominator population for a consistent application to the other data sources analyzed here. Other than TESS, it is the only source for an estimate of poisoning events for which direct medical care was not received (telephone consultation only). Its main weakness is that it is respondent based (i.e., poisoning episodes are not confirmed by medical records) and thus subject to recall and nonresponse biases. In addition, poisoning events are relatively uncommon in the dataset; because only 2 years of data are used, some of the estimates are not as robust as desired.

NATIONAL AMBULATORY MEDICAL CARE SURVEY

This source is a national probability sample survey of visits made in the United States to the offices of nonfederally employed physicians classified as working in settings that are "office-based care." Visits to private, non-hospital-based clinics and health maintenance organizations are included, but those that occur in federally operated clinics are not. Visits from sampled physicians are sampled systematically for abstraction to a form subsequently completed by the physician or the physician's staff. Sample data are weighted to produce annual national estimates for the noninstitutionalized, civilian population. Data include only actual visits for patient care; telephone calls or visits to pay bills are excluded.

NAMCS includes three groups of data items that can be used to ascertain office-based visits to physicians for poisoning. The first is the "reason for visit" variables, which are up to three of the patient's complains, symptoms, or reasons for visit in the patient's own words, listed in order from most to least important. These verbatim responses are then coded later by NCHS staff using a coding scheme that differs from the ICD-9. The second group of items consists of up to three causes of injury, poisoning, or adverse event that resulted in the visit; these are subsequently provided ICD-9 E-codes by NCHS staff. The third group consists of up to three diagnoses, representing the physician's best judgment at the time, and they are coded to ICD-9-CM codes by NCHS staff. In 2001, a question was added concerning whether this was the first or subsequent visit for a particular problem.

A strength of this system as it pertains to poisoning epidemiology is that it is based on medical records. In addition, the "reason for visit" code provides an additional source of case capture beyond ICD-9-CM coding. Because it is not linked to emergency department or hospital care, it is likely that case selection effects (e.g., a poisoning or toxic exposure event not perceived by the patient as threatening enough to bypass a physician's office and go directly to an emergency department) are prominent in the mix of poisonings captured by this survey, which may present a limitation. NAMCS has not been featured in previously published estimates of poisoning incidence.

NATIONAL HOSPITAL AMBULATORY MEDICAL CARE SURVEY

This is the hospital ambulatory care complement to NAMCS. Both the outpatient and emergency files contain the "reason-for-visit," ICD-9 E-code, and ICD-9-CM diagnosis fields contained in the NAMCS data. Thus, NHAMCS has all of the strengths of NAMCS, but captures visits in the hospital outpatient and emergency department settings. Because of the disposition information contained in the dataset, it is also possible to estimate hospital admissions preceded by emergency department care, which can be presumed to be a major route of poisoning admissions to the hospital. The NHAMCS outpatient (non-ED) component may be particularly relevant to low-income or elderly patients who may be more likely to receive care in hospital-based clinics rather than in private practice settings. It is possible to combine these three sources to obtain estimates for all ambulatory visits made in the United States (outpatient NAMCS, outpatient clinic NHAMCS, and emergency department visits from NHAMCS), another strength of these surveys.

NATIONAL HOSPITAL DISCHARGE SURVEY

This source covers discharges from a sample of short-stay hospital visits that are noninstitutional and nonfederal. Up to seven ICD-9 diagnosis codes (including ICD-9 CM and E-codes) can be provided. Because this system includes disposition, it provides an independent source for estimating poisoning-related fatalities for patients who died in a hospital. Because the system incorporated source of admission in 2001, it also provides an estimate of hospital-to-hospital transfers of poisoning cases, a factor that must be taken into account as a "discounting" measure in incidence estimates to prevent double counting of cases. The potential for underutilization of E-codes and for the impact of ranking of poisoning in a severely ill patient with multiple-organ failure (where the underlying poisoning event may be obscured if it falls down the rank order list of diagnostic codes included) may lead to underestimation in this system.

METHODOLOGY

Coding Poisoning and Drug Overdose

The dominant coding scheme used is the *International Classification of Diseases* (World Health Organization, 1992–1994). Although it is now in its 10th revision (ICD-10), in the United States this latest revision is currently applied only to mortality data. The ninth revision is generally applied to other morbidity and survey data (World Health Organization,

1989). Certain datasets are not coded by either ICD-9 or ICD-10 criteria, such as the TESS poison control center data system (see Chapter 7). Within the ICD coding scheme, provision has been made to differentiate among different types of poisoning and drug overdose events and among different categories of intent. There is no defined grouping of ICD codes that establishes a single category subsuming all poisoning events.

Some of the problems described may be magnified in the ICD-10, in which injury coding (including for poisoning) has changed substantially. Specifically, in ICD-10 it may not be possible for "intent" codes to separate out ethanol-related toxicity from toxic syndromes caused by alcohol substitutes such as methanol. Added instructions for coding deaths further impact definitions by requiring the principal cause of death to be categorized as due to a selected group of mental and behavioral disorders if such a disorder appears among contributing causes in a poisoning death. Thus an acute acetaminophen fatality in a chronic ethanol abuser (if this was listed as a contributing cause) would be coded in ICD-10 as a primary alcohol-related death (World Health Organization, 1992–1994). CDC recently added a series of special "U" codes (allowed for in the ICD-10 scheme) to capture terrorism-related fatalities. Some of these new codes also could be relevant to poisoning, such as U01.7 for terrorism involving chemical weapons (http://www.cdc.gov/nchs).

Data Management

As a precursor to pooling data for multiple years for each source, the proportion of poisoning observations for each year was examined to ensure that poisoning estimates were relatively stable during the period. Each source was analyzed separately, but demographic variable recoding was done on each to create consistent categorical variables for gender, age, race, and region. Estimates associated with relative standard errors (ratio of the standard error to its estimate) >.3 or based on a small sample size were retained, but are noted. Sampling weights included in the data files were rescaled to an annual timeframe and used to create population estimates. SUDAAN, the standard computerized statistical package that calculates estimated rates taking into account the sampling weights built into the design of each survey, was used whenever possible to adjust for the multistage sampling design of the surveys.

4

Historical Context of Poison Control

At the beginning of the 20th century, there was no modern specialty of toxicology, no poison control centers, no oversight of pharmaceutical manufacturing or drug labeling, and little knowledge regarding the treatment of poisonings in the United States. Household and occupational toxic hazards were poorly understood. As public health concepts were in their formative stages, surveillance of toxic exposures and the morbidity and mortality associated with these exposures were virtually unknown. Similarly, emergency response personnel and systems of emergency care that could respond to poisonings in the home or workplace were virtually nonexistent.

This chapter summarizes some of the key historical determinants of poison control and management as a modern health care service. Through a review of these developments, we can better understand the origin of our current "system" of poison control and management. Understanding this history can enhance efforts to advance this vital health care service by identifying potential barriers to the evolution of an optimal poison control and management system.

BRIEF OVERVIEW

Issues of misbranding, mislabeling, and adulteration of food and drugs concerned those who were involved in public health as well as health care providers and led to the founding of the United States Pharmacopeia (1820), the American Medical Association (1847), the American

Pharmacists Association (1852), and other organizations. Concerns about food and drug safety, beginning in the 19th century, provided the impetus for the first human clinical food additive trials to demonstrate safety and efficacy in December 1902 (Hurt, 1985). These efforts led to the Pure Food and Drug Act of 1906 (Lewis, 2002). This legislation ("Wiley Act") created the Food, Drug, and Insecticide Administration, which in 1930 became the Food and Drug Administration (FDA). The Act required approval for foods and drugs meant for human consumption. Subsequently, the federal government passed a number of laws, created regulations, and proposed other controls and management of poisoning. These efforts are summarized in Box 4-1.

In the 1930s, childhood poisoning was recognized as a significant component of pediatric practice and patient morbidity. Unfortunately, little information existed regarding the toxicity of household products and management recommendations. Jay Arena, M.D., a pediatrician at Duke University, began to systematically collect information regarding toxic hazards in the early 1930s and provided advice to physicians on poisoning cases in the surrounding area. He provided one of the first reports on the hazards of household products to children (Martin and Arena, 1939). Louis Gdalman, R.Ph., a pharmacist in Chicago, collected information during World War II. He developed a toxicological information system using index cards and eventually converted to microfiche. This system eventually covered more than 9,000 commercial and consumer products. Moreover, Gdalman established the precursor to the modern poison control center by personally taking telephone calls 24 hours per day (Botticelli and Pierpaoli, 1992; Burda and Burda, 1997).

Although recognized as a growing problem during this period, the magnitude of childhood poisoning was not appreciated until a 1949–1950 epidemiological study focusing on children under 5 years of age reported a significant number of poisoning deaths (Bain, 1954). In 1950, the American Academy of Pediatrics (AAP), which was founded in 1930, established its Accident Prevention Committee, chaired by George M. Wheatley, M.D. That committee surveyed the 3,000 members of AAP and found that 49 percent of reported "accidents" treated by AAP members involved poisoning (Wheatley, 1953).

In 1953, Edward Press, M.D., and Gdalman developed the first formal poison control center in Chicago. Their center provided professional telephone advice and included a standard data collection form (Botticelli and Pierpaoli, 1992; Burda and Burda, 1997). These centers rapidly developed, with as many as 265 by 1958 and 661 by 1978 (Scherz and Robertson, 1978).

Provision of timely information to physicians regarding drugs and the toxicity of other agents was the driving force for poison control center

BOX 4-1
Key Legislative and Regulatory Activities Related to Poison Prevention and Management

1906 *Pure Food and Drug Act ("Wiley Act")* Created the pre-FDA (Food, Drug, and Insecticide Administration). Required federal approval for sale of all foods and drugs meant for human consumption. Most patent medicines, after testing, were no longer approved for human consumption.

1927 *The Caustic Poison Act* Required labels to warn parents and protect children from lye (e.g., soap making) and 10 other caustic chemicals.

1930 Food, Drug, and Insecticide Administration becomes the Food and Drug Administration (FDA).

1938 *Food, Drug, and Cosmetic Act* Requires demonstration of new drug safety of non-narcotic prescription drugs.

1951 FDA defines and restricts further drugs to prescription.

1957 *National Clearinghouse for Poison Control Centers (NCHPCC) established in FDA* Mandate for poison control centers to collect data and provide to FDA. Used 1/3-page carbon-copy data forms. Provided 5" by 8" drug information cards. Published a bulletin on drug overdose management. Published *The Clinical Toxicology of Commercial Products.*

1961 *Poison Prevention Week* established by Pub. L. No. 87–319 (75 Stat. 681). Third week in March designated as National Poison Prevention Week.

1966 *Child Protection Act* Bans toys and other articles so hazardous that warning labels cannot be written.

1966 "Baby" aspirin packaging regulation limited to 36 tablets of 81 mg to protect children.

1970 FDA requires first patient package insert (oral contraceptives).

1970 *Poison Prevention Packaging Act (FDA responsibility until 1973)* Establishes Consumer Product Safety Commission (CPSC).

1972 FDA releases report on over-the-counter drugs. Reviews safety, effectiveness, and appropriate labeling.

1973 *Emergency Medical Services Systems Act of 1973 (Pub. L. No. 93–154), Department of Health and Human Services* Develops regional trauma centers, burn centers, and others, and establishes poison control centers as one of seven priorities. Funds some new poison control centers for 3 years.

1982 CPSC provides tamper-resistant packaging regulations.

1987 NCHPCC program terminated.

1995 FDA adopted "hierarchic" imprint coding of medications. Imprinting used alphanumeric codes and logos Imprinting yields. ~43 percent accuracy rate of unknown tablet and capsule identification.

1997 FDA Modernization Act.

1997 FDA passes iron medication regulations. Requires unit-dose packaging for products with 30 mg or more per dosage unit.

2000 *Poison Control Center Enhancement and Awareness Act 106–174* (February 25, 2000) intended to stabilize poison control center funding.

2002 Homeland Security Act of 2002 (Sec. 505).

operations. These early leaders also recognized that effective telephone triage could avert unneeded medical visits or lead to early treatment at home. Hence, centers soon began to provide advice directly to laypersons and nonphysician care providers. This feature distinguishes poison control centers in the United States from similar centers in other countries, where the task of giving advice remains largely restricted to physicians.

In 1957, the Surgeon General established the National Clearinghouse for Poison Control Centers (NCHPCC) within the FDA. At the time, the FDA and the U.S. Department of Agriculture represented the only federal agencies related to consumers with jurisdiction over drugs and chemicals. Product ingredient information was provided and poison exposures were tracked through NCHPCC. Funding was also provided to develop the text *Clinical Toxicology of Commercial Products*, authored by Robert Gosselin, M.D., Harold Hodge, M.D., and Marion Gleason at the University of Rochester.

At the 1958 AAP annual meeting, the American Association of Poison Control Centers (AAPCC) was founded (Mofenson, 1975) (see Box 4-2 for the AAPCC statement of objectives). AAPCC continues to serve as the voluntary association for poison control centers. As the lead professional organization regarding poison control and management, AAPCC—along with other toxicology groups—continues to host medical toxicology scientific presentations and continuing education sessions at its annual meeting in combination with several other societies. In 1968, both the American Academy of Clinical Toxicology and the American College of Emergency Physicians were founded. One impact of both organizations was to take the focus of poisoning beyond pediatric exposures. The American Board of Medical Toxicology gave its first examination for physician toxicologists in 1974 and fellowship training programs were instituted at about the same time. Emergency medicine was recognized as a specialty in the

BOX 4-2
Statement of Objectives of the American Association of Poison Control Centers

To provide a forum for poison centers and interested individuals to promote the reduction of morbidity and mortality from poisonings through public and professional education and scientific research.

To set voluntary standards for poison center operations.

SOURCE: http://www.aapcc.org//aapcc.htm.

United States in 1979 when it received conjoint status from the American Board of Medical Specialists (ABMS) and became a primary specialty 10 years later. Medical toxicology began administering examinations in 1974 and was recognized by the ABMS as a certificate of added qualification in 1994. Although AAP has remained active in the area of poisonings with its Section on Injury and Poison Prevention, founded in 1990, the organization no longer sponsors AAPCC meetings.

During the late 1970s, systems of emergency care were developed following passage of the Emergency Medical Services (EMS) Systems Act of 1973 (Pub. L. No. 93–154). The application of technology and centralized public service (communication) access points produced the opportunity for integration of poison control centers within EMS systems. Concurrent steps to enhance home safety (e.g., product labeling, smaller quantities of over-the-counter medications per package, prescription drug safety caps, childproof cabinet locks) coincided with a shift in awareness of mortality risk to include adult poisoning as a major emphasis of care. Given a growing emphasis on adult poisoning management and poisoning response with EMS services, an increasing number of leaders in medical toxicology, including the poison control center medical directors, began to come from a background in emergency medicine followed by a fellowship in clinical toxicology (see Chapter 5). Professional activities by medical toxicologists, pharmacists, and nurses also have grown dramatically in both the management and operation of poison control centers. Centers no longer use clerical personnel or sanitarians to manage exposures.

Certain aspects of poisoning prevention in the past 30 years have been independent of poison control center clinical functions. For example, the introduction by the FDA of imprint codes on tablets and capsules provided a much improved method of tablet and capsule identification. Although the system has major drawbacks (e.g., use of logos that are difficult to categorize, describe, and list), there has been faster determination of potential medication exposures (Marder et al., 2001; Symonds and Robertson, 1967).

Similarly, the use of safety caps on medications and chemical compounds has reduced the number of exposures in children and may have been the single most important reason for reduction in morbidity and mortality since the early 1970s (Anonymous, 1982; Arena, 1959; Palmisano, 1981; Rodgers, 1996; Walton, 1982). Although there have been various public campaigns to use other child safety devices, the only data that exist are related to safety caps. Child-resistant packaging was required for various prescription drugs beginning in 1974. Looking at the period from 1974 to 1992 and comparing it with a previous period, estimates show a reduction of 460 child deaths and a mortality rate reduction of 45 percent as a result of this packaging (Rodgers, 1996). Figure 4-1 from that study

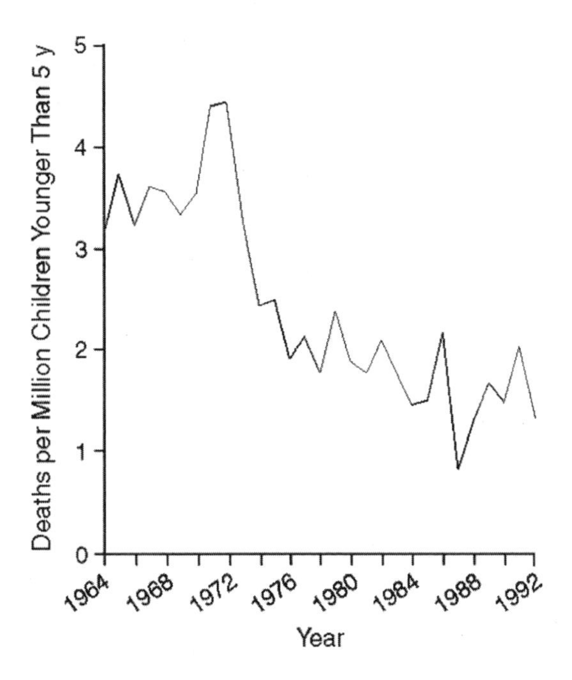

FIGURE 4-1 Child mortality rates due to the unintentional ingestion of oral prescription drugs, 1964–1992.
SOURCE: Rodgers (1996).

shows the drop in child mortality rates over the time period; Figure 4-2 shows the difference between the predicted rates with and without child-resistant packaging.

Events both before and after September 11, 2001, have heightened national concerns regarding homeland security and the threat of radiologic, biological, and chemical weapon exposure. The Health Resources and Services Administration (HRSA) and the Centers for Disease Control and Prevention (CDC) have recognized the importance of poison control centers as a component of an all-hazards emergency planning and response system that is integrated with state health departments and supports regional and hospital-based emergency service efforts. As discussed in Chapters 5 and 9, the incorporation of poison control centers in this manner has been variable and remains underdeveloped. This new potential role—combined with substantive changes in funding and federal oversight—clearly marks a break with the past and the beginning of a new period for poison control centers.

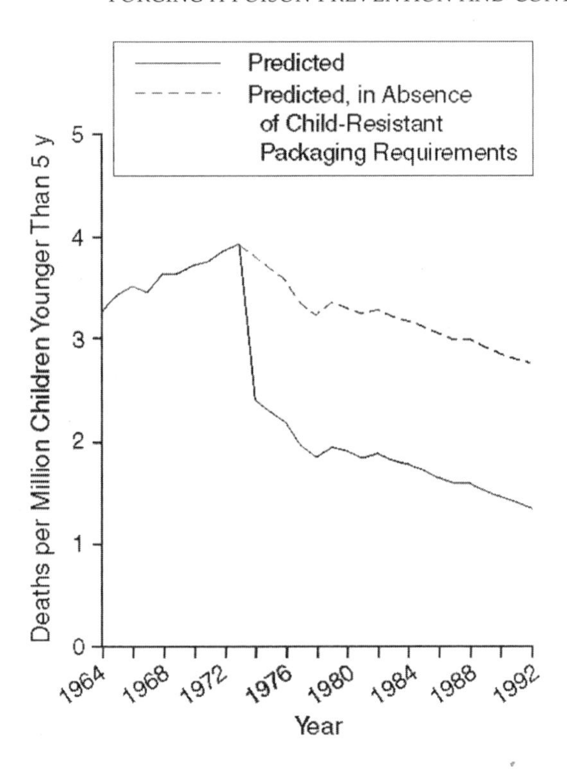

FIGURE 4-2 Fitted model for predicted child mortality rates due to unintentional ingestion of oral prescription drugs, 1964–1992.
SOURCE: Rodgers (1996).

POISON CONTROL CENTERS

Origins

Although the preceding overview provides a brief context for poison control centers, the detailed history of their development provides further insight into their current status and function.

Edward Press, with the support of the AAP Illinois Chapter, the Illinois Department of Public Health, seven hospitals, five Chicago medical schools, the American Medical Association, the FDA, and others, formed a committee on April 1, 1953, to begin development of the first poison control center (Botticelli and Pierpaoli, 1992; Burda and Burda, 1997). By 1954, 11 centers had been established in the city of Chicago alone, with the objective of providing information to physicians for treatment of children exposed to toxic agents. During this time, visiting nurses from the board of health also visited the homes of poisoning victims in Chicago.

In 1961, the advisory committee of the poison control center in Chicago consolidated the 11 poison control centers into one information center at Presbyterian–St. Luke's Hospital. In 1962, the Master Poison Control Center was established with Joseph R. Christian, M.D., as medical director, and Chicago pharmacist Gdalman as director responsible for operations. The advisory committee also resolved its concerns about professional liability exposure and agreed to allow direct calls to the poison control center from the public.

Embraced as a lifesaving idea by the pediatric community, the number of centers rapidly increased nationally from 1953 to 1958, when 265 poison control centers were reported to exist. This expansion occurred with no consistent funding or formal organizational structure (Arena, 1983). It was not until the 1970s that emergency medicine became a potent force creating professional demand for improved poison information. The result was the extension of standardized poison and drug information and consistent access to toxicologists. By that time, more than 660 poison control centers had developed (Scherz and Robertson, 1978). Recognizing the changing epidemiological trends, poison control centers began in the 1970s to expand their efforts beyond a primarily pediatric focus to serving the full population. However, the prevention education in centers has continued to emphasize pediatric poisoning prevention.

Evolution of Structure and Function

As noted previously, poison control centers were established to provide drug and chemical toxicity information and patient management guidance to physicians. These services were expanded to handle telephone calls from laypersons in the 1960s. Initially, most centers consisted simply of a telephone and a designated individual to answer that telephone. The individual responding to the calls at times was a clerical person, pediatric house officer (physician in training, pharmacist), or other interested (or designated) person. Neither training nor educational materials were standardized.

In 1957, the first efforts to standardize poison information were undertaken within the FDA by NCHPCC. These included (1) index cards containing information on drugs, chemicals, household products, and plants, and (2) a monthly newsletter summarizing the poisoning literature. NCHPCC also funded the publication of a book, *The Clinical Toxicology of Commercial Products*, with the first edition in 1957 (Food and Drug Administration).

In the 1970s, poison control centers began to offer clinical toxicology fellowship programs for physicians and other scientists. A medical subspecialty certificate of added qualification in medical toxicology became possible in 1994 for diplomates of the American Board of Emergency Medicine, the American Board of Pediatrics, or the American Board of Preventive Medicine through the American Board of Medical Specialties. Currently, candidates for this subspecialty certification must complete a residency in one of the sponsoring boards or other boards by petition and a medical toxicology fellowship affiliated with a poison control center.

Professional activities by pharmacists and nurses have grown dramatically in both the management and operation of poison control centers. Centers no longer use clerical personnel to manage exposures. Beginning in the early 1980s, AAPCC developed and promoted criteria enabling nonphysicians to become specialists in poison information. As an adjunct to their extensively trained personnel, nonphysician, pharmacist, and nursing personnel could serve as a supervised poison information provider.

Designations of poison control centers were made by each state health department, peaking in 1978 with 649 sites in the 50 states and 12 more in U.S. territories and the Virgin Islands (Scherz and Robertson, 1978) (Table 4-1). However, there were few large poison control centers and the number receiving more than 1,000 calls per year never exceeded 80. In 1970, less than 6 percent of poison control centers received more than 9 to 10 calls per day, or 3,285 to 3,650 per year (Manoguerra, 1976).

In the absence of a federal certifying body or federal poison control center regulations for staffing and operations, AAPCC developed certification systems for both the centers and their personnel (Lovejoy et al., 1994). As a result of an increasing expectation for center certification, tenuous financial support, and economically driven service cutbacks by hospitals and teaching institutions, the number of centers dropped rapidly over three decades.

In 1983, the number of centers had dropped to 395 and in 1994 to 87. By 2002, there were 64 centers reporting to the AAPCC Toxic Exposure Surveillance System (TESS) data collection system, with coverage of 99.8 percent of the U.S. population, a dramatic increase from the 52 percent of the population covered in 1993 (Lovejoy et al., 1994; Watson et al., 2003). Following introduction of the national toll-free number, 100 percent of the U.S. population is currently covered by 64 centers. The surviving poison control centers developed from the consolidation and expansion of the early centers (Rumack et al., 1978). Such centralization of information and treatment was shown early to reduce poisoning mortality; for example, in one hospital it declined from 8 to 4 percent (Teitelbaum, 1968). Reorganization of poison control centers largely has been driven by local economic mandates rather than by public health initiatives. Currently, several cen-

TABLE 4-1 Poison Information Centers: States, Territories, and Virgin Islands, 1978

Alabama	8	New Jersey	34
Alaska	5	New Mexico	6
Arizona	17	New York	21
Arkansas	8	North Carolina	8
California	9	North Dakota	7
Colorado	9	Ohio	13
Connecticut	10	Oklahoma	8
Delaware	1	Oregon	1
Florida	32	Pennsylvania	74
Georgia	11	Rhode Island	4
Hawaii	1	South Carolina	2
Idaho	3	South Dakota	2
Illinois	102	Tennessee	8
Indiana	33	Texas	21
Iowa	4	Utah	1
Kansas	14	Vermont	0
Kentucky	10	Virginia	19
Louisiana	5	Washington	11
Maine	1	West Virginia	16
Maryland	6	Wisconsin	5
Massachusetts	6	Wyoming	2
Michigan	28	Canal Zone	1
Minnesota	27	District of Columbia	1
Mississippi	13	Guam	1
Missouri	15	Puerto Rico	5
Montana	3	Virgin Islands	4
Nebraska	2	Total 50 States and the	
Nevada	2	District of Columbia = 650	
New Hampshire	1	Total = 661	

SOURCE: Adapted from Scherz and Robertson (1978).

ters supply parts or all of the service needs of other states. Some centers serve areas at a great distance (e.g., the Oregon Poison Center serves Alaska and the Rocky Mountain Poison and Drug Center in Colorado serves Hawaii). States that do not have onsite centers provide services such as education and outreach through community organizations and state public health agencies.

Although concern has been expressed that states without poison control centers would be less well served than states with centers, there is no evidence to show decreased call rates in states without centers physically present, such as Alaska, Hawaii, Wyoming, Montana, North Dakota, and Idaho.

Evolution of Data Collection and Poisoning Surveillance

Collection of data by early poison control centers was fragmented and generally nonstandardized. In 1957, NCHPCC collected limited data from centers and published a yearly statistical report for the aggregate of reporting centers approximately 24 months after the end of the data collection year.

AAPCC developed the Toxic Exposure Surveillance System, a data collection system, in 1983. TESS is the data source for AAPCC annual reports (Watson et al., 2003). TESS was developed in part to supply marketable, comprehensive data for pharmaceutical companies and federal agencies (e.g., Consumer Product Safety Commission, or CPSC). TESS also provides poison control centers with a standardized poisoning exposure record. This system initially used "mark sense" forms (to be described later in this chapter), but has been advanced to digital format. Annual TESS summary data for reporting poison control centers are published annually in the *American Journal of Emergency Medicine* (Watson et al., 2003).

During the past several years, changes to TESS have been funded by CDC and its Agency for Toxic Substances and Disease Registry. Funding in excess of $6 million has largely supported enhancement of the proprietary software underlying TESS and has not subsidized data collection by poison control centers.

TESS provides useful exposure data, but captures only a fraction of the most seriously poisoned patients (Blanc et al., 1995; Hoppe-Roberts et al., 2000). From a historical perspective, a key shortcoming of TESS is that it did not develop as a public access database comparable to governmental sources of other vital statistics.

The TESS data collection program is proprietary to AAPCC and thus is not managed by any public health or government agency. Consistent with this fact, the underlying software for both data collection and analysis was developed and remains owned by a private company with ties to AAPCC. As new data fields have been added to TESS over time, individual poison control centers must provide the additional time and personnel required to acquire the data without full compensating revenue. However, approximately $250,000 from the sale of data is returned to individual centers annually based on the number of cases submitted as partial compensation. This compensation from the AAPCC central office ranges from $3,000 to $7,000 per center each year. Individual centers reporting to the Committee estimated the net cost of providing such data (i.e., beyond that compensated) ranges from $50,000 to $100,000 per year. Federal agencies such as CPSC and the FDA, as well as most state agencies, also purchase TESS data reports.

Technology

Poison Information

Drug information and poisoning management cards used by the early poison control centers were replaced by microfiche as the required database grew. The first major commercial product for this purpose was the Poisindex® (founded in 1973 by Barry H. Rumack). Although there were several other databases such as ToxiFile from Illinois and a compilation of NCHPCC cards from Detroit, they did not publish for more than a few years. Poisindex was published on microfiche in August 1974 and contained a compilation of consumer and commercial products coded to treatment algorithms (Rumack, 1975). These "managements" were written by an editorial board and covered care of exposed patients at home or in a health care facility. Poisindex was provided electronically for mainframe computers beginning in 1981. In 1985, Poisindex was published on CD-ROM for the first time and coupled with a personal computer. Although CD-ROM continues to be its major method of distribution, the software is also available through Internet subscription and over private intranets. Poisindex is used by all U.S. poison control centers and the majority of centers around the world. Validity of the Poisindex database has been independently verified (Wan et al., 1993). Data contained within Poisindex is provided voluntarily by consumer product, industrial, and other manufacturers and repackagers. A company called Micromedex employs an internal staff to obtain and code products, as well as to prepare the management documents for review by an outside editorial board.

Data Acquisition

Although NCHPCC summarized data from many poison control centers to provide estimates of poisoning in the United States, the process suffered from limited standardization of data collection and definition and from its voluntary submission nature. AAPCC began centralized collection of poison exposures with TESS in 1983 using "mark sense" paper forms that were scanned and converted to digital data. Current data collection occurs using a computerized program that allows data capture during a poison call. The embedded product codes in Poisindex were used by AAPCC to enable connection of each case with an appropriate product or products. Currently, four computer-based data collection products interface with TESS and are used for data collection (see Chapter 8).

Drug Recognition

One useful computerized drug identification database is Drugdex. It interfaces with Poisindex and is published by Micromedex (http://www.micromedex.com) as part of a suite of databases used by many poison control centers. Electronic matching between the computerized case reports filled out by center specialists and the product and ingredient indexes of Poisindex is instantaneous. This drug identification software works seamlessly with the four electronic data collection programs currently used by centers for uploading data to AAPCC as part of TESS.

Communications

The enhancement of satellite, microwave, and cable communications systems has improved the ability of laypersons and professionals to contact poison control centers. This development of a sophisticated communications network has allowed specialists in one center or region to provide backup support for another center or region. A nationwide telephone number, 1-800-222-1222, was introduced in 2002 and allows triage of poison control center exposure or information calls to a center anywhere in the United States. Before this telephone number was introduced, poison control centers used a combination of local telephone numbers and state or regionwide toll-free numbers.

Funding of Poison Control Centers

Initially, funding for work performed by the poison control centers was borne by the host institution and by the pediatricians and pharmacists who were involved. There is no indication that any of the indirect federal funding through Medicare supplements to teaching hospitals was used in centers. The sponsoring institution sometimes provided consulting personnel from the laboratory and other areas of the hospital without direct charge. There was no federal funding beyond the money used within NCHPCC to generate epidemiological poisoning summaries, treatment cards, and the book, *The Clinical Toxicology of Commercial Products* (Food and Drug Administration, 1957). It has been long held that poison control centers save money for the health care system by avoiding the need for emergency department visits as well as permitting treatment at home (Food and Drug Administration, 1973; Harrison et al., 1995, 1996; Krenzelok, 1998; Morton, 1998; Olson et al., 1999; Woolf et al., 1997; Zuvekas et al., 1997).

Poison control centers have not been included as specific requirements in the key public health block grants to state health departments that were established in the early 1980s, so there has been no national strategic or tactical plan for their development or linkage with public health departments. This has resulted in center function variation over time, with each center evolving its own culture within its own institutional and community structure. The variability between the current poison control centers in terms of both geographic location and institutional support is largely a product of each center's funding history and functional role for its sponsoring institution. For example, those centers located in a children's hospital often have served as part of the institution's community outreach program, while those located in a school of pharmacy often have served primarily as part of the school's education and training environment. In both examples, the institutions have been willing to absorb some costs, although in one instance the poison control center provides community goodwill and raises the recognition of the institution, whereas in the other the center primarily enriches a training program.

One poison control center director wrote in a pharmacy journal in 1976: "The development of poison control centers over the past 20 years has been haphazard. Each state was allowed to organize its own system resulting in very little uniformity of services and quality control" (Manoguerra, 1976:p. 382). Historically and recently many mandates, such as an expectation to participate in emergency planning and response, have been without accompanying financial support. Until the early 1970s, most medical directors received salaries from other sources and not from the poison control center. Sources of center funding included hospitals (primarily children's hospitals for community outreach), states, counties, cities, health insurance firms, universities, schools of pharmacy, public health agencies, and contracts with pharmaceutical firms, chemical companies, and others (see Chapter 5). Poison control centers have rarely been considered central to a community's health care, and many have closed as institutions have faced financial difficulties. Hospital and other institutional administrators have not seen the centers as a revenue source, but instead as public relations, education, or training cost centers. Only a handful of the centers have developed affiliated clinics or inpatient services that create direct hospital revenue. Many of the centers that have closed did not meet the needs of their community. Regionalization resulted in consolidation and closure of some centers, producing a more focused use of resources. As Table 4-1 indicates, some states had a large number of centers, and their closure was more likely to have benefited the system than hurt it. There has not been a clear understanding about how many centers should exist. When the first centers were designated by various state health departments starting in 1958, there was no concept of regionalization comparable to what exists today.

FDA Report on Poison Control Management

A 1973 FDA report (Food and Drug Administration, 1973) evaluated a variety of central and regional models for providing poison control services (see Box 4-3). These proposals were intended to bring about economies of scope and scale. One proposal suggested that a single national poison control center "with a single national center utilizing inward Wide Area Telephone Service (inward WATS) to answer calls on a nationwide basis" (p. 77) be established to replace the existing collection of centers (at that time there were 597). It was estimated that "45 persons would be required for a presumed call volume of 320,000 using 13 phone lines, but if the call volume doubled, it would require 22 phone lines and 78 people. The cost of 13 lines was estimated at $296,000 per year and at an average cost of $15,000 it was estimated that 45 personnel would cost $675,000 per year." It was suggested that expert staff were needed to "back-up para-medicals answering phones" and that they would be "pharmacists or physicians" (p. 88). The report noted possible drawbacks to establishing a national center, including (1) persuading the public to call a toll-free number, (2) following up with patients at long distances, and (3) the required increases in federal funding.

Several other recommendations in the report (see Box 4-3) were instituted over time by poison control centers, although the FDA, which commissioned the report, ceased to be involved in center activities in 1987.

Other key findings contained in the 1973 FDA report characterized the poisoning rates of the time:

- The mortality rate from accidental poisoning combined across all ages was increasing 4.4 percent annually, while the mortality rate from accidental poisoning among children under age 5 was decreasing 4 percent annually. (Although exposures in children under the age of 5 are still considered "accidental" exposures, in older children and adults they are considered either "unintentional" or "intentional.")
- The average number of poisoning incidents reported to NCHPCC per poison control center reporting between 1965 and 1971 increased for all ages combined, as well as for each age group separately, except for children under 5 years of age. For the latter group, the number of incidents decreased.

In addition, other events were believed to have a favorable impact on the incidence of poisoning at that time:

- Enactment of and initial regulations created under the Poison Prevention Packaging Act (the Act was being administered by the FDA at that time prior to the formation of CPSC).

BOX 4-3
Conclusions and Recommendations from *Evaluation of the Poison Control System* (1973)

1. Major strengths of the poison control *system*
   a. The treatment of accidental poisonings is enhanced as more knowledge is disseminated
   b. The reporting system provides human experience information
   c. The total estimated cost (federal and nonfederal) for 1972 was $7,486,776
   d. The estimate of annual cost savings to the public was $4,107,050 for 1972
2. Major deficiencies of the poison control *system* and recommendations for improvement
   a. The uneven distribution of poison centers geographically and by population could be eliminated by a national poison center
   b. The services of the poison control system are poorly advertised to the public, but advertising could readily be improved
   c. Many poison centers do not submit case histories to the Clearinghouse [NCHPCC], but providing technical assistance to centers could be an inducement to increase center cooperation
   d. The poison control system is not putting enough emphasis on prevention
   e. Summary poison statistics may not accurately reflect national incidence, morbidity, and mortality, but this potential deficiency can be removed by increasing the number of poison centers submitting case histories and improving reporting procedures
   f. Summary statistics are not geared to examining the operations of individual poison centers and are of marginal use to the centers; these needs can be met by modifying the format of the annual statistics and increasing feedback of center-related data to poison centers
   g. The poison control system has no mechanism for performing ongoing evaluation of its activities and should implement an evaluation system using indicators matched to a program structure
3. Major strengths of *federal* poison control activities
   a. Clearinghouse cards
   b. Poison prevention activities
   c. Enforcing and regulating hazardous substances and poison prevention packaging
   d. Publication of the book *Clinical Toxicology of Commercial Products*
   e. Total federal expenditure in 1972 of $1,300,000
4. Major deficiencies of *federal* poison control center activities
   a. Acquisition of product information from manufacturers could be improved by developing and disseminating to manufacturers a recommended format with which to supply product information and a timetable for such submissions
   b. Insufficient advertising of services provided by poison centers could be improved by modifying Poison Prevention Week activities to include advertising of poison center services

continued

BOX 4-3 Continued

   c. Clearinghouse poisoning report forms and procedures are difficult for poison centers to use, but could be improved by redesigning the forms and routing sampling poison center opinion about clearinghouse services
   d. Clearinghouse procedures for disseminating product cards are irregular
   e. The federal poison control activities have no mechanism for performing ongoing evaluation and should implement an evaluation system using indicators matched to a program structure
5. Major strengths of poison *centers*
   a. 90 percent of poison centers are able to provide service on a 24-hour basis
   b. All but one of the centers provides treatment information over the telephone
   c. 94 percent of the centers accept phone calls from the public
   d. The operation of a single poison center is estimated to cost an average of $10,860 per year
6. Major deficiencies of poison *centers*
   a. The majority of poison centers are devoting no staff time for poison prevention activities and should increase their prevention efforts
   b. The majority of poison centers do not have their own special telephone number
   c. Only half of the poison centers are following up on cases and should make use of local government social services agencies to follow up cases
   d. Only half of the poison centers are keeping a record of phone calls

NOTE: The report addresses the federal system and then individual centers.
SOURCE: Food and Drug Administration (1973, pp. 107–131).

• Enactment and initial regulations created under the Hazardous Substances Labeling Act.
• Establishment of Poison Prevention Week.
• Legislation passed in 1967 limiting the number of aspirin tablets in baby aspirin bottles.
• Growth in the number of poison control centers and to a lesser extent the establishment of NCHPCC and publication of *The Clinical Toxicology of Commercial Products* (released during the period 1957–1959); periodically updated.

The report estimated that in 1972 consumers saved $4 million and avoided 400,000 emergency department visits by receiving free treatment information by telephone from poison control centers.

PUBLIC OVERSIGHT

FDA

In 1979, the FDA's NCHPCC reached its zenith with 45 staff members. The center provided support services but no direct funding to poison control centers. Participation in NCHPCC statistical reports was a voluntary center activity. The development of a taxonomy of poisoning by NCHPCC for standardized reporting was incomplete and no standardization of center organization, service delivery, or clinical outcomes assessment was achieved. As NCHPCC's emphasis evolved into a monthly compilation of toxicology literature, the FDA became less committed to management of NCHPCC and finally ceased to work in this area in 1987.

HRSA and Other Organizations

Poison control centers do not operate within any single federal mandate having regulatory or reporting authority. Furthermore, there is no requirement that emergency departments, critical care units, or any other care facilities or providers contact or report poisoning cases to any center. Although some states, such as New York, have requirements for reporting of poisoning cases as well as food poisoning cases, the discrepancies among various databases suggest that capture of all cases does not occur. This may be due to definitional issues addressed in one of the Committee's recommendations in Chapter 10. The center's role is generally unclear to most health care providers; a center is generally regarded as a place that the public and professionals can call about poison exposures.

Since 2000, more than $60 million have been infused into the AAPCC and poison control centers through the Poison Control Center Enhancement and Awareness Act (Pub. L. No. 106–174), which was enacted in 2000 and reauthorized in 2003 (see Appendix 4-A for the Reauthorization). HRSA and CDC were given responsibility by the Secretary of Health and Human Services for implementing the Act. HRSA administers stabilization and incentive grants to the poison control centers. These grants are reviewed each year for goals, progress, and financial accounting. Interestingly, the related HRSA grant guidelines have discouraged the use of such funds to support delivery of existing services (i.e., to provide fiscal stabilization); instead, they have been earmarked for "enhancement of services." Thus, the Act has done little to stabilize poison control centers in dire financial need of support for basic service delivery.

CDC administers funds to the AAPCC under the Act for upgrading TESS, developing a national poison control telephone number, and developing new nationwide media campaigns. AAPCC has received up to

$5 million per year to support projects such as revisions of a proprietary data collection instrument and development of a real-time surveillance tool that collects data from poison control centers. To date, no real-time data feedback are provided by AAPCC to the centers nor do the centers have real-time access to the electronic database; rather, they are notified by telephone about suspicious activities identified through this real-time tool. Furthermore, centers have only limited access to their own electronic and national data upon request. AAPCC also receives federal income from the Environmental Protection Agency and CPSC for access to TESS surveillance data.

PUBLIC HEALTH LINKAGES

As noted previously, public health agencies, for the most part, had little involvement with poison control centers until 2001, when bioterrorism and related activities created interest in poison control center activities. Few public health leaders were involved during the formative years of poison control centers. This may have been related in part to NCHPCC being located within a regulatory agency. A few centers have developed relationships with public health agencies, but only rarely have they received significant funding or other support. Furthermore, a few states have provided Maternal and Child Health Bureau funding to poison control centers (see Chapter 9).

An Institute of Medicine report on the role of the public and private sectors in injury prevention mentions poisoning only briefly, despite listing it as the third leading cause of injury-related death. Furthermore, the report does not include AAPCC data collection and TESS in its list of more than 30 databases (Bonnie et al., 1999).

Some public health agencies, particularly at the state level, have maintained close ties with their state's poison control centers, but lack of funding has limited such participation. Public health authorities have indicated interest in drugs of abuse and other issues under their purview but, for example, have rarely been interested in unintentional drug ingestion in the home. The lack of collaboration with public health agencies also may be related to the observation that poison control centers have owned their data and most have required compensation from other agencies to share those data.

State activities in poison control centers have been quite variable. New York had a particularly focused development in this area in coordination with the larger picture of injury control (Fisher, 1986; Fisher et al., 1986). A review of poison control statutes that existed by the early 1980s was conducted and some early strategic planning was suggested at about that time (Fisher, 1981; Russell and Czajka, 1984).

Integration of poison control centers and drug information centers was addressed at a number of sites in the late 1970s and early 1980s as pharmacists became more involved with their development (Czajka et al., 1979; Troutman and Wanke, 1983; Wanke et al., 1988).

SUMMARY

The following key messages can be drawn from the discussion provided in this chapter:

- Poison control centers and the work of dedicated poison specialists have had a significant impact on U.S. health care. Key achievements include:
  —Development and implementation of medication safety caps.
  —Establishment of limits on the number of children's aspirin tablets and subsequently other over-the-counter medications in a bottle.
  —Development of imprint code regulations to help speed identification of medications.
  —Use of TESS data to encourage passage of federal regulations in 1997 to reduce the number of iron tablets in a container.
  —Demonstration that nearly 80 percent of human exposures can be managed in the home using poison control center personnel guidance, thus reducing the burden on the health care system and providing reassurance to parents (Watson et al., 2003).
  —Demonstration of the ability to provide an immediate response to public health exposure concerns, such as anthrax, and subsequent participation in bioterrorism responses.
- The current structure of poison control centers is quite variable and developed as a result of historical factors that may be irrelevant to current functional needs.
- Poison prevention efforts have historically focused on children, despite more recent recognition of greater risk for morbidity and mortality in adults.
- More emphasis has been placed on treating patients with drug abuse and alcohol problems as the role of poison control centers has broadened to include adults. Although medical toxicologists see such patients regularly as part of their management of critically ill patients, further integration of these aspects into poison control center services is warranted as part of the spectrum of poisoning treatment.
- Attention to the special problems of the elderly, along with the important contributions of pharmacists in reducing adverse reactions in this population, deserves attention as an aspect of development of poison control centers.

- Funding for poison control centers is piecemeal, and centers often receive unfunded mandates for data provision and other services. A current example is the expectation of active participation in regional emergency planning and response and the provision of additional data for all-hazards emergency preparedness and response surveillance without dedicated resources. Furthermore, federal grants earmarked for poison control center enhancement have done little to stabilize centers in need of financial support for basic service delivery.
- There is considerable opportunity for coordination and cooperation between poison control centers and public health agencies at federal, state, and county levels. However, without federal or state points of accountability, many poison control center oversight roles have been assumed by the American Association of Poison Control Centers. These factors have led to a lack of integration of center data with the public health system.

Appendix 4-A

Poison Control Center Enhancement and Awareness Act Amendments of 2003

Public Law 108–194
108th Congress

An Act

Dec. 19, 2003
[S. 686]

To provide assistance for poison prevention and to stabilize the funding of regional poison control centers.

Be it enacted by the Senate and House of Representatives of the United States of America in Congress assembled,

Poison Control Center Enhancement and Awareness Act Amendments of 2003.
42 USC 201 note.
42 USC 300d–71 note.

SECTION 1. SHORT TITLE.

This Act may be cited as the "Poison Control Center Enhancement and Awareness Act Amendments of 2003".

SEC. 2. FINDINGS.

The Congress finds the following:

(1) Poison control centers are our Nation's primary defense against injury and deaths from poisoning. Twenty-four hours a day, the general public as well as health care practitioners contact their local poison centers for help in diagnosing and treating victims of poisoning and other toxic exposures.

(2) Poisoning is the third most common form of unintentional death in the United States. In any given year, there will be between 2,000,000 and 4,000,000 poison exposures. More than 50 percent of these exposures will involve children under the age of 6 who are exposed to toxic substances in their home. Poisoning accounts for 285,000 hospitalizations, 1,200,000 days of acute hospital care, and 13,000 fatalities annually.

(3) Stabilizing the funding structure and increasing accessibility to poison control centers will promote the utilization of poison control centers, and reduce the inappropriate use of emergency medical services and other more costly health care services.

(4) The tragic events of September 11, 2001, and the anthrax cases of October 2001, have dramatically changed our Nation. During this time period, poison centers in many areas of the country were answering thousands of additional calls from concerned residents. Many poison centers were relied upon as a source for accurate medical information about the disease and the complications resulting from prophylactic antibiotic therapy.

(5) The 2001 Presidential Task Force on Citizen Preparedness in the War on Terrorism recommended that the Poison Control Centers be used as a source of public information and public education regarding potential biological, chemical, and nuclear domestic terrorism.

(6) The increased demand placed upon poison centers to provide emergency information in the event of a terrorist event

involving a biological, chemical, or nuclear toxin will dramatically increase call volume.

SEC. 3. AMENDMENT TO PUBLIC HEALTH SERVICE ACT.

Title XII of the Public Health Service Act (42 U.S.C. 300d et seq.) is amended by adding at the end the following:

"PART G—POISON CONTROL

"SEC. 1271. MAINTENANCE OF A NATIONAL TOLL-FREE NUMBER.

42 USC 300d–71.

"(a) IN GENERAL.—The Secretary shall provide coordination and assistance to regional poison control centers for the establishment of a nationwide toll-free phone number to be used to access such centers.

"(b) RULE OF CONSTRUCTION.—Nothing in this section shall be construed as prohibiting the establishment or continued operation of any privately funded nationwide toll-free phone number used to provide advice and other assistance for poisonings or accidental exposures.

"(c) AUTHORIZATION OF APPROPRIATIONS.—There is authorized to be appropriated to carry out this section $2,000,000 for each of the fiscal years 2000 through 2009. Funds appropriated under this subsection shall not be used to fund any toll-free phone number described in subsection (b).

"SEC. 1272. NATIONWIDE MEDIA CAMPAIGN TO PROMOTE POISON CONTROL CENTER UTILIZATION.

42 USC 300d–72.

"(a) IN GENERAL.—The Secretary shall establish a national media campaign to educate the public and health care providers about poison prevention and the availability of poison control resources in local communities and to conduct advertising campaigns concerning the nationwide toll-free number established under section 1271.

"(b) CONTRACT WITH ENTITY.—The Secretary may carry out subsection (a) by entering into contracts with one or more nationally recognized media firms for the development and distribution of monthly television, radio, and newspaper public service announcements.

"(c) EVALUATION.—The Secretary shall—

"(1) establish baseline measures and benchmarks to quantitatively evaluate the impact of the nationwide media campaign established under this section; and

"(2) prepare and submit to the appropriate congressional committees an evaluation of the nationwide media campaign on an annual basis.

"(d) AUTHORIZATION OF APPROPRIATIONS.—There are authorized to be appropriated to carry out this section $600,000 for each of fiscal years 2000 through 2005 and such sums as may be necessary for each of fiscal years 2006 through 2009.

"SEC. 1273. MAINTENANCE OF THE POISON CONTROL CENTER GRANT PROGRAM.

42 USC 300d–73.

"(a) REGIONAL POISON CONTROL CENTERS.—The Secretary shall award grants to certified regional poison control centers for the purposes of achieving the financial stability of such centers, and for preventing and providing treatment recommendations for poisonings.

"(b) OTHER IMPROVEMENTS.—The Secretary shall also use amounts received under this section to—

"(1) develop standardized poison prevention and poison control promotion programs;

"(2) develop standard patient management guidelines for commonly encountered toxic exposures;

"(3) improve and expand the poison control data collection systems, including, at the Secretary's discretion, by assisting the poison control centers to improve data collection activities;

"(4) improve national toxic exposure surveillance by enhancing activities at the Centers for Disease Control and Prevention and the Agency for Toxic Substances and Disease Registry;

"(5) expand the toxicologic expertise within poison control centers; and

"(6) improve the capacity of poison control centers to answer high volumes of calls during times of national crisis.

"(c) CERTIFICATION.—Except as provided in subsection (d), the Secretary may make a grant to a center under subsection (a) only if—

"(1) the center has been certified by a professional organization in the field of poison control, and the Secretary has approved the organization as having in effect standards for certification that reasonably provide for the protection of the public health with respect to poisoning; or

"(2) the center has been certified by a State government, and the Secretary has approved the State government as having in effect standards for certification that reasonably provide for the protection of the public health with respect to poisoning.

"(d) WAIVER OF CERTIFICATION REQUIREMENTS.—

"(1) IN GENERAL.—The Secretary may grant a waiver of the certification requirement of subsection (c) with respect to a noncertified poison control center or a newly established center that applies for a grant under this section if such center can reasonably demonstrate that the center will obtain such a certification within a reasonable period of time as determined appropriate by the Secretary.

"(2) RENEWAL.—The Secretary may renew a waiver under paragraph (1).

"(3) LIMITATION.—In no instance may the sum of the number of years for a waiver under paragraph (1) and a renewal under paragraph (2) exceed 5 years. The preceding sentence shall take effect as if enacted on February 25, 2000.

Effective date.

"(e) SUPPLEMENT NOT SUPPLANT.—Amounts made available to a poison control center under this section shall be used to supplement and not supplant other Federal, State, or local funds provided for such center.

"(f) MAINTENANCE OF EFFORT.—A poison control center, in utilizing the proceeds of a grant under this section, shall maintain the expenditures of the center for activities of the center at a level that is not less than the level of such expenditures maintained by the center for the fiscal year preceding the fiscal year for which the grant is received.

"(g) MATCHING REQUIREMENT.—The Secretary may impose a matching requirement with respect to amounts provided under a grant under this section if the Secretary determines appropriate.

"(h) AUTHORIZATION OF APPROPRIATIONS.—There are authorized to be appropriated to carry out this section $25,000,000 for each of the fiscal years 2000 through 2004 and $27,500,000 for each of fiscal years 2005 through 2009.

"SEC. 1274. RULE OF CONSTRUCTION.

42 USC 300d–74.

"Nothing in this part may be construed to ease any restriction in Federal law applicable to the amount or percentage of funds appropriated to carry out this part that may be used to prepare or submit a report.".

SEC. 4. CONFORMING AMENDMENT.

42 USC 14801 note.

The Poison Control Center Enhancement and Awareness Act (42 U.S.C. 14801 et seq.) is hereby repealed.

Approved December 19, 2003.

LEGISLATIVE HISTORY—S. 686:

SENATE REPORTS: No. 108–68 (Comm. on Health, Education, Labor, and Pensions).

CONGRESSIONAL RECORD, Vol. 149 (2003):

June 20, considered and passed Senate.

Nov. 19, 20, considered and passed House, amended.

Dec. 9, Senate concurred in House amendment.

○

5

Poison Control Center Activities, Personnel, and Quality Assurance

Poison control centers were developed as sources of information about the management of pediatric poisonings, but have evolved to encompass a wide range of additional functions (Hoffman, 2002; Leikin and Krenzelok, 2001; Youniss et al., 2000; Zuvekas et al., 1997). Here we provide an overview of poison control center activities (Box 5-1) and the personnel who perform them (American Association of Poison Control Centers, 2002a; Poison Center Annual Reports, 2000–2002). Some of these activities are discussed only briefly here and covered more thoroughly in other chapters (e.g., public education, data collection and surveillance systems, and sentinel events).

Most activities discussed in this chapter are provided by all poison control centers, such as responding to telephone calls from the public or health care providers regarding poison exposures. Other activities are offered by only some poison control centers, depending on their interests, capabilities, affiliations, and the funding initiatives of sponsoring agencies. For example, centers affiliated with medical toxicology training programs serve as clinical training sites for medical toxicology residents. Collaboration between poison control centers and appropriate public health agencies, health care providers and provider groups, government agencies, and academic resources is embedded within the activities to be described, and is assumed throughout this discussion.

The first half of this chapter reviews the service and administrative/research activities of the poison control centers. The professional education role of the centers, including the poison specialists who work for them, is emphasized in the latter half of the chapter.

BOX 5-1
Current Activities of Poison Control Centers

Phone consultation regarding individual exposures, potential exposures, or information
- Public callers
- Health care professionals

Capacity to respond to mass exposures or potential exposures
- HAZMAT/occupational
- Bioterrorism
- Nonpoisoning exposures for which the poison control center may be the default information source

Contract services
- Industry support

Data collection and reporting
- Systematic reporting of exposures to a collective poison prevention and control database
- Sentinel event reporting

Research
- Toxicology
- Health care delivery

Program evaluation
- Quality improvement
- Outcomes

Public education, in collaboration with public health agencies
- Prevention
- Use of the health care system

Professional education
- Specialist in poison information and poison information provider training
- Medical toxicologists
- Continuing education of health care professionals
- Other health care trainees

ACTIVITIES

Telephone Consultation

The primary activity of poison control centers in the United States is telephone consultation, providing direct services to the public, emergency medical services personnel, health care providers, and public health agencies (Zuvekas et al., 1997). These services depend on the rapid, efficient call handling by specialists with training in clinical toxicology, supported by medical toxicologists, consultants in subspecialty areas, and poison information databases. The general process of call management is illustrated in Figure 5-1 and will be discussed. The skill sets and training of personnel responding to telephone calls are reviewed later in this chapter under the section on health care professionals.

Table 5-1 presents an overview of poison control center direct services to patients; emergency medical services (EMS) personnel; medical staff in offices, clinics, and emergency departments; in-hospital staff; and local and state public health officials. Examples include providing (1) comfort to caregivers or patients; (2) advance notification and care guidance to EMS, emergency department, and in-hospital personnel; and (3) consultation on syndromal case clustering with public health officials.

Calls from the Public

Poison control centers provide information to the public regarding poisoning exposures and respond to requests for information about poisons—defined as calls in which there is no actual exposure discussed (Hoffman, 2002). Calls to poison control centers can be made directly by the public using a toll-free telephone number or through referrals to centers via 911 or other emergency numbers. Human exposures or suspected

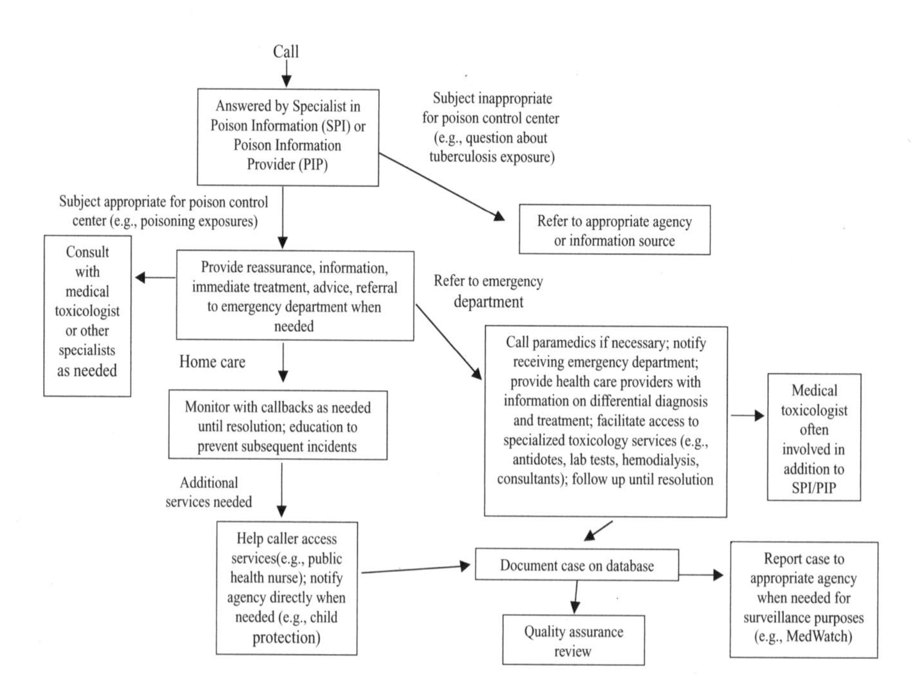

FIGURE 5-1 Typical sequence of events at poison control center handling a call from the public.

TABLE 5-1 Poison Control Center Direct Service Matrix

	Service Recipient				
Service Type	Patient/ Family Member	EMS Personnel	Office, Clinic, or Emergency Department Personnel	In-Hospital Personnel	County and Regional Public Health
Comfort and reassure	yes	—	—	—	—
Identify products	yes	yes	yes	yes	—
Determine potential toxicity	yes	yes	yes	yes	—
Assist 911 response	yes	yes		—	—
Direct poisoning first aid	yes	yes	yes	—	—
Guide prehospital triage decisions	yes	yes	—	—	—
Provide advance notification if care transferred	—	yes	yes	—	—
Offer continued care guidance and monitoring	yes	yes	yes	yes	—
Provide poison prevention education	yes	—	—	—	—
Guide advanced patient management and diagnostics	—	—	yes	yes	—
Guide caregiver protective actions	—	yes	yes	yes	—
Supply reference materials related to patient management	—	—	yes	yes	—
Link provider with special toxicology resources	—	—	yes	yes	—
Guide hazardous materials management	—	yes	yes	yes	yes
Consult on syndromal case clustering	—	—	yes	yes	yes
Participate in public health notifications as health threats emerge	—	—	—	yes	yes

exposures represent 68 percent of all calls, and requests for information represent 32 percent (Watson et al., 2003). In 2002, there were 2.38 million human exposure calls and 1.1 million information calls to poison control centers across the country (see Table 5-2 for the content of these calls). Approximately 85 percent of the human exposure calls were categorized as unintentional, 11.7 percent as intentional, and 2.4 percent as adverse reaction. Furthermore, most calls were judged to have little or minor effect, with 74 percent managed in the home (Watson et al., 2003).

Assistance to the public regarding human exposures includes assessment of the type and severity of poisoning, suggestions for at-home care when appropriate, reassurance to the caller, and referral to a health care provider when necessary. Follow-up is provided with callbacks as needed to assure satisfactory resolution of the episode. Preventive measures are suggested to avoid similar poisoning episodes, such as removing certain items from the home or placing them out of reach of children. When callers are referred to a health care facility, that facility is notified and information regarding the case and the relevant toxicology of the poison involved is made available. Access to the public is provided via around-the-clock toll-free telephone lines staffed by poison control center personnel. Most calls are answered by specialists in poison information (SPIs) or poison information providers (PIPs) (Zuvekas et al., 1997). Medical toxicologists are available at all times by telephone for backup consultation. Poison control centers also have consultants on call with specific areas of expertise, such as mycologists and herpetologists. Tables 5-2 and 5-3 present the most common reasons for exposure and information calls, respectively.

Poison control centers also respond to calls from the public regarding animal exposures to poisons; exposures to chemicals at work or in the home; requests to evaluate symptoms that the caller believes may be related to poisoning; questions about environmental pollutants, plants, herbal medicines, drug interactions, or envenomations; or requests for general information about poisoning topics (e.g., first aid) (Leikin and Krenzelok, 2001). The range of topics is broad in part because people call the poison control center when they are not sure who else to call. Such calls may be referred to more appropriate agencies, for example, the local public health department for possible food poisoning. Nearly all of the poison control centers maintain websites with information on selected topics and links to other information sources, and callers may be referred to those websites when appropriate.

Calls from Health Care Professionals

Poison control centers provide information to a wide variety of health care providers regarding exposures and potential exposures. As the focus

TABLE 5-2 Exposure Call Content (most common agents involved in 2,380,028 exposure calls* reported to the TESS data collection system in 2002)

Exposure Type	Percentage
Analgesics	10.8
Cleaning substances	9.5
Cosmetics and personal care products	9.2
Foreign bodies	5.0
Sedatives/hypnotics/antipsychotics	4.7
Topical	4.4
Cough and cold products	4.4
Antidepressants	4.2
Bites/envenomations	4.1
Pesticides	4.0
Plants	3.6
Food products, food poisoning	3.2
Alcohol	2.9
Antihistamines	2.9
Antimicrobials	2.7
Cardiovascular drugs	2.6
Hydrocarbons	2.5
Chemicals	2.3

*Actual or suspected contact with any substance that has been ingested, inhaled, absorbed, applied to, or injected into the body. Calls involving the therapeutic use of a medication without adverse effects or toxicity are not considered exposures.
SOURCE: Adapted from Watson et al. (2003).

TABLE 5-3 Information Call Content to Poison Control Centers (most common topics involved in 1,110,635 information calls* reported to the TESS data collection system in 2002)

Topic	Percentage
Drug identification	50.0
Drug information	16.0
Poison information	11.0
Prevention/safety/education	6.0
Environmental	3.7
Substance abuse	0.9
Teratogenicity	0.6

*Requests for information that lack an identifiable exposed person.
SOURCE: Adapted from Watson et al. (2003).

of poison control centers has shifted from pediatric exposures to include adult exposures, which are far more likely to result in significant morbidity and mortality, consultation with health care providers has become an increasingly important function. Users of this service include emergency medical technicians (EMTs) who obtain prehospital treatment guidance; first responders such as firefighters or HAZMAT teams; nurses and physicians in hospitals and clinics; nurses and social workers who staff advice lines; and physicians, nurses, or other professionals with public health agencies (Hoffman, 2001; Leikin and Krenzelok, 2001). Consultation with emergency department and intensive care unit staff has become an important role for poison control center staff. Emergency or critical care physicians or nurses managing poisonings in their clinic are given guidance related to poison identification, anticipated toxic effects, initial intervention, and whether transfer to an emergency department or inpatient facility is needed. Emergency physicians and nurses may receive suggestions regarding differential diagnosis, initial intervention, selection of laboratory tests, where to send nonstandard laboratory test requests, the need for consultation and the identification of appropriate consultants, and disposition of the patient.

Poison control center staff provide guidance regarding the indications for specific procedures such as hemodialysis, the availability and use of specific antidotes, or the need for transfer when the necessary facilities or expertise needed for patient management are not available. Staff assist with arranging patient transfers or putting physicians in contact with appropriate consultants. Specific reference materials or pertinent literature are provided. Calls from health care professionals are handled initially by SPIs or PIPs, but are often referred to medical toxicologists. Approximately 20 percent of the human exposure calls made to poison control centers come from health care facilities (Poison Control Center Annual Reports, 2000–2002).

Follow-up calls from the poison control center are used to ascertain patient status, symptom resolution, compliance with recommended therapy, and where appropriate, status after discharge. For both public and health care professional callers, health care agencies are contacted when they can assist with patient care or when reporting is required or appropriate. The adverse effect of Health Insurance Portability and Accountability Act regulations, or hospitals' interpretations of these guidelines, on the ability to obtain follow-up information is an issue that poison control centers are currently addressing (http://fpicn.org/HIPAA_Compliant.htm).

Technology Support for Telephone Consultation

Since 2001, the American Association of Poison Control Centers (AAPCC) has been responsible for establishing and implementing a national toll-free telephone number (1-800-222-1212) that is advertised by all U.S. poison control centers. Individuals (either from the public or health care professionals) who wish to report a poisoning exposure or ask for information regarding a product or procedure can call this number and be routed to the local poison control center for a response. Centers may continue to maintain their own local and toll-free numbers because people in their service area are familiar with these numbers and have them posted at home. California maintains a separate toll-free number for use by health care professionals. Specialized telecommunication devices for communication with deaf individuals are used, and there are links to agencies that specialize in communicating with them. Translation services are available to ensure rapid and accurate communication with individuals who do not speak English or who speak English as a second language. Generally, these services are provided through a three-way conference call with a translation service equipped to translate several languages; however, four centers have bilingual speakers onsite 24 hours a day, 7 days a week (http://www.aapcc.org/pccsurveyresults/2002/2002Table4.pdf).

The primary information aid in exposure analysis and treatment is Poisindex® published by Micromedex. As noted in Chapter 4, Poisindex provides (1) information on the composition of most commercial and natural products; (2) a description of the toxicity of the products; and (3) suggested treatment options. Micromedex employs an internal staff to obtain and code products as well as to prepare the management documents for review by an outside editorial board. The embedded product codes are used by AAPCC to connect each poison control center case to a product or category.

As the call progresses, the certified specialist in poison information (CSPI) or SPI collects and records demographic information about the caller (e.g., age, gender, geographic location), notes the product code associated with the exposure, judges severity, and provides treatment recommendations using one of four commercially available data collection systems (discussed in Chapter 8) that automatically upload data every 4 to 10 minutes to the Toxic Exposure Surveillance System (TESS) housed at AAPCC headquarters in Washington, D.C. Electronic matching between the computerized case reports filled out by the specialists and the product and ingredient indexes of Poisindex® is automatic and works seamlessly with all four electronic data collection programs.

Capacity to Respond to Mass Exposures or Potential Exposures

HAZMAT, Occupational, or Environmental Exposures

Poison control centers are an integral part of most local HAZMAT (hazardous materials) response protocols, such as a chemical leak from an overturned tanker truck (Burgess et al., 1997; Mrvos et al., 1988). In this capacity, poison control centers work in close collaboration with local public health agencies and other agencies with responsibilities in this area, such as fire and police departments. Centers may be accessed in these situations by the public, responders at the incident scene, health care providers caring for exposed individuals, or public health agencies. The role of the poison control center includes providing information to individuals with exposures or potential exposures, assisting in triaging injured patients and notifying the receiving health care facility, providing information regarding the toxicology of the chemicals involved in the incident and the management of exposed patients, and gathering data regarding exposures (e.g., locations, types of injuries) that may be useful for managing the incident or for surveillance and follow-up. Poison control centers serve a similar role in acute or chronic occupational exposures, providing initial information regarding the toxicology of the exposure, assistance with initial triage and management, and coordination with responsible public health agencies (Blanc and Olson, 1986). Centers also have been accessed following natural disasters such as earthquakes to provide information on air and water quality (Nathan et al., 1992).

All-Hazards and Other Public Health Emergencies

Poison control centers can play an important role in preparedness and response to acts of bioterrorism, chemical terrorism, or other public health emergencies. The efforts involved in the rapidly evolving, present-day building of capacity in the areas of bioterrorism and chemical terrorism preparedness and response is likely to strengthen the ability of centers to respond to natural disasters and other threats to public health. In 2001, 3,395 exposure calls regarding agents classified in TESS as "weapons of mass destruction" (WMD) were received and acted on by poison control centers. Examples include reports of anthrax, suspicious powder, chemical weapons, and other suspicious substances.

State plans for public health and hospital emergency preparedness are beginning to acknowledge roles for poison control centers. These state programs, funded by cooperative agreement grants from the Centers for Disease Control and Prevention (CDC) and the Health Resources and Services Administration (HRSA), help the public health system and hospitals prepare for acts of bioterrorism, outbreaks of infectious disease, and other public health threats and emergencies. For a further discussion of state involvement, see Chapter 9.

To help describe the relevant services that poison control centers can provide, the following framework is adapted from the Haddon Matrix: *Preevent*, including planning, education, and surveillance; *Event and Response*, including detection, investigation, dissemination of information to the public, antidote distribution, and communication with agencies; and *Recovery*, including serving as an information resource for the public and health care providers.

Preevent activities include the following:

- *Preparedness planning:* Poison control center personnel, particularly toxicologists and pharmacists, serve in multiple capacities related to local, regional, and national emergency preparedness planning; examples include service on state and local preparedness committees and advisory groups. CDC and HRSA guidance in 2003 to state and local applicants for cooperative agreement funding for public health preparedness and response to bioterrorism specifies that poison control centers be represented on advisory committees for these cooperative agreements. Centers also assist with Strategic National Stockpile planning; help with local and regional capacity assessments related to bioterrorism preparedness; and develop diagnostic and treatment protocols and associated standardized staff education programs. Poison control centers also maintain antidote supplies and/or facilitate local/regional inventory tracking for essential medicines and other supplies.
- *Surveillance:* Poison control centers participate in toxicosurveillance for the identification of sentinel events that may represent intentional bioterrorist events or natural toxin exposures. Every 4 to 10 minutes, poison control centers upload case data to TESS. To assist in improving public health surveillance, CDC's National Center for Environmental Public Health, the Agency for Toxic Substances and Disease Registry (ATSDR), and AAPCC are working to convert TESS into a real-time public health surveillance system. This conversion will generate more immediate and appropriate responses to public health threats that may be related to toxins or chemicals in the environment. In addition to a national toxicosurveillance effort performed in conjunction with AAPCC, poison control centers participate in local syndromic surveillance and report notifiable conditions affecting multiple individuals to local and/or state health officials.
- *Continuing education and research and preparation for bioterrorism through effective training programs:* Poison control centers provide training to various groups (e.g., the public, emergency medicine residents, medical toxicology

fellows, emergency physicians and nurses, and public health officials) on the medical effects of WMD. Centers also serve as a repository for specialized databases regarding agents of bioterrorism and chemical terrorism.

Event and Response activities include the following:

• Poison control centers can provide assistance with early recognition and notification of bioterrorism and chemical terrorism events; coordinate antidote distribution and guide appropriate antidote use; assist health care professionals with management of exposed patients and rescue personnel; disseminate threat and preventive/therapeutic information to the public; and provide consultative support to public health and law enforcement authorities.

Recovery activities include the following:

• Poison control centers provide a single source (i.e., one telephone number) for coordinating exposure and treatment information; serve as an information resource to the public, media, and medical practitioners regarding existing health effects in the aftermath of a terrorist act; reassure these same groups to help minimize panic; and provide information to public health agencies about the long-term effects associated with a terrorist attack and how to respond to and treat those effects.

Case vignettes illustrative of how poison control centers contribute to the investigation and management of bioterrorism events are presented in Appendix 5-A. The case study of the Rocky Mountain Poison and Drug Center illustrates the integral involvement of a center into a public health system. The Texas Poison Control Center's experience illustrates the role a poison control center can play in communicating with the public and provider community when an urgent health issue emerges, in this instance the receipt of hundreds of calls during the 2001 anthrax attacks. The Northern New England Poison Center vignette describes how a center works in the early identification of (i.e., toxicosurveillance) and response to a communitywide cluster of illness (i.e., arsenic exposure).

Nonpoisoning Exposure Information Requests to Poison Control Centers

Because of their well-publicized availability, poison control centers are often the default choice for the public to call regarding exposures to agents other than chemicals. For example, centers have received calls about Severe Acute Respiratory Syndrome and must be able to respond with appropriate information or referrals. In addition, incidents can occur in which the nature of the offending agent is unknown and could be an infectious agent or a chemical. In this case, the public, health care professionals, and public health agencies may all access the poison control center for information (Geller and Lopez, 1999).

Industry Contracts

Poison control centers may contract with industry to provide poisoning information to the public and health care providers for a specific commercial product or group of products. For example, the toll-free number listed on a commercial product for information about poisoning may be answered by center staff, usually through a separate, dedicated telephone line. Summary information regarding the number, types, and outcomes of these exposures can assist companies with reporting requirements regarding adverse events associated with their products and guide reduction of the hazard. Poison control centers may also contract to provide material safety data sheets needed by an employer (Krenzelok and Dean, 1988). The number and extent of poison control center contracts with industry is not well documented.

Data Collection and Reporting

Data collection and reporting are critical activities of poison control centers. This topic is examined in detail in Chapter 7 along with the contribution of other data sources in developing an overall understanding of poisonings in the United States.

Systematic Reporting of Exposures to a Collective Poison Control Center Database

Poison control centers currently report all exposure data to the TESS database (Watson et al., 2003). These data are used to document the spectrum of exposures causing poisonings and their consequences. While this database represents only the fraction of all exposures that generate calls to centers (Blanc et al., 1995), the data may be used to identify trends and potential targets for education, surveillance, public health measures, or research. This type of database represents a unique opportunity to collect detailed data regarding certain types of poisoning exposures.

Sentinel Event Reporting

Sentinel events are initial cases or events indicating a more widespread problem. Poison control centers, because of their telephone consultation function, may be the first part of the health care system to become aware of new types of poisonings in the community, such as previously unknown adverse reactions to a medication, the use of a new type of drug of abuse (e.g., scopolamine poisoning among heroin users [*Morbidity and Mortality Weekly Report*, 1996]), or an unusual illness suggestive of bioterrorism. Centers may also recognize clusters of cases that would otherwise go unnoticed as isolated events. Prompt reporting of these events to the appropriate public health agencies and in the medical literature can help provide an early warning and offer the option of generating a timely response. Poison control centers also notify public health agencies of reportable diseases or cases or incidents that may be of interest, such as suspected food poisoning.

Research

AAPCC (http://www.aapcc.org) considers research to be a part of the overall mission of poison control centers. Center research may be broadly divided into toxicology research, which focuses on the mechanisms, treatment, or prevention of poisonings; and poison control center clinical services research, which focuses on the role and contribution of centers to studying and managing these problems (North American Congress of Clinical Toxicology, 2003). Examples of poison control center clinical services research include the evaluation of different poison control center service models with regard to costs and outcomes; comparison of different data collection tools or models; efficacy of using center data for real-time surveillance of previously unrecognized toxic effects of medications or mass poisonings; effectiveness of education strategies; comparisons of treatments; or strategies for monitoring herbal medicine toxicities. Although any investigator, regardless of location or affiliation, could perform such research, poison control center staff are particularly well positioned to conduct research in these areas. Their involvement in the management of exposure and information calls provides a unique opportunity to identify important or emerging issues. Access to center data and familiarity with the health care delivery issues facing poison control centers provide staff with a unique perspective on issues surrounding the delivery of center services, and the means to study strategies for improving these services.

Despite these opportunities, the research output of poison control centers is generally modest. One survey of center research efforts, which included all types of research except for "bench science," found that 5 percent of poison control center staff time (primarily non-SPI staff) was devoted to research and that centers published a mean of three journal articles yearly (Zuvekas et al., 1997). Some centers, such as those affiliated with medical toxicology training programs, are more active in research. However, as noted in Chapter 6, they have little specific funding to conduct this research. Hence, presentations at the AAPCC annual meeting, the largest annual gathering of clinical toxicologists, tend to emphasize clinical observations. The abstracts accepted for the AAPCC meetings during the years 2001–2003 were tallied (Table 5-4), and results showed the vast majority of presentations were case reports and descriptive database summaries (some of the latter included prospective observational studies). True clinical trials, randomized or not, made up about 2 percent of the abstracts presented. This latter figure does not include a small number of human pharmacokinetic studies, which were generally observational and included only one to two subjects (these are listed separately). Laboratory-based or animal/tissue studies were the subject of about 11 percent of the abstracts. Clearly, the presentation of prospective interventional and laboratory-based research in poisoning mechanisms and management at this meeting was limited. Even research abstracts focusing on poison system impact made up less than 10 percent of the abstracts. Furthermore, only 1.5 percent of the abstracts addressed public/professional poison education programs.

TABLE 5-4 Summary of Research Abstract Contents, AAPCC Meetings, 2001–2003

Types of Studies	2001	2002	2003	Total	(%)
Case series/report	110	93	93	296	40.1%
Descriptive database review	38	58	64	160	21.7%
Pharmacokinetic	8	2	3	13	1.8%
Survey	0	2	11	13	1.8%
Cohort analysis	26	19	19	64	8.7%
Case-control study	4	5	5	14	1.9%
Uncontrolled clinical trial	2	0	3	5	0.7%
Historically controlled clinical trial	0	1	1	2	0.3%
Nonrandomized controlled clinical trial	2	1	3	6	0.8%
Randomized controlled clinical trial	1	1	0	2	0.3%
Survey or administration	24	22	28	74	10.0%
Education	1	5	5	11	1.5%
Cell culture/biochemical	5	2	1	8	1.1%
Whole organ or whole animal study	12	13	9	34	4.6%
Laboratory research/product analysis	4	12	20	36	4.9%
Total	237	236	265		
(%)	32.1%	32.0%	35.9%		

SOURCES: *Journal of Toxicology, Clinical Toxicology* (2001, 2002, 2003).

Several factors may limit research activity. First, the common sources of funding for center services (federal, state, or local funds) neither support nor require research, and there is no research requirement for AAPCC certification of poison control centers (American Association of Poison Control Centers, 1998). Second, federal funding for poison control center-based research is limited. Research focusing on toxicologic mechanisms or specific poisons can be funded via programs with a related focus, such as drug abuse (e.g., phencyclidine toxicity [Hardin et al., 2002, p. 1642]), mental health (e.g., antidepressant overdose [Pentel et al., 1995, p. 817]), or environmental safety (e.g., lead poisoning [Osterloh and Kelly, 1999, p. 1980]).

By contrast, research focusing on poison control center health services delivery has not been a focus of programs within the National Institutes of Health (NIH). A search of NIH program announcements for 2003 using the keywords "poison system," "poison control," or "poison center" revealed only one announcement specifically relevant to poison control center services, and that announcement involves bioterrorism preparedness (http://grants1.nih.gov/grants/guide/index.html). A search using the term "poison" revealed 18 other announcements, none of these dealing with poison control center services delivery but rather focusing on specific areas such as drug abuse (alcohol, inhalants), environmental toxicology, or alternative medicines (chelation therapy). A search of the Computer Retrieval of Information on Scientific Projects (CRISP) database of current NIH grants for 2003 using the same terms revealed none dealing with general poison control center services issues, one each involving the utilization of poison control center surveillance to monitor gamma-hydroxybutyrate toxicity or foodborne illnesses, and one regarding antidote design for venomous bites (http://crisp.cit.nih.gov/). This indicates that sources of funding for research about poison control center services or general management issues in toxicology (such as the use of activated charcoal for drug ingestions) are limited.

A third possible reason for limited research is that poison control center staff have extensive clinical or administrative responsibilities that reduce the time they have available for research. Moreover, the training of poison control center staff is predominantly clinical.

Public Education

The following is a general statement of the types of public education currently provided by the poison control centers. A detailed discussion is provided in Chapter 8.

Primary Prevention: Public Education

Poison control centers provide public education on poisoning prevention (Zuvekas et al., 1997). This education may be offered in collaboration with public health agencies, nongovernment organizations, or industry. The role of the centers is to provide expertise in poisonings; the collaborator provides expertise in injury prevention, as well as offering the infrastructure for public education (see discussion in Chapter 8).

Secondary Prevention: Utilization of Health Care Resources

A potential benefit of having poison control centers available to answer poisoning calls is the appropriate triage of patients to health care facilities when needed (improving patient care), and avoiding the use of health care facilities when not needed (reducing costs). Poison control centers provide public education focused on facilitating these outcomes, such as calling the center for poisoning questions or exposures rather than calling physicians' offices.

HEALTH CARE PROFESSIONALS AND THEIR TRAINING BY POISON CONTROL CENTERS

Poison control centers provide education for many categories of health care professionals. A critical component of this education is the training of the health care professionals who actually work in or with centers. To better describe how this training is provided, we begin with a description of the personnel who staff centers and the range of activities they perform. Activities include medical direction, center management, telephone consultation, professional training, public education, and research. A discussion of the specific contributions of poison control centers to their training follows.

Health Care Professionals

A wide variety of health care and public health professionals contribute to the recognition, prevention, and management of poisonings. This section focuses on those who staff or interact directly with poison control centers, recognizing that many other types of individuals contribute to poison control efforts in other capacities.

The term *toxicologist* is a general description of an individual dealing with any aspect of acute or chronic poisonings, and it does not have a specific definition or implication with regard to training or job description. For example, this term may be used to describe individuals whose activities range from molecular biology to epidemiology, as long as they deal in some way with the toxic effects of chemicals. The term *clinical toxicologist* implies a more clinical orientation, but likewise has no specific definition or implications. *Medical toxicologists* are physicians with specific training and board certification in the subspecialty of medical toxicology, which focuses on the care of poisoned patients.

Specialists in poison information are health care professionals (primarily nurses or pharmacists) who serve as poison control center staff with the primary responsibility of responding to telephone calls regarding poisoning exposures or requests for information. *Poison information providers* are individuals who may lack training in nursing, pharmacy, or medicine but serve in a similar capacity to SPIs within poison control centers, but with supervision by an SPI or a medical or managing director.

Medical Toxicologists

Medical toxicology is a subspecialty for physicians defined by the Accreditation Council for Graduate Medical Education (ACGME) as a "clinical specialty that includes the monitoring, prevention, evaluation and treatment of injury and illness due to occupational and environmental exposures, pharmaceutical agents, as well as unintentional and intentional poisoning in all age groups" (Accreditation Council for Graduate Medical Education, 2003). Medical toxicologists first complete training in any primary medical specialty (usually emergency medicine, occupational medicine, pediatrics, internal medicine, or pathology), and then an additional 2-year fellowship in medical toxicology (Wax and Donovan, 2000).

AAPCC requires a board-certified medical toxicologist as medical director as a condition of poison control center certification. In this capacity, medical toxicologists provide overall clinical supervision and medical backup of poison control center personnel and contribute to teaching, administrative, and research efforts within the center. Medical toxicologists also serve as consultants to centers, providing medical backup, teaching, or research expertise.

Common roles for medical toxicologists outside poison control centers include direct care or consultation regarding poisoned patients, teaching of medical toxicology fellows and other health care professionals, toxicology research, and medicolegal consultation. A smaller number of medical toxicologists work in various capacities in public health or government agencies, including CDC and the U.S. Food and Drug Administration, or in industry (Wax and Donovan, 2000).

Managing Direction

According to the AAPCC criteria for certification, the managing director of a certified poison control center provides "direct supervision of poison center staff, strategic planning, and oversight of administrative functions of programs (e.g., staff training, quality assurance, budgeting, etc.)" (American Association of Poison Control Centers, 1998, p. 6). A managing director with responsibilities for toxicological supervision must be board certified or board prepared in applied toxicology. This position can be filled by a physician, a pharmacist, or a nurse. In some centers the managing director is also the medical director.

Specialists in Poison Information

Specialists in poison information are the primary poison control center staff who answer telephones and respond to callers wanting information regarding poisonings. A 1998 survey of centers in the United States reported that SPIs were nurses (53 percent), pharmacists (40 percent), or physicians (3.5 percent) or had other backgrounds (2 percent) (Youniss et al., 2000). A certification process for SPIs is offered by AAPCC, allowing them to become *certified specialists in poison information* (these procedures are discussed in the section on quality assurance).

In addition to their primary role in answering poison control center telephone calls from the public and professionals, SPIs may supervise PIPs (staff with less advanced training in health care), serve administrative functions, conduct public or professional education, or participate in research projects. However, most SPI time is spent responding to telephone calls (75 percent) or handling administrative tasks (20 percent), with little time devoted to education (3 percent) or research (1 percent) (Zuvekas et al., 1997).

Poison Information Providers

Like SPIs, the primary role of PIPs is answering poison control center telephone calls from the public and health care professionals. PIPs differ from SPIs in that they lack training as nurses, pharmacists, or physicians, but rather have a variety of other health-related backgrounds. Because they have less advanced training in health care, PIPs working in AAPCC-certified poison control centers are required to do so under the onsite supervision of a CSPI or the managing or medical director of the center. The addition of PIPs to the more traditional use of SPIs to answer center telephones serves to enlarge the pool of providers available for such employment and to reduce costs.

Health Educators

Most poison control centers have a full- or part-time health educator whose primary role is education of the public regarding poisoning prevention and promotion of poison control center use. These efforts are often focused on the pediatric age group. In some centers, SPIs contribute to public education (Zuvekas et al., 1997). The role of health educators is discussed in more detail in Chapter 8.

Consultants

A wide variety of consultants provide additional capacity and expertise to poison control centers. Centers have medical toxicologists who serve as consultants to provide additional medical backup to SPIs and PIPs, to share on-call responsibilities with the medical director, or to contribute to teaching or research. In addition, poison control centers have consultants with expertise in specialized areas, such as veterinary medicine, herpetology, or mycology. These individuals are important to centers because it is not possible for their staff to have the detailed knowledge needed to manage the wide variety of poisonings encountered. Consultants typically volunteer their services and represent backgrounds as diverse as their subject matter.

Role of Poison Control Centers in Training Their Personnel and Other Health Care Providers

All poison control centers provide training for their own SPI and PIP staff because no other means of training is available. Clinical experience in the setting of a center is required for the training of medical toxicologists, and centers affiliated with such training programs are involved extensively in the education of these physicians. Thus, poison control centers play an essential role in the education of each of the major categories of center personnel and they also provide continuing medical education in their service area to health care professionals and to groups such as fire or police departments. Additionally, centers with affiliations or geographic proximity to training programs for physicians, nurses, dentists, pharmacists, or emergency medical technicians also provide some experience in toxicology as part of their training.

SPIs and PIPs

The training of both SPIs and PIPs is accomplished by the poison control center via lectures, assigned readings, observing center staff, participating in hospital rounds, and responding to poison information or exposure telephone calls under supervision. There is no standard curriculum or published description for such training. AAPCC requires of certified poison control centers only that such training be under the supervision of the center medical director. SPI training takes from 3 months to a year to allow the SPI to answer center telephones independently (Committee briefings from Dart, 2003; Heard, 2003; and Trestrail, 2003).

Medical Toxicologists

Poison control centers provide an essential part of the training of medical toxicology fellows. ACGME specifically requires that accredited fellowship programs have an affiliation and geographic proximity to a poison control center for the clinical portion of their training. This center-based training includes answering telephones, providing medical backup to SPIs and PIPs, providing direct consultation to health care professionals calling the center, interacting with health care agencies that collaborate with or use the center, understanding center data collection and reporting, and gaining experience in center administration. A survey of medical toxicologists indicated that 46 percent of their clinical experience during training involved center-based activities (Wax and Donovan, 2000).

Currently, 12 poison control centers (ACGME-accredited programs) offer medical toxicology fellowships. In some cases, medical toxicology fellows spend time at a second center that serves a different geographic area or patient population or has staff with different expertise in order to gain a broader range of experience.

Clinical Toxicology Fellowships

Several poison control centers offer 1- or 2-year clinical toxicology fellowship training programs for clinical pharmacists (e.g., see http://www.pharmacy.umaryland.edu/pps/residents/toxicology.htm and http://www.hscj.ufl.edu/pharmacy/residency/index.html). There are no uniform criteria, curricula, or accreditation processes for such fellowships. The purpose of these fellowships is to prepare clinical pharmacists for careers focusing on the management of the poisoned patient, but their content may vary depending on the institution and the interests of the trainee.

Professional Continuing Education

Poison control centers provide professional continuing education in their service area. The type and extent varies greatly, reflecting the needs of the community, the availability and funding of center staff, and the particular expertise of the staff (Zuvekas et al., 1997). Courses or individual lectures may be located at or near the poison control center, or in communities throughout the service area. The target audience may include physicians, nurses, pharmacists, emergency medical technicians, veterinarians, dentists, public health workers, or government employees who share responsibility for some aspect of poisoning prevention or management, such as HAZMAT responders. Poison control center staff often share the podium with other professionals at such conferences because the range and scope of toxicology is broad and collaboration with other professionals and agencies is an integral part of both center service and educational efforts.

Poison control center staff also contribute to professional continuing education via publications, including editing toxicology textbooks, writing book chapters or review articles, contributing sections to references such as Poisindex®, and preparing review or teaching documents for government agencies such as ATSDR.

Education of Other Health Care Professionals

Poison control centers contribute to the education of a wide variety of health care professionals, including medical, nursing, pharmacy, and dental students; medical and pharmacy residents and fellows; and EMTs. Training may include rotations through the centers where they observe or assist with call management or information retrieval, lectures in formal courses, or the participation of poison control center personnel in hospital rounds. Education of these trainees is important as a means of disseminating standards of care for the poisoned patient, and in creating interest in this field among younger trainees who might later choose to obtain further training in toxicology or careers involving poison control centers.

Poison control center education of trainees also contributes to the scope of their education. However, there are many more training programs for health care professionals than there are poison control centers, and not all training programs are located near a center. Thus the extent to which poison control centers contribute to the education of health care professionals varies greatly (Zuvekas et al., 1997). For example, a survey of emergency medical technician training programs showed that 81 percent of such programs had access to a regional poison control center, and 11 percent offered rotations in centers to trainees (Davis et al., 1999). Where such relationships are established, poison control centers can contribute to the education of health care professional students, residents, or fellows through lectures incorporated into their required courses (e.g., a lecture on management of overdose as part of a pharmacology course for medical students), as clinical rotations in which students spend time in the center, or as participants in center research projects. This form of poison control center educational involvement is not mandated for any health care professional training (other than that of medical toxicologists).

QUALITY ASSURANCE

There are both external and internal evaluation mechanisms for assuring the quality of poison control centers. Externally, a certification process is offered by AAPCC for poison control centers or systems (see Box 5-2), defined as two or more poison control centers functionally and electronically linked to provide services (http://www.aapcc.org/MEMBERS/center.htm). Fifty-two of the existing 63 poison control centers in the United States are currently certified. Although certification is voluntary, some funding sources (e.g., states, HRSA) require AAPCC certification as a condition of funding. The key elements of certification are as follows:

1. Providing free 24-hour-a-day, 365-days-a-year telephone service to a defined geographic area designated by the involved state(s) in order to respond to calls from the public and health professionals regarding poisoning exposures or information;
2. Offering access to hearing-impaired and non-English-speaking callers;
3. Providing staffing by qualified SPIs, PIPs, a managing director, and a medical director, and other medical backup as needed;
4. Achieving a minimum penetrance of 7 human poison exposure calls per 1,000 population served each year;
5. Developing comprehensive public and professional education plans;
6. Submitting all exposure data to the TESS database;
7. Having written operational guidelines and a disaster plan;
8. Providing quality assurance monitoring; and
9. Maintaining current institutional membership in AAPCC.

In addition to certifying centers, AAPCC provides the certification process for SPIs, allowing them to become certified specialists in poison information. To be eligible for this certification, a SPI must be trained as a nurse, pharmacist, or physician, and must acquire experience at a poison control center, consisting of 2,000 hours answering poison information or exposure calls, and 2,000 calls answered. One AAPCC requirement of certified poison control centers is that all SPIs acquire CSPI certification within 2 years of achieving eligibility. AAPCC considers certification by the American Board of Applied Toxicology to be an acceptable alternative

BOX 5-2
Requirements for Certified Poison Center or Poison Center System

Provide evidence that the center/system adequately serves its entire region.
Where multiple states are involved, designation from each state is necessary.
- Where a state declines to designate any poison center/system, designation by other political or health jurisdictions may be an alternative.
- In instances where more than one center or system is designated to serve the same area, evidence of cooperative arrangements must be provided.

Direct incoming telephone system that is extensively publicized throughout the region to both health professionals and the public.
- Must use AAPCC nationwide toll-free number (1-800-222-1222).
- May not impose a direct fee to individual members of the lay public (either by direct billing or pay-for-call services) for emergency calls received within its region.
- Must be able to respond to inquiries in languages other than English as appropriate to the region.
- Access for hearing-impaired individuals must be provided.

A plan to provide poison center services in response to natural and technological disasters must be in place.
Provide a Medical Director certified by the American Board of Medical Toxicology.
Medical Director full-time equivalent to be determined by call volume.
Staff certification and roles are as follows:
- For certification of center, 50 percent of staff must have passed exam.
- Provide triage and treatment for individual calls.
- Provide follow-up at appropriate intervals, validate cases, provide caller education.
- Obtain specific data for data collection and poison prevention.
- Provide education for other health care providers.
- Provide poison prevention education for public.
- Collect prehospital care and triage data; some hospitalized data may be obtained, but this is primarily the role of the Medical Director.

A managing director with toxicological supervision must be board prepared or board certified. For physicians the board can be ABMT or ABMS; for non-physicians the board must be ABAT.

SOURCE: American Association of Poison Control Centers (2003a, pp. 16–23).

to CSPI certification to satisfy this requirement or, for physicians, certification in medical toxicology by the American Board of Medical Toxicology or American Board of Medical Specialists.

Internally, poison control centers have instituted programs consisting of case reviews, call monitoring, and other measures designed to evaluate the quality of service provided (McGuigan, 1997). A quality improvement program focusing on "high risk, high volume or problem-prone cases" is required for AAPCC certification of poison control centers (American Association of Poison Control Centers, 1998). Centers collect data through callbacks to individuals or to the health care facility caring for them to assess outcomes of cases (Zuvekas et al., 1997).

Data are limited regarding the evaluation of poison control center performance for the purposes of ongoing quality assurance of case management, assessing the impact of the larger poison control system, or assessing various poison control strategies. Most such data (reviewed in Chapter 6) focus on poison center utilization or cost-effectiveness, specifically the role of poison control centers in reducing health care costs by reducing unneeded emergency department or primary care visits. Some studies have examined the performance of poison control center personnel in managing structured simulated cases (e.g., Thompson et al., 1983). This mechanism potentially could be used for ongoing quality control of poison control centers, but has not been specifically studied for this purpose. Essentially no data are available regarding the impact of poison control centers, individually or in aggregate, on health outcomes (morbidity or mortality). The lack of such data is partly because of the following factors:

1. The incompatibility of existing databases (nonuniform methods of data collection, disease definitions, and reporting) that makes population-based data collection and evaluation difficult;
2. Low rates of some outcomes such as mortality from poisonings, as well as the wide variety of types of poisonings managed by poison control centers, such that very large samples are needed to examine trends or make comparisons; and
3. A lack of funding for outcomes-based quality assurance activities or research, and the resulting lack of data on whether these are feasible or how to accomplish them.

Data regarding the effectiveness of poison education by poison control centers are discussed in Chapter 8. As with poisoning management data, these are largely limited to intermediate outcomes, such as the retention of the education messages or short-term changes in behaviors (e.g., safe storage of hazardous household products), rather than to health outcomes, such as a reduction in incidence of poisoning, morbidity, or mortality.

OPPORTUNITIES

1. Poison control centers have developed an infrastructure to respond to calls from the public or professionals regarding poisoning exposures or to obtain information about poisonings. This infrastructure has been adapted successfully to respond to HAZMAT incidents, and is now being further developed to contribute to all-hazards emergency preparedness and response, including biological or chemical terrorism. The latter role represents a significant opportunity to use the unique capabilities and expertise of poison control centers to support our national capacity to prevent, recognize, or respond to such incidents.

2. There are few specific funding opportunities available for research regarding poison control center services delivery. Dedicated funding for such studies, and for analysis of the data generated by poison control centers, would provide important and much needed information for the further development of poison prevention and control programs.

3. A certification process exists for poison control centers and for SPIs to become CSPIs; however, there is no certification for PIPs. Development of certification for PIPs could help to assure the competence of this expanding role. Certification of SPIs is currently the responsibility of the professional organization (AAPCC) to which these same individuals belong and which represents their interests. Certification of poison control centers is also the responsibility of AAPCC. Poison control centers are in fact required to join AAPCC to become certified. AAPCC assumed this dual role of professional organization and certifying body out of necessity because no other agency expressed an interest in acquiring the required expertise or developing such a process. A more common model for certification of health care professionals or programs is for certification to be the responsibility of an independent agency, rather than an organization to which the applicants belong. For example, medical toxicologists are certified by the American Board of Medical Specialties rather than by a toxicology organization. With the continued development of poison control centers and their increased integration into the public health system, alternative certification processes may offer advantages over the current system. Certification of centers, SPIs, and PIPs by an independent organization would offer greater independence of the process from the participants, wider input from the health care community, and wider recognition of their skills and contributions.

SUMMARY

Poison control centers perform a wide variety of activities related to the prevention, recognition, and management of poisonings. The types of poisonings addressed include commercial products, medications, drugs of abuse, plants, venomous animals, industrial chemicals, and potential agents of bioterrorism. Because accessibility is widely publicized and available around the clock and free, centers are used extensively as sources of information or advice regarding actual or potential poisoning exposures. Poison control centers may also be accessed when the nature of the exposure (e.g., chemical versus infectious agent) is uncertain, and perform a triage function by either providing information or referring the call to an appropriate agency. Poison control centers have evolved a common set of activities to deal with these needs, primarily centered around telephone lines staffed by specialists with training in clinical toxicology and backed up by medical toxicologists, a wide range of consultants, and extensive collaborations with public health agencies and first responders. Complementary educational efforts are also offered by most poison control centers. Some centers provide training of medical toxicology residents, research, surveillance to detect emerging syndromes or bioterrorism events, or contracts to provide information services to industry.

Both the core activities shared by most poison control centers and the additional activities offered by some centers fill distinct needs in the nation's public health system. Opportunities for enhancing poison control center effectiveness exist in the potential for further expansion of efforts in the area of emergency preparedness and response. There is a need for funding to support data analysis and research regarding center services delivery and further development of the certification process for SPIs and PIPs.

Appendix 5-A

Case Studies of Poison Control Centers in Emerging Health Situations

Three brief examples are provided. The first focuses on the development of cooperative arrangements between a poison control center and other local and state organizations involved in preparedness. The second provides general steps taken by a poison control center in response to calls about anthrax. The third illustrates the interaction between a poison control center and other health care agencies in diagnosing a threat and providing a response.

ROCKY MOUNTAIN POISON AND DRUG CENTER: INTEGRATION INTO PUBLIC HEALTH SYSTEM

The Rocky Mountain Poison and Drug Center (RMPDC) is the designated regional poison control center for Colorado; Montana; Las Vegas, Nevada; Idaho; and Hawaii. RMPDC is the coordinator for the Metropolitan Medical Response System (MMRS) for Denver County and works with the Denver Health Center for Public Health Preparedness. RMPDC's ability to bring groups together as a neutral third party within a competitive health care market led to its selection as the coordinator for the MMRS in 1997. RMPDC personnel are members of the HRSA-supported hospital preparedness advisory committee and work closely with the activities established via the CDC's Public Health Preparedness and Response for Bioterrorism Cooperative Agreement with the state of Colorado.

Specific public health preparedness activities undertaken by RMPDC include:

- Establishment of a set of emergency response procedures for call center workers, located on workstation desktops.
- Use of the nurse advice line to assist Denver Health with syndromic surveillance.
- Use of the nurse advice line to assist Denver Health with handling calls from the public regarding public health topics, including emerging problems, smallpox vaccination, and West Nile virus.
- Creation of stockpiles of antidotes and ability to survey for inventory levels of local/regional supplies of antidotes.
- Involvement in drills within Colorado (but not with other states served by RMPDC), including TOPOFF (top officials).

- Provision of training, including development of Web-based training modules in emergency preparedness and response.

TEXAS: ANTHRAX CALLS

(Note: The following is based on material provided to committee by Doug Borys, Director, Central Texas Poison Control Center; also, an article on anthrax calls to the Texas Poison Control Center is forthcoming, which will be authored by M. Forrester.)

The Texas Poison Control Center's (TPCC's) response to bioterrorism is focused on three areas: incident response, emergency preparedness, and professional and public education. Under incident response, TPCC reported receipt of hundreds of calls during the 2001 anthrax attack. Medical staff at TPCC wrote a public education piece on anthrax that was distributed on state and other Internet sites, and copies were distributed to health care providers and the public throughout Texas. Local emergency preparedness activities include the participation of TPCC staff in emergency planning committees of their own host institution, the city in which they are located, and in groups located throughout the state of Texas. At the state level, TPCC staff are members of the following organizations:

- Texas Department of Health, Hospital Preparedness Planning Committee
- Texas Department of Health Preparedness Coordinating Council
- Texas Institute for Health Policy Research, Policy Advisory Committee, Disaster Response Project
- Texas Medical Association, Bioterrorism Task Force

Also, the West Texas Poison Center serves as an active member of the El Paso/Ciudad Juarez-binational local emergency planning committee.

In terms of professional and public education, staff members of the six TPCCs are involved in a variety of terrorism-related panels, programs, and lectures, including:

- Chaired and lectured on a panel on biochemical weapons and the Latino community in Washington, D.C., at a national Latino health leadership conference.
- Served as regional director of advanced HAZMAT life support course (Texas, Louisiana, Oklahoma, and Arkansas) that included chemical terrorism preparedness; medical director for several courses.
- Participated in dozens of training sessions related to weapons of mass destruction (non-Advanced HAZMAT Life Support).

NORTHERN NEW ENGLAND POISON CENTER: ARSENIC EXPOSURE[1]

The Northern New England Poison Center (NNEPC) received a call from a rural hospital in the northeast corner of Maine requesting epidemiological assistance with what hospital staff thought was an outbreak of infectious disease or foodborne illness. One hour later, the hospital called the NNEPC staff on the poison control center hotline for information regarding the possibility that this incident involved poisoning—not an infectious disease or foodborne illness outbreak.

The staff reviewed potential causes and requested that the beverage and foods potentially linked to the exposures be secured for later evaluation, and paged the center's two boarded toxicologists for consultation. More patients had presented to the rural emergency department for treatment of severe symptoms of dehydration and hypotension. The most serious cases were transferred to a larger hospital. A differential diagnosis of possible toxins, including arsenic, paraquat, and ricin, was discussed and laboratory evaluation arranged with the state Environmental Testing Lab. A diagnosis of arsenic was confirmed and supported. AAPCC called NNEPC to investigate because all NNEPC cases had been automatically uploaded to TESS in Washington, DC. AAPCC notified CDC and the Department of Homeland Security.

NNEPC and the Maine Medical Center, the poison control center's host institution, had developed mechanisms for locating and delivering antidotes throughout the state. After September 11, 2001, NNEPC developed a state antidote stockpile to provide initial doses of antidotes for attacks with weapons of mass destruction. The state police and a plane normally used to transport critically ill patients transported the antidotes. NNEPC staff provided antidote administration and dosing guidelines to the clinicians involved in treatment, and worked at the treating hospital regarding emergency use of unapproved drugs. The center developed information materials for the health care professionals involved in treatment to help explain the therapy to patients and to obtain legal consent, as well as fact sheets for patients and their families to explain the effects of acute arsenic poisoning. At the request of treating health care professionals, NNEPC drafted a letter outlining appropriate monitoring and admission/discharge information. This information was used to guide physician management and satisfy insurers, who were calling the hospitals daily.

[1]Based on a letter to the Institute of Medicine dated May 22, 2003, from Maine Medical Center and interviews with CDC/ATSDR on August 8, 2003.

Multiple media requests were made of NNEPC. Toxicologists provided more than 20 interviews discussing arsenic toxicity in general and used the opportunity to raise awareness of unrelated but regionally significant chronic arsenic exposure through well water. The center staff followed these patients over subsequent weeks.

During this time, the poison control center continued to manage approximately 100 other hotline calls daily, and the publicity of the cases led to an increased number of calls about food poisoning and potential chronic heavy metal exposure. Information will be shared with the Maine Bureau of Health and the involved hospitals to furnish data that will be useful to others treating patients poisoned with arsenic in the future.

6

Current Costs, Funding, and Organizational Structures

The previous chapter describes the primary activities of poison control centers in the United States as well as the staffing, financial, and operational characteristics of these centers. The purpose of this chapter is to examine the costs and organizational structures of these centers in an effort to identify those characteristics that contribute to efficient operation and service delivery.

An earlier study of these factors was conducted in 1997 by Zuvekas et al. in response to a set of recommendations from the Poison Control Center Advisory Work Group (1996). The study's purpose was to identify approaches to more efficient service delivery. The methods included six in-depth case studies examining the time and costs associated with poison control center activities and a written survey mailed to all 75 centers that posed a series of general questions regarding size, location, activities, and organizational affiliations. The results showed that the majority of staff time and expense, regardless of center size or penetrance, was associated with providing telephone advice to the public and health care professionals. Their analysis of costs is discussed later in this chapter (see potential economies of scale).

The major recommendations were based on the use of communication technology to link the centers, thereby more effectively distributing the calls across them and allowing for consolidation. It was estimated that 50 centers might be a target number based on the reasoning that some small centers might be easily combined; technology would allow more efficient communication; and funding by the states might be more easily stabilized. No data were presented to directly support the recommendation for 50 poison prevention and control centers. Furthermore, no recommendations were presented regarding specific internal organizational structures, modes of operation, or the need to develop service quality measures. Finally, it was outside the scope of the study to compare the cost of poison control center service delivery with other delivery mechanisms such as emergency departments.

The data and analysis presented in this chapter are an effort to further explore and clarify these issues. The first section of this chapter focuses on a review of the economic evaluations of services delivered by poison control centers and the direct and indirect cost savings gained by using them. The second section describes the staffing and operational characteristics of centers, evaluates their economies of scale, and compares their organizational characteristics that exhibit contrasting values on size and efficiency.

In conducting these analyses, we used a variety of data sources, including the Toxic Exposure Surveillance System (TESS); nonaudited, self-reported survey data provided by the American Association of Poison Control Centers (AAPCC); statistical analysis of secondary data to explain variation in efficiency of poison control centers; and an analysis of qualitative interview data obtained from a sample of 10 poison control centers. These centers were a stratified, nonprobability sample based on cost per human exposure call handled in 2001, population served, and penetrance.

COST-EFFECTIVENESS AND COST-BENEFIT ANALYSES

While poison control centers perform a number of activities (see Chapter 5), as Phillips and colleagues state: "The primary benefit of poison control centers is that they provide advice that allows poisonings to be appropriately handled at home or triaged to a health care facility, thereby avoiding unnecessary visits to health care facilities or inappropriate and potentially harmful home treatments." They also serve as a free resource for those without primary care or with limited access to primary care. In 2002, Watson et al. (2003), using TESS data, found that public calls to a poison control center were managed in a non-health-care facility—usually in the patient's home (74 percent); were treated in a health care facility (23 percent); and were referred to a health care facility but the patient did not go (2 percent). Indeed, it is the benefits of this triage role, as well as better health outcomes from the center's interfacing with emergency departments, that are the focus of the peer-reviewed literature on economic costs.

A number of published studies provide cost-effectiveness and cost-benefit analyses of various aspects of the poison control center system and some take account of the potential morbidity and mortality benefits.[1] In many of these, the lack of necessary or appropriate data presents a challenge. Nonetheless, taken as a whole, this literature makes a convincing case that, at least in terms of treatment management guidance for the public, poison control centers save the health care system economic resources and save members of the public time, lost wages, and anxiety. Indeed, the literature supports the proposition that for every dollar spent on treatment management activities, multiple dollars are saved by the health care system as a whole. These studies do not examine the cost-effectiveness among poison control centers, but rather compare the centers with other health care providers such as emergency departments.

The focus of the published literature on the economics of poison control centers also accords with the fact that treatment management guidance is the dominant activity of centers in terms of expenditures, representing on average some 70 percent of those expenditures (Zuvekas et al., 1997). Still, insofar as the peer-reviewed literature on economic costs has examined only their role in treatment management guidance responding to direct calls from the public, conclusions about the cost-effectiveness of other center activities are not available, including providing consultation for patients in intensive care units.

One instructive study of the cost-effectiveness of poison control centers is King and Palmisano (1991). Louisiana discontinued its center during the late 1980s; King and colleagues analyzed a natural experiment based on the resultant experiences. The researchers compared various outcomes during the discontinuance of the center to the period just before the discontinuance, as well as to the outcomes in neighboring Alabama during the period of Louisiana's discontinued service (a period during which the poison control center service in Alabama remained). During the discontinuance, it was estimated that self-referrals to the emergency department increased by a factor of more than fourfold and the number of home management cases declined to less than half. Before the closure of the Louisiana poison control center, the triage patterns in Alabama and Louisiana were nearly identical. During the closure, the rates of poison-

[1]Cost-effectiveness analysis (CEA) captures the cost, per unit of specific outcome, of competing interventions. CEAs include cost per averted emergency department visit, cost per saved snail darter, and so forth. CEA is often the best way to rank competing interventions that would divide a rather fixed budgetary pie to achieve the same objectives. CEA is also valuable in ranking interventions whose outcomes are hard to value in monetary terms. Cost-utility analysis examines cost per unit of health outcome, operationalized as quality-adjusted life-years. Cost-benefit analysis assigns monetary values to all health and nonhealth outcomes.

ing self-referrals to the emergency department in Alabama did not increase.

The projected incremental costs to Louisiana during the discontinuance of excess health care facilities visits were estimated at $1.4 million. This can be compared with the $400,000 in savings to the state from closing its center. In short, had the state spent the $400,000 to keep its poison control center open, it would have saved the system $1.4 million, for a net savings of $1 million. Phrased differently, this implies a savings to the health care system of more than $3 for every dollar invested in the center. This is an underestimate of the benefits of the poison control center insofar as it does not take into account a reduction in mortality and morbidity, or in anxiety and time to the public.

Phillips et al. (1998) used the results of another "natural experiment" to examine cost savings of poison control centers. Between 1993 and 1994, a single county in California lost funding for its center. Public callers to the center received a recorded announcement advising them to dial 911 for poisoning exposures and information. If they called 911, they were patched into a neighboring poison control center to which they had prior direct access. An analysis was done of individuals who called the center during this interruption of service compared with a matched set who called subsequently after service was resumed. The outcomes during the period of blockage were substantially different than during the control period, even though the disruption did not involve lack of access to the center, but only patching into one through 911. Fourteen percent of callers with restricted access were treated in an inappropriate location (e.g., treated by an emergency department when they might have been managed at home), compared with 2 percent who had direct access to a poison control center. In a further analysis of the costs associated with the same blocked-caller episode, it was found that restricting access resulted in an additional $10.98 per case in net societal costs (all costs and benefits, including patient time and transportation and marginal costs for resources used as a result of the block) and an additional $33.14 per case in health care purchaser costs (Olson et al., 1999).

Harrison et al. (1996), in one of the most thorough of the existing analyses, adopted a decision theoretic analysis to evaluate treatment management guidance for the public. In addition to secondary data on costs such as emergency department visits, ambulances, and other factors, Harrison and colleagues used data assembled from an expert panel of toxicologists to estimate probabilities of morbidity outcomes, mortality outcomes, and adverse treatment impacts of cases coming into an emergency department. Thus the researchers were able to consider not only differences in direct costs to the health care system, but also differences in morbidity and mortality due to the provision of poison control center services. They did this in the context of four typical poison exposures (e.g., acute cough or cold preparation overdose in children younger than 13 years of age).

For cough or cold preparation overdose, they concluded that when calls go through a poison control center first, the costs per case average $414 (in 1995 dollars), with a probability of .004 of morbidity and .000006 of mortality. In contrast, without a poison control center, the costs per case are $664, with a probability of .01 of morbidity and .00002 of mortality. The authors concluded that for cold preparation overdoses, and under the assumptions of their model, poison control centers lead to lower costs to the health care system and better outcomes in terms of morbidity and mortality. The cost savings result both from the centers triaging visits to the emergency departments and from the cost savings associated with better health outcomes for those cases going to the emergency departments that have already gone through the centers. These cost savings amount to an average of $250 per case ($664 minus $414). This compares with a cost per call to the poison control center in the $25-to-$30 range (Zuvekas et al., 1997).

Applying the same methodology to acetaminophen overdoses, the cost savings are $343 per case. For antidepressant overdoses, the cost savings are estimated to be $347 per case. For a standard cleaning substance exposure in children, the cost savings are estimated to be an average of $297 per case. In each of these cases as well, each dollar of poison control center expenditure on treatment management guidance results in a cost savings to the health care system of $10 or more. This is a lower bound estimate insofar as it does not take into account the benefits of poison control centers in terms of time and anxiety to the public, nor to their substantial positive impact on morbidity and mortality outcomes.

Miller and Lestina (1997) provided an analysis of cost savings from poison control centers that has been widely cited in legislation, congressional testimony, and many popular venues. It concluded that for every dollar invested in poison control centers, there are savings of about $6.50 to the health care system as a whole. While this magnitude of cost savings is not wildly different from other convincing analyses, there are significant limitations to the Miller and Lestina analysis.

Miller and Lestina estimated that the total societal costs of poisonings would be reduced from $3,315 million if there were no poison control centers to $2,905 million if the whole population of the country had access to a center, a savings of $310 million. They compared this with the cost of centers, which they indicated to be between $60 and $80 million. These data form the basis of their conclusion that there is a 6.5-to-1 cost savings for each dollar invested in poison control centers.

The $390 million in savings, an amount greater than Miller and

Lestina's total estimate of net savings, comes from increased costs of hospitalization. One assumption that leads to these savings is problematic. This assumption concerns individuals who, had there been a poison control center, would have been treated at home but, absent a center, go to the emergency department. It is assumed that these individuals will have the same probability of hospitalization and the same hospitalization costs as experienced by the whole population of individuals who are hospitalized for poisoning. If this assumption is not valid, the resulting analysis will overestimate the savings attributable to poison control centers.

On the other hand, there are some cost-saving features of poison control center systems not taken into account by Miller and Lestina. For example, Miller and Lestina used total center costs (e.g., including education), not costs associated with telephone-based case management. Also, better health outcomes for cases that did need to go to the emergency department (e.g., Harrison et al., 1996) were not taken into account. Thus, while the widely quoted figure from Miller and Lestina (1997) of a 6.5-to-1 cost savings for investments in poison control centers is not outside the bounds found in other studies, their methodology makes their particular conclusion problematic.

All of the above analyses focus on tangible cost savings associated with poison control centers. In such analyses, intangible psychological benefits to the public of such centers are not considered. Yet parents and caregivers often experience lowered levels of anxiety if they are able to call the centers and be reassured, when warranted, that a trip an emergency department is not necessary, and they are subsequently advised about how to treat the situation at home. There is also the comfort of knowing that this service exists even if one does not use it. These intangible benefits are hard to quantify.

One study that considers these psychological benefits is Phillips et al. (1997). The researchers asked individuals who had called a poison control center and members of the public what they would be willing to pay to have a center to which they could have access. A wide range of methodological concerns can be raised about hypothetical answers to willingness to pay that are given by members of the public without the benefit of deep reflection and thoughtful calculation of intangible and tangible benefits; nonetheless, the results are informative. For those who had called a poison control center, the average willingness to pay to have a center was $6 to $7 per month, or $72 to $84 per year; for members of the public, the results were an average of $2.55 per month, or about $30 per year. The willingness to pay these figures can be compared with the actual cost per person in service area of a treatment management guidance function, which Zuvekas et al. (1997) estimated to range from 22 to 58 cents per year. Thus, Phillips and colleagues (1997) found a difference of at least 50 to 1 between the lower estimate of willingness to pay and the upper estimate of cost incurred. Again, this provides additional evidence of the benefits of poison control centers compared with their cost.

CURRENT COSTS, ORGANIZATION, AND STAFFING

While the previous review of the existing literature suggests cost savings resulting from the activities of poison control centers, the samples on which many of these studies are based are limited and may not represent the population of centers as a whole. Furthermore, this literature does not attempt to account for potential variation in operating efficiency of centers or explain the sources of that variability. Below we analyze available data from AAPCC to address these issues. Because these data were collected for other purposes and are largely self-reported by the individual poison control centers, they are not ideally suited to such an analysis. However, given the paucity of research in the area of center operations and financial performance, even these limited data can provide valuable additional information and help inform recommendations relating to their size, structure, and consolidation. Results of these analyses may also highlight potentially important organizational or policy issues that require further investigation with better data and more rigorous research methods.

Preliminary Characterization

This section provides a description of the population of poison control centers in terms of their staffing, population served, revenue sources, and other operating characteristics. In 2001 (American Association of Poison Control Centers, 2002b) there were 65 poison control centers, located in 42 states plus the District of Columbia. The number of centers located in each state ranged from zero (Alaska, Idaho, Montana, Nevada, Rhode Island, South Dakota, and Vermont) to six (Texas).[2] Regional distribution of poison control centers in the United States is depicted in Figure 6-1. The majority of centers are located in the South, while the remainder are distributed roughly evenly among the Northeast, Midwest, and West. Most poison control centers were certified by the AAPCC (N = 51; 78 percent). Descriptive data from the AAPCC survey administered to the poison control centers in 2001 shows the following characteristics (see Table 6-1 for the mean, range, and standard deviation of each characteristic[3]).

[2]Data describing poison control center operations in 2002 were not available for analysis.

[3]California is treated as one center because of the form in which the revenue and expense data were provided to the Committee.

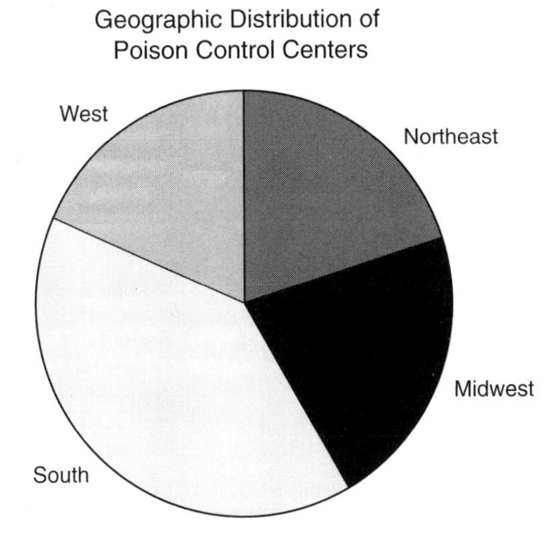

FIGURE 6-1 Geographic distribution of poison control centers.

Population Served and Call Volume

Figure 6-2 shows the distribution of population size served across all poison control centers. As can be seen from the figure, the size ranges from 634,000 to nearly 35 million. Because the largest poison control center is California's, which operates as a four-region poison control system, average population size served by centers may be highly skewed; therefore, the median population served of 3.8 million is probably a better indicator of the average for all poison control centers. Centers handle three types of calls: human exposure, information, and animal exposure. Total call volume for human exposure and information calls combined averaged 55,687 per center in 2001, although there was great variability from center to center (range 4,716 to 300,321).

Staffing

The majority of poison control centers (92 percent) had 24-hour staffing by specialists in poison information (SPIs). Compared with SPIs and certified specialists in poison information (CSPIs), poison information providers (PIPs) made up a small proportion of the center staff (8 percent on average, range 0 to 67 percent). PIPs tend to have different backgrounds and have less training than SPIs or CSPIs. They are typically drawn from

TABLE 6-1 General Characteristics of Poison Control Centers

	N	Median	Mean	Standard Deviation	Minimum	Maximum
Population Served	62	3,765,293	4,593,562	4,492,049	634,448	34,501,130
Calls:						
Human poison exposure calls	62	31,514	37,155	31,125	3,150	230,438
Information calls	62	11,928	16,569	13,066	1,491	62,003
Animal poison exposure calls	62	1,043	1,850	2,380	0	12,118
Nonexposure calls	62	89	117.7	97.23	0	450.0
All calls	62	47,272	55,687	41,767	4,716	300,321
Human exposure calls: % total calls	62	68.5%	66%	11.2%	37.7%	88.%
Penetrance:						
Human exposure calls per 1,000 population	62	8.291	8.611	2.502	4.946	16.79
Staffing:						
Managing director full-time equivalents (FTEs)	61	1.00	0.971	0.455	0	3.500
FTEs of medical director funded	62	.50	0.667	0.498	0	3.150
FTEs of medical director	62	.60	0.751	0.559	0.0125	3.500
FTEs: Administrative staff	62	2.00	2.100	1.976	0	10.73
Health educator FTEs	62	1.00	1.151	1.048	0	7.500
Total PIP and CSPI/ SPI FTEs	62	9.85	10.94	7.511	0	55.90
Total PIP FTEs	62	0	1.060	2.366	0	14.50
Total CSPI plus SPI FTEs	62	9.30	9.877	5.915	0	41.40
FTEs: SPI plus CSPI : % total FTE	61	100.0%	91.8%	15.5%	33.3%	100.0%
Expenses:						
Total expenses	61	1.2E6	1.38E6	968,595	116,579	6.89E6
Personnel expenses	61	1.0E6	1.1E6	775,509	101,579	5.7E6
All nonpersonnel expenses	61	181,071	276.431	254,894	15,000	1.05E6
Expenses per 1,000 population	61	303.2	336.8	131.9	82.76	723.8

TABLE 6-1 Continued

	N	Median	Mean	Standard Deviation	Minimum	Maximum
Expenses per human exposure call[a]	61	37.18	39.98	13.54	13.72	76.78
Expenses per call (all)	61	24.83	26.31	8.509	9.222	54.00
Personnel expenses per 1,000 population	61	257.1	272.8	112.7	70.60	605.0
Personnel expenses per human exposure call	61	30.81	32.15	11.08	11.71	65.29
Personnel expenses all calls	61	20.82	21.23	7.124	7.868	42.09
Nonpersonnel expenses per 1,000 population	61	55.82	63.99	39.64	3.303	183.7
Nonpersonnel expenses per human exposure call	61	7.11	7.834	5.076	0.661	23.20
Nonpersonnel expenses all calls	61	4.52	5.083	3.256	0.454	15.83
Total revenue	61	1.2E6	1.4E6	1.02E6	116,579	7.07E6

[a]Includes industry calls which represent less than 1 percent.
SOURCE: American Association of Poison Control Centers, 2001 Survey (2002a).

non-health-services backgrounds and, like SPIs, they are not certified in poison information. One poison control center (North Dakota) had neither SPIs nor PIPs, although they did field 4,641 calls from a population of 634,448.

The median number of health educator full-time equivalents (FTEs) was 1 (mean 1.1, range 0 to 7.5); 4 centers (6 percent) had no health educators; and 17 (29 percent) had more than one health educator FTE. There was usually a managing director (average FTE 0.97, median 1.0, range 0 to 2) and a medical director (average funded medical director FTE 0.67, median 0.5, range 0 to 3.1). In nine poison control centers (14 percent), the managing director was also the medical director. Two poison control centers had no managing director and four had no medical director. While most (77 percent) poison control centers had one or more administrative staff FTEs (mean 2.1, range 0 to 10.7), seven centers (11 percent) had none. The median number of FTEs for administrative staff was two.

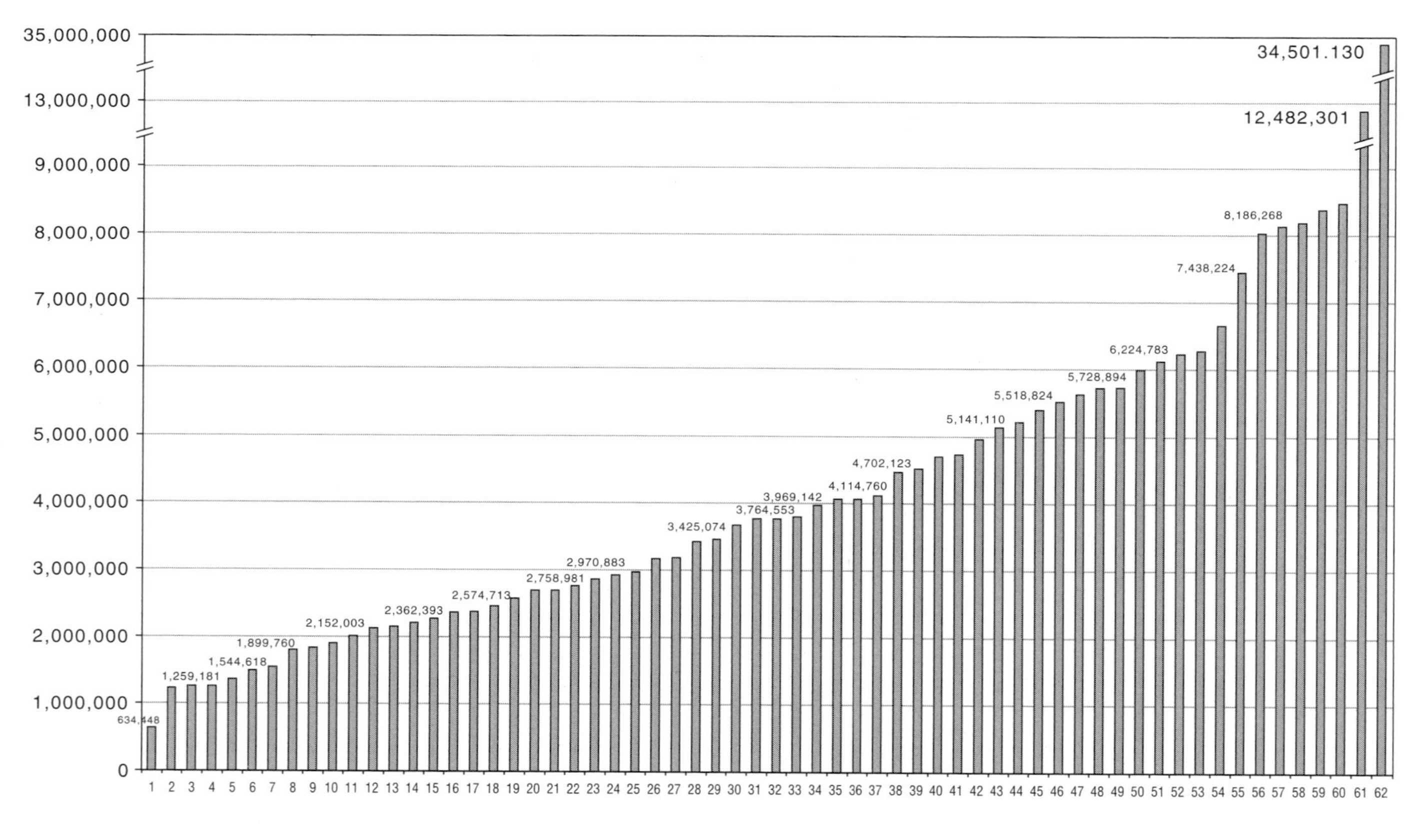

FIGURE 6-2 Population served at poison control centers in 2001 (N = 62).

Funding and Expenses

Reported revenues received ranged from a low of $116,000 to a high of $7.07 million[4] (median $1.27 million, mean $1.40 million). As Table 6-2 (http://www.aapcc/org/surveyresults2001.htm) indicates, the funding of poison control centers is fragmented. Across the states there are more than 30 separate funding sources. Approximately 6 percent of total poison control center funding comes from federal and state Medicaid programs; 3 percent from federal block grants; and 8 percent from other federal programs, for a total of 17 percent from federally associated programs. Approximately 44 percent of total funding comes from states, with many different approaches to state funding, ranging from line-item appropriation to state-funded universities to telephone surcharges. Hospitals represent 15 percent of total funding (either as host institutions or network members). Another 3 percent of funding comes from a wide range of donations and grant sources and 20 percent from myriad other sources, some of which are itemized in Table 6-2.

This fragmented pattern of funding is evident within as well as among states. The mean number of funders per center is 5.5 (mean 5.5, SD 2.7, range 2–12). In most cases, the three top funding sources provided the bulk of the funding (proportion derived from the top three funders ranged from 57 to 100 percent, with a mean of 91 percent).

Because of the lack of regular funding sources, poison control centers report that significant time is spent raising revenues and that there has been substantial instability in funding. As financial pressures on state governments and health systems have risen, the willingness of traditional funders to continue to provide revenues has diminished, leaving many centers facing great uncertainty, budget pressures, and cutbacks.

For example, in the past year, two centers have been forced to close due to changing priorities and budget cuts. Furthermore, at least one center lost its institutional funding and spent several months convincing the state to provide the needed support; during this time center staff were operating under extreme uncertainty and much program planning was suspended.

Total expenses for personnel plus telecommunications and equipment costs averaged $1.38 million (range $117,000 to $6.89 million in 2001 dollars). In terms of personnel costs, expenses ran about $40 per human exposure call (range 13 to 77); this was equivalent to 33 cents per capita (range 8.2 to 72 cents). Figures 6-3 and 6-4 display the distributions of personnel and nonpersonnel costs per human exposure call across all

[4]California system.

TABLE 6-2 FY2001 U.S. Poison Control Center Funding, All Centers Providing Data (N = 64)[a]

Funding Source	Amount ($)	Percentage of Total	Number of Centers with Funding Source
Federal (excluding Medicaid and Block grants)			
HRSA stabilization grant	6,679,259	6.39%	52
HRSA incentive grant	567,875	0.54%	11
Other	1,049,619	1.00%	6
Total Federal	8,296,753	7.94%	69
Medicaid			
Federal	3,593,737	3.44%	5
State	2,456,235	2.35%	7
Unknown	311,529	0.30%	1
Total Medicaid	6,361,501	6.09%	13
Block grants:			
Maternal Child Health	3,157,555	3.02%	6
Preventive Health	80,000	0.08%	1
Other	270,000	0.26%	2
Total block grants	3,507,555	3.36%	9
State:			
Line item appropriation	19,385,768	18.54%	27
Through state-funded university:			
Direct line item appropriation from state	8,084,711	7.73%	7
University designated funds	5,529,715	5.29%	17
Total through state-funded university	13,614,426	13.02%	24
Telephone surcharge	9,018,608	8.63%	8
911 fees	50,000	0.05%	1
License fees	0	0.00%	0
Other	4,444,372	4.25%	19
Total state	46,513,174	44.49%	79
City	665,352	0.64%	5
County	896,301	0.86%	13
Hospital (other than host):			
Member hospital network	5,162,071	4.94%	12
Donations/fees from area hospitals	350,737	0.34%	8
Total hospital (other than host)	5,512,808	5.27%	20
Host institution:			
Host hospital	10,841,344	10.37%	38
Other host institution	378,696	0.36%	3
Total host institution	11,220,040	10.73%	42
Donations/grants:			
Children's Miracle Network	225,528	0.22%	5
Community service organizations	70,225	0.07%	4
Corporations	486,064	0.46%	13
Events	23,815	0.02%	1

TABLE 6-2 Continued

Funding Source	Amount ($)	Percentage of Total	Number of Centers with Funding Source
Foundations	1,459,935	1.40%	17
Individuals	418,995	0.40%	19
United Way/Federated campaigns	869,139	0.83%	10
Total donations/grants	3,553,701	3.40%	69
Health insurers/HMOs (excluding HMO hospitals)	24,400	0.02%	3
Other business sources[b]	2,169,625	2.08%	20
Other	2,491,469	2.38%	14
Total direct funding	91,212,677	87.25%	64
Estimated additional in-kind and subsidized support[c]	13,327,026	12.75%	51
Total funding	104,539,703	100.00%	64

[a]One noncertified center serving a population of 1,224,398 submitted a survey with incomplete funding information.
[b]E.g., portion of industry contract funding used to provide general poison control services in the center's region.
[c]Estimated as 15 percent of total identified expenses.
SOURCE: http://www.aapcc.org/surveyresults2001.htm.

centers. Note that the figures suggest a relatively smooth distribution on both measures of expenses, and that it is therefore unlikely that a few "outlying" centers are disproportionately influencing the averages on these measures. Taken as a whole, AAPCC data indicate that poison control centers display wide variability in virtually all aspects of staffing, operations, and costs. The population of centers can therefore best be characterized as highly heterogeneous, with few common structural or operational characteristics to form the basis for characterizing a "typical" center.

Chapters 2 and 5 discussed the core functions of poison control centers. Table 6-2 provides a rough estimate of the national costs of supporting the core activities, approximately $100 million per year. This figure is slightly less than the sum of reported expenditures in Table 6-2, with the reduction providing some consideration of the cost of noncore activities (e.g., public education; education of health care professionals not working toward a career in medical or clinical toxicology). Conversely, if an average total cost of approximately $40 per human exposure call is used (this amount represents poison control center expenses divided by the number of human exposure calls) and this amount is multiplied by the call vol-

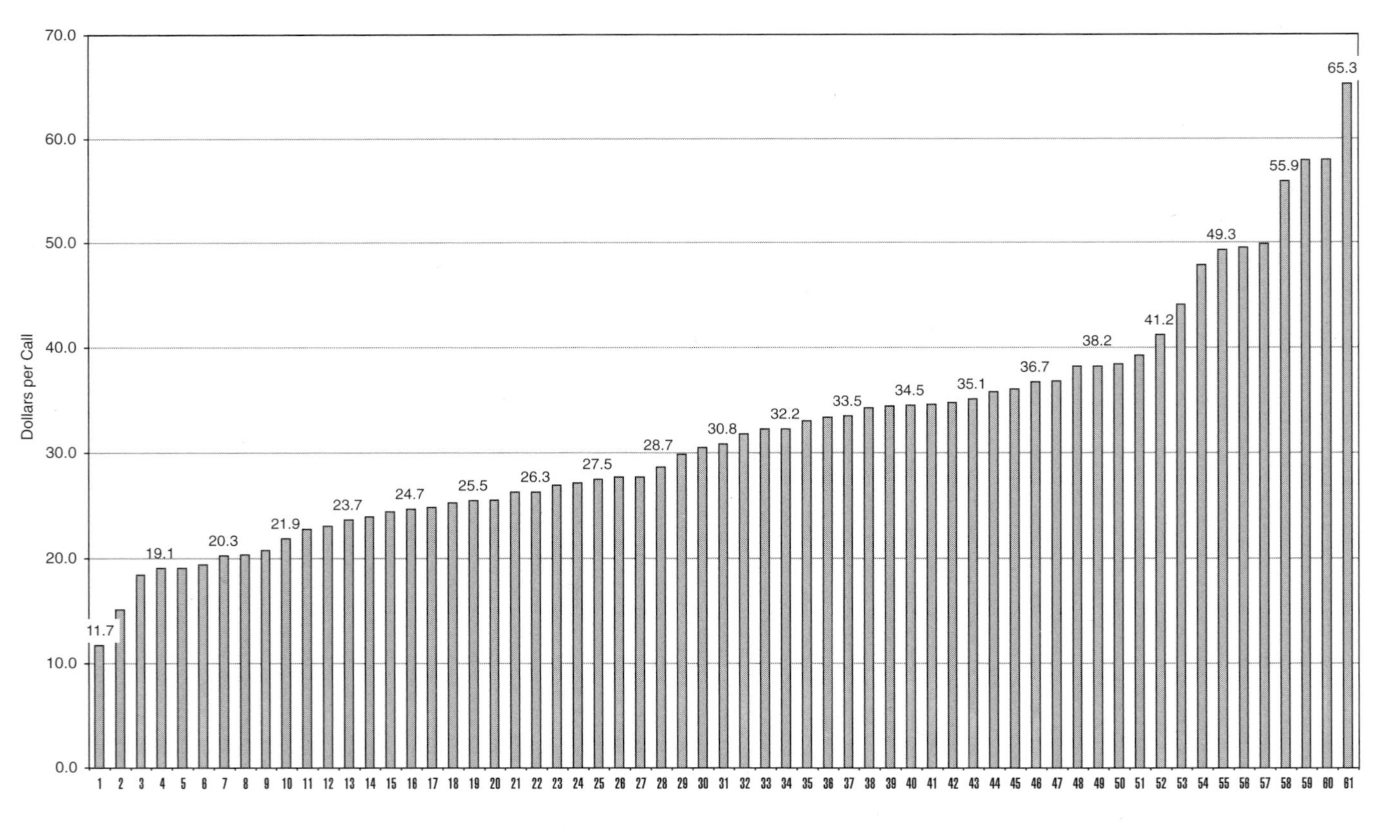

FIGURE 6-3 Total personnel expenses per human poison exposure, 2001 (N = 62).

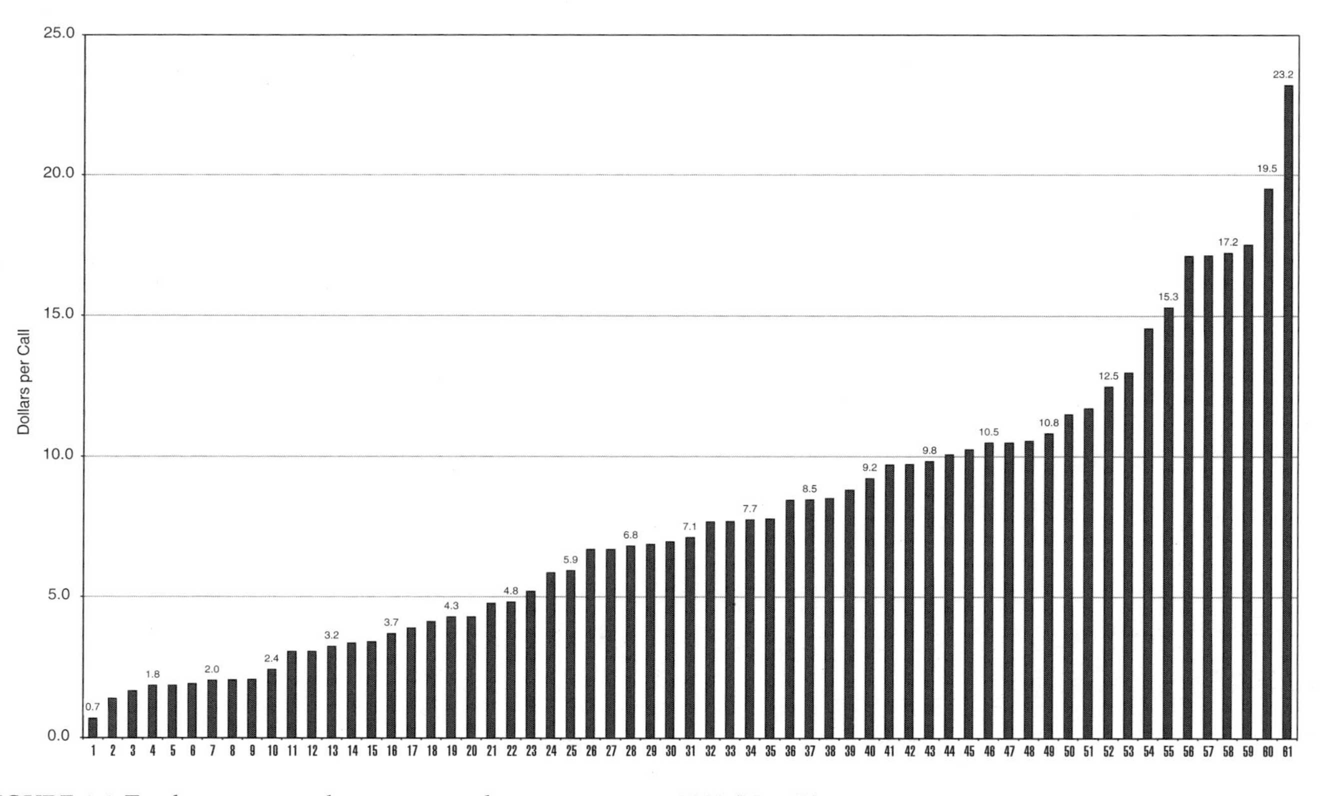

FIGURE 6-4 Total nonpersonnel expenses per human exposure, 2001 (N = 62).

ume reported in the 2002 data (approximately 2.4 million calls) a figure with an order of magnitude of approximately $100 million per year is obtained.

Potential Economies of Scale Through Consolidation or Networking

An important issue in the economics of poison control centers is whether there are economies of scale in the cost of operating the centers as a function of the number of individuals in the centers' catchment area. As mentioned earlier, the population served by a poison control center ranged from 634,000 to 35 million in 2001. Are centers serving larger populations able to provide the same service at reduced costs per call or per person in the population? If so, that would be one argument for consolidating centers into fewer and larger centers.

In an unpublished study, Zuvekas et al. (1997) conducted a detailed examination of the activities of six poison control centers and allocated all activity costs among eight functions: treatment management guidance, public prevention and promotion, professional education, protocol assessment and poison surveillance and data collection, interaction with local and regional public health and safety officials, and research. They found that the costs of treatment management guidance ranged from $21 (in 1995 dollars) to $43 per human exposure call, with four of the six centers in the $21-to-$27 range. They also found that the lowest estimated cost per person in service area ($.22/year, in 1995 dollars) and the lowest cost per human exposure call ($21.14) occur in the largest poison control centers (with a service area population of 8 million) studied, whereas the highest of these pair of costs ($.58 and $43.18, respectively) occurs in the smallest center (with a service area population of 1.3 million). Additionally, the second smallest poison control center studied, with a service area population of 2.8 million, had costs slightly higher than most of the larger centers. It is important to note that these analyses also do not consider other benefits and tradeoffs associated with increased or reduced costs. For example, while high-cost centers serving relatively small populations may appear inefficient from a purely financial standpoint, they may be offsetting benefits from those increased costs, including more diverse services offered or more innovation in the delivery of services (e.g., use of promotoras, or community health workers, in predominantly Hispanic regions).

Even considering these caveats, the data are not definitive. The costs per call and per person served do not appear in this very limited sample to increase once a center is above 2 million served. Additionally, there are no measures available about quality of service. It may be that a center that happens to be small compared with other centers hires more senior and more professionally capable staff. The result could be higher costs per call served, but also better outcomes in terms of triage or interfacing with emergency departments. There is no evidence in the literature about the cost-effectiveness of more experienced (and more expensive) or better trained (and more expensive) staff, nor are data available on whether the poison control centers with higher costs per call experience higher costs because of these considerations.

To address this issue further, we used 2001 survey data on poison control center operations (American Association of Poison Control Centers, 2002a), described elsewhere in this chapter, to examine economies of scale. If economies of scale were present, one would expect to observe a decline in expenses per call as the population served increases, followed by a plateauing of expenses per call at some breakpoint in the population served distribution. The data are not well suited for fine-grained analysis of economies of scale, but insofar as reported data on costs can be used to construct costs per human exposure call, and insofar as no differences in terms of quality of service provided by the centers are assumed, they can be informative. Based on Figure 6-5, only a very weak relationship is found between size of population served and costs per call handled, albeit this relationship is in the expected direction. Furthermore, population served explains only about 10 percent of the variation in cost per call, suggesting that cost savings may be gained through changes other than consolidation. Moreover, even if there are economies of scale in handling calls from the public, this is only one of the core functions of a poison control center, and the question of the optimal size of a center must consider these other functions as well.

Explanatory Models of Poison Control Center Costs

In an attempt to explain the wide variation in costs incurred by poison control centers, expenses per human exposure call (a call fielded about a person who may or may not have been poisoned) were modeled as a function of a limited set of explanatory variables available from the AAPCC annual survey. In general terms, these explanatory variables captured centers' regional location, source of funding, size of catchment area, and staffing characteristics. The primary benefit of this analysis was to allow us to capture the independent effects of these explanatory variables, holding constant the effects of the other variables in the model. Results were obtained through a set of regression models run for the 58 to 59 centers with complete data on the predictors in the models: population served, educator FTEs per PIP/SPI/CSPI FTEs, percentage of SPI/CSPI FTEs that were certified (CSPI), having 24-hour coverage, percentage of revenue from Health Resources and Services Administration stabilization grants, percentage of revenue from host hospital, percentage of revenues

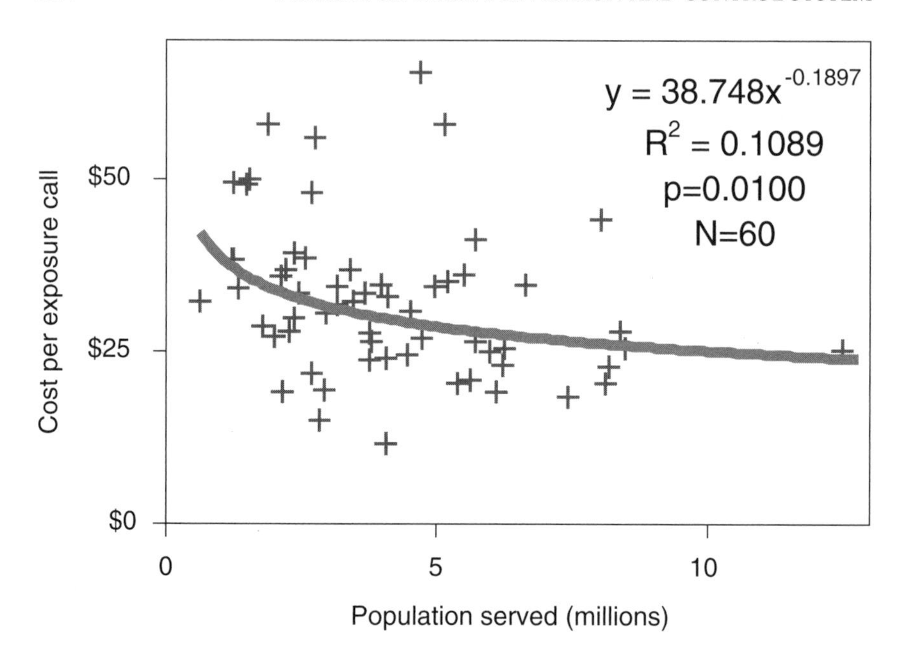

Data Considered	Mean ± SD, Range [Minimum, Maximum] (N)
Population Served (millions)	4.38 ± 2.55, 3.80 [0.63, 12.8] (65)
Cost per Exposure Call	32.3 ± 11.1, 21.3 [11.7, 65.3] (60)

Relationships Examined	Relationship	R^2	p-value
Linear (2 parameters, N = 60)	$y = -1.44 \times 38.2$	0.0893	0.0204
Power (2 parameters, N = 60)	$y = 38.7 \times ^{-0.1896}$	0.1089	0.0100

FIGURE 6-5 Cost per human exposure call versus population served.
NOTE: Regression line shows least-squared fit of log(cost per call) to log(population served).

from area hospitals (other than the host hospital), location in the Northeast region of the United States, and location in the Southern region of the United States.

Three models were run, predicting Total Expenses per Human Exposure Call, Personnel Expenses per Human Exposure Call, and Nonpersonnel Expenses per Human Exposure Call.

Total expenses per human exposure call was negatively associated with a larger population served. Factors positively associated with this dependent variable were location in the Northeast region, 24-hour coverage by SPIs, and more health educator FTEs.

Results for personnel expenses per human exposure call were virtually identical to the results for total expenses per human exposure call. This dependent variable was negatively associated with a larger population served and positively associated with location in the Northeast region, 24-hour coverage by SPIs, and more health educator FTEs. This pattern of results suggests that total expenses are driven largely by factors related to personnel costs. This finding is further supported by the fact that the model for *nonpersonnel cost per human exposure call* was not significant overall and none of the predictors reached significance at the $p<.05$ level. This suggests that compared with personnel expenses, expenses related to nonpersonnel items (e.g., telecommunications) play a relatively minor role in total center cost differentials.

Nonpersonnel expenses per human exposure call, including telecommunications, increased with subsidization and area hospital funding, but decreased with host hospital funding. Nonpersonnel expenses were positively associated with 24-hour SPI coverage, and two were negatively associated with greater percentage of human exposures calls.

The most important predictor of expenses per call was 24-hour coverage, which accounted for a $19 difference in cost per call. Whereas the vast majority of poison control centers have such 24-hour coverage as a requirement of accreditation by AAPCC, these results further reinforce the importance of personnel costs and staffing patterns as key determinants of center costs and efficiency. Of nearly equal importance in the model was center location in the Northeast region of the county. Relative to centers located in the West and Midwest, these poison control centers incurred $11 to $13 more per call, even holding constant other variables such as population served and funding source. Again, these differences are likely to reflect higher wage rates in this part of the country, a factor that drives up the cost of providing services. Finally, it should be noted that these models account for at best 42 to 45 percent of the total variation in cost per call. This means that a substantial portion of cost differentials across centers is not accounted for by variables in the models and that other unmeasured factors may be contributing to such differentials.

Qualitative Analysis of Organizational Characteristics

Sample Selection

The analysis of the 2001 AAPCC survey data shows significant variability among poison control centers on a number of dimensions, including total costs, personnel costs, and nonpersonnel costs per (1) population served, (2) total calls, and (3) human exposure calls. The regression models described earlier were limited insofar as they contained only variables available from the AAPCC survey and on average explained less than half of the variance in expenses per human exposure call. Notably absent from the set of explanatory variables were organizational characteristics of the poison control centers that might account for differences in costs as well as other outcomes of interest, such as staff turnover and retention or service increases and declines.

To describe and compare the organizational characteristics of poison control centers in depth, a stratified nonprobability sample of 10 centers was selected for further qualitative study based on semistructured interviews with key informants at each site. The two principal strata of selection were based on cost per human exposure call handled in 2001 and total population served combined with total human exposure calls per population per 1,000 (penetrance).

The first stratum was defined by being the highest or lowest quartile for at least one of three defined categories of expense per exposure call: total, personnel, or nonpersonnel expense. Six of the 10 centers were in the highest quartile by at least one of these measures. Four were in the lowest quartile. The second stratum was defined by population and penetrance. Centers based on high or low quartile were also selected here, but these could have been discordant. Five centers were in the highest quartile of population served and three were in the lowest, while four were in the highest quartile of penetrance and four were in the lowest. All 10 were in an extreme quartile for at least one of the two. Five were discordant, either highest-lowest or lowest-highest. Either the managing director or the medical director of each center participated in a one-hour interview. The questions are included in Appendix 6-A.

The distribution of the operating characteristics among the centers included in the interview sample is provided in Table 6-3.

TABLE 6-3 Distribution of Operating Characteristics Sample for Qualitative Study

Poison Control Center	Total Cost	Personnel Cost	Population	Penetrance
A		Low	High	High
B	High		High	Low
C	High		High	Low
D		High	High	Low
E			High	
F	Low		Low	High
G	High	High	Low	High
H		High	Low	
I	Low	Low		Low
J	Low	Low		High

Findings

Tables 6-4 and 6-5 present an overview of the differences between the lower (high-cost) and higher (low-cost) efficiency centers and between centers serving large and small populations. The following sections provide a brief narrative description of survey response in each area of inquiry.

TABLE 6-4 Differences Between Low-Efficiency and High-Efficiency Centers

Survey Variables	High Efficiency	Low Efficiency
Affiliation	Most are private not-for-profit	Most are public not-for-profit
Staffing	Slightly lower staff turnover rate Less likely to mention low pay as a contributor to staff turnover More staff hours spent weekly on all poison control center activities; more staff hours spent on poison prevention	Slightly higher staff turnover rate More likely to mention low pay as a contributor to staff turnover Fewer staff hours spent weekly on all poison control center activities; fewer staff hours spent on poison prevention other than direct client response
Services and activities		
• Current practices	More full-time equivalents (FTEs) devoted to education and outreach	Fewer FTEs devoted to education and outreach
• Areas of growth	No differences	No differences
• Areas of decline	Industry contracts, general call volume, resident training	Professional education, fellowship training, general call volume
Interorganizational relationships	More likely to have partnership or joint venture arrangement with another organization Less shared staff and shared information technology (IT) Lower proportion of calls referred to outside providers	Less likely to have partnership or joint venture arrangement with another organization More shared staff and shared IT Higher proportion of calls referred to outside providers

Continued

TABLE 6-4 Continued

Survey Variables	High Efficiency	Low Efficiency
Quality improvement and assurance	No differences	No differences
Research and training	No differences	No differences
Future organizational challenges	More likely to have a strategic plan specific to poison control center Less likely to cite problems related to complex reporting and accountability Less likely to cite balancing core poison control functions with other activities such as research and bioterrorism response and preparedness	Less likely to have a strategic plan specific to poison control center More likely to cite problems related to complex reporting and accountability More likely to cite balancing core poison control functions with other activities such as research and bioterrorism response and preparedness

TABLE 6-5 Differences Between Centers Serving Large and Small Populations

Survey Variables	Large Population	Small Population
Affiliation	No differences	No differences
Staffing	More likely to use paid consultants other than medical director Less likely to cite low pay as reason for turnover Employ more full-time equivalents (FTEs) More hours spent on all poison control center activities and nonclient response activity	Less likely to use paid consultants other than medical director More likely to cite low pay as reason for turnover Employ fewer FTEs Fewer hours spent on all poison control center activities and nonclient response activity
Services and activities		
• Current practices	More extensive involvement in professional education, public education, and outreach	Less extensive involvement in professional education, public education, and outreach

TABLE 6-5 Continued

Survey Variables	Large Population	Small Population
• Areas of growth	Professional education, public education, and outreach	
• Areas of decline	Services eliminated because of inadequate funding or high costs and declines in exposure calls and overall call volume	Services eliminated because of inadequate funding or high costs; public and professional education most likely to be reduced
Interorganizational relationships	More likely to partner or joint venture with other organizations	Less likely to partner or joint venture with other organizations
Quality improvement and assurance	No differences	No differences
Research and training	More likely to have fellowship funding	Less likely to have fellowship funding
Future organizational challenges	More likely to have a formal strategic plan Staff recruitment and retention somewhat less likely to be an issue	Less likely to have a formal strategic plan Staff recruitment and retention a significant issue

General Eight of the ten centers surveyed were established during the 1950s. The governance of most of the centers falls under the board of the organization with which they are affiliated. One center is governed by the city department of health, and the one independent center has its own board of trustees. Four centers either have or are considering forming an advisory board.

The organizational structure of most of the centers surveyed is relatively flat, with a managing director and medical director comprising the senior staff, with poison information personnel and educators reporting directly to them. Two centers have additional midlevel personnel, such as associate directors or education or administrative coordinators.

Most centers have undertaken fairly significant organizational change in the past 10 years. In some cases, these changes were precipitated by a consolidation of service areas (usually because one or more other centers had closed). In other cases, the changes were in response to a fiscal crisis, the resolution of which led to new funding and organizational arrangements. At least one center indicated making staffing changes in order to seek AAPCC certification. Many of the centers also underwent functional changes (e.g., adding "extended services" such as bilingual services, additional education and outreach, research, other hotlines, and bioterrorism/public health emergency response services).

Three centers serve smaller populations (i.e., below the mean population served by all poison control centers), while the remaining seven serve larger populations (above the mean). Not surprisingly, "low-population" centers tend to employ a smaller number of staff than "high-population" centers (FTEs of 10 to 14.5 versus FTEs of 14 to 29). Those centers that have midlevel personnel, such as associate directors or education or administrative coordinators, are centers that serve larger populations.

Five centers have lower costs per human exposure call (i.e., below the mean cost for all poison control centers), while the remaining five have higher costs (above the mean). All of the lower-cost centers serve larger populations; none of the centers surveyed serving smaller populations fall in the lower-cost category. Centers with higher costs employ similar numbers of FTEs as those with lower costs. FTEs at lower-cost centers range from 14 to 26, while the number at higher-cost centers ranges from about 10 to 29.

Affiliation The 10 centers surveyed are evenly divided between private not-for-profits (5) and public not-for-profits (5). Eight centers identify themselves as "owned" by a hospital, university, or similar entity; for one, ownership status is unclear, and for another ownership is wholly independent. Three identify themselves as part of a university, five as part of a hospital, one as part of a hospital association, and one as independent. In all cases but the independent center, respondents indicated that major management decisions need to be approved by the larger entity of which they are a part. In some cases, even more minor decisions (such as the allocation of all expenditures) require such approval. Some funding agencies also exercised decision approval over various aspects of their expenditures (e.g., use of grant funds).

Most hospital-based centers report and are accountable to a high-level vice president or the president/chief executive officer, and this is also true of university-based centers (e.g., reporting to a dean). Again, some centers have accountability relationships to outside entities, usually governmental agencies such as the state department of health.

There are few differences in nonprofit status, reporting, and accountability among centers serving larger and smaller populations or among those classified as high and low cost. The one notable exception is that four of five higher-cost centers described themselves as public not-for-profits, while four of five lower-cost centers described themselves as private not-for-profits.

Staffing Most centers indicate using at least some volunteers for activities such as education and outreach or mailings. Others count unpaid medical toxicologist backup and students on clinical rotations as volunteers. All centers use consultants as backup medical toxicologists and other medical experts, and these are often not paid. At three centers, the medical director is a contracted, paid consultant. Other consultants include an education coordinator, a "financial person," a computer maintenance/repair service, and a language line used for translation of calls from non-English speakers.

The overall number of FTEs employed at each of the 10 centers ranges from approximately 10 to 29, and estimates of average annual turnover range from "very low" to 20 percent; the higher turnover is concentrated among SPIs. Reasons for turnover include new hires not successfully making it through orientation, personnel moving, and staff getting better job offers elsewhere. SPI pay is seen as noncompetitive with most other pharmacist and nursing jobs, despite the fact that 24/7 coverage and heavy job responsibilities are seen as more taxing.

The total number of estimated weekly hours that center personnel spend on all poison control center activities ranges from a low of 370 to a high of 1,095. Estimated hours that center personnel spend on poison prevention and control activities other than direct client response range from 60 to 500.

High- and low-population centers do not seem to differ in use of volunteers and they seem no more (or less) likely to use consultant arrangements for their medical directors. However, centers serving high-population areas are more likely to use other types of paid consultants. Estimates of average annual staff turnover also do not differ between centers serving large populations versus centers serving smaller populations. Centers serving smaller populations are somewhat more likely to mention low pay as contributing to turnover rates.

Centers serving smaller populations estimate lower numbers of hours spent weekly on all poison control center activities (370 to 490 versus 670 to 1,095 hours), as well as lower numbers of weekly hours spent on poison prevention and control activities other than direct client response (60 to 120 versus 100 to 500 hours). This is not surprising in view of the lower number of overall FTEs they employ.

An examination of high- and low-cost centers shows no difference in use of volunteers and no difference in the use of consultant arrangements for their medical directors. However, lower-cost centers are somewhat more likely to use other types of paid consultants. Furthermore, average annual staff turnover does not differ greatly between higher- and lower-cost centers, although estimates of turnover were perhaps slightly lower among lower-cost centers. Lower-cost centers, however, are much less likely to mention low pay as contributing to turnover rates.

With one exception, higher-cost centers estimate fewer hours spent weekly on all poison control center activities (370 to 690 versus 670 to 1,000 hours). The exception is one higher-cost center reporting 1,095 total hours weekly. Higher-cost centers also tend to spend fewer weekly hours on poison prevention and control activities other than direct client response (60 to 195 versus 70 to 500 hours). These results are perhaps surprising given that the two types of centers do not differ greatly on total number of FTEs employed.

Services and Activities

Besides handling exposure and nonexposure calls, all 10 centers conduct some professional education and/or residency training and some public education and outreach. Most sites also conduct research. Other activities include writing grants, writing guidelines, providing bilingual services, collecting data, and carrying out industry contract services.

The centers' primary target population is the public. Some sites cover their entire state, while others cover a regional area. All 10 centers in the survey sample handle calls from health care professionals and health service organizations. One site also offers bilingual services and covers all Spanish speakers in two states. The number of FTEs estimated for this activity range from 6 to 21. In centers serving smaller populations, the range is 5.5 to 10.75 versus 6 to 21 FTEs for centers serving larger populations. No differences were found between high- and low-cost centers.

Professional education and residency training is targeted both at medical toxicology residents and fellows as well as residents and students in pediatrics, emergency medicine, and pharmacy. The centers vary in the number and intensity of these educational activities, with some hosting a number of toxicology fellows, for example, while others provide shorter-term training (i.e., rotations) for students and residents from four or five different departments. Nearly all centers also provide education for practicing professionals, through grand rounds and lectures at area (or their own) hospitals and other health care institutions, as well as special conferences and seminars. Again, the number and intensity of such activities varies. Center directors have a difficult time estimating the FTEs devoted to such training, which is often conducted by the medical director (and other medical personnel, if the center has any). Estimates range from 0.25 to 4 FTEs, for professional education and resident/student training combined. Low-population centers averaged 0.5 FTEs while high-population centers ranged from 0.67 to 4 FTEs. Low-cost centers employed 1or 2 FTEs for professional education; high-cost centers ranged from 0.5 to 4 FTEs for this activity.

Public education and outreach activities include public workshops, promotional activities conducted in partnership with local organizations (such as health fairs), bilingual training for volunteer peer educators (promotoras), newsletters, and prevention education at schools, adult education programs, and Supplemental Women, Infants, and Children program sites. The number of FTEs estimated for these types of activities varies greatly, from 0.75 to nearly 5. Centers serving smaller populations employ from 0.5 to 1 FTE in education and outreach, while centers serving larger populations range from 1 to nearly 5. Centers with lower costs employ from 1.25 to nearly 5 FTEs in education and outreach, while the estimates for centers with higher costs range from 0.5 to 2.

Nine of the 10 sites surveyed conduct research, very broadly defined. For some this involves case reports and series of literature reviews and includes research conducted by onsite fellows. One center mentioned internal research for the purposes of quality assurance, while two others mentioned joint research conducted with other centers. Again, directors find it difficult to estimate FTEs for research because the research effort tends to be split among numerous staff, many of whom devote only a small amount of their time to this activity. Estimates range from 0.1 to 6 FTEs. With one exception, research FTEs are no more limited at centers serving smaller populations than they are in centers serving larger populations. Low-population center research FTEs range from minimal to 1.75; for high-population centers this ranges from 0.1 to 1.25 (with one outlier site having 6 research FTEs, predominantly fellows). Research FTEs at lower-cost centers appear similar to higher-cost centers. Lower-cost center research FTEs range from 0.1 to 1.25; for higher-cost centers, this ranges from a low of 0.1 FTE to 1.75 FTEs (again with one outlier site having 6 research FTEs, predominantly fellows).

Finally, the centers report other activities, including data collection (two), writing guidelines (one), grant writing (one), and industry service contracts (two). The level of effort for these other activities ranges from 0.2 to 1 FTE. Centers serving smaller populations seem no less likely to engage in other activities than centers serving larger populations, with one low-population center engaging in more additional activities than any other center surveyed. No differences were found between higher- and lower-cost centers in the interview sample.

Respondents mentioned six main areas of growth in services and six main areas of decline. Growth areas described include the following:

- Six centers indicated they have experienced the most growth in the area of public education and outreach, with some mentioning an increase in grant funding as the reason.
- Three centers indicated growth in professional education.
- Three mentioned increases in call volume.
- Two described an increase in administrative burden as a result of Health Insurance Portability and Accountability Act (HIPAA) requirements and funding issues (e.g., grant writing and administration).
- One center mentioned research.
- One center mentioned bilingual services.

Centers experienced similar areas of growth in services regardless of whether they serve larger or smaller populations, but to differing degrees. More centers serving larger populations reported growth in professional education (three centers), and growth in public education and outreach (five centers). Only one center serving a smaller population reported growth in either of these areas of service (education and outreach). Two high-population and one low-population center reported increased call volume. One of each kind of center described an increase in administrative burden. One center serving smaller populations mentioned research and another mentioned bilingual services as top growth areas.

Both higher- and lower-cost centers experienced growth in public education and outreach (three centers of each type), professional or resident education (two higher-cost centers, one lower-cost center), and call volume (two lower-cost centers, one higher-cost center). One of each kind of center also described an increase in administrative burden. One higher-cost center mentioned research and another mentioned bilingual services as top growth areas.

Areas of decline described include the following:

- Four centers reported professional education as the activity area in which they have experienced the most decline, with one center indicating that the problem was not demand, but rather the center's decreasing ability to meet the demand.
- Three centers reported decreasing call volume in general.
- Two centers reported a decrease in human exposure calls as a percentage of all calls.
- Two centers reported a decrease in drug information calls (in both cases because this service had been discontinued).
- Three centers said the number of industry or other contracts for special services has decreased.
- One center reported a decline in the amount of marketing the center has been doing due to a lack of funding.

Centers serving smaller populations reported declines in service all related to the inability to fund activities—either a direct loss of funding (e.g., loss of grant) or discontinuation of services due to unacceptably high costs (e.g., decision to discontinue accepting drug information calls). Two of these centers also reduced education activities, in one case an unfilled fellowship position and in another a decrease in professional education. Centers serving larger populations, while also experiencing declines due to funding issues, also reported declines in service due to other factors—two of these centers reported declines in exposure calls as a percentage of all calls, while three others reported declines in overall call volume. Also mentioned were decreases in industry contracts, professional or resident training, and other types of calls (animal exposures, drug decoding, pesticide calls). One center turned down requests for various services due to lack of staff.

Declines in service are somewhat different between higher- and lower-cost centers. Three higher-cost centers experienced declines in professional education or fellowship training, while one lower-cost center suffered a decrease in resident training. Two lower-cost centers indicated declines in industry or other contracts. Both types of centers, however, mentioned declines in call volumes, with one higher-cost and two lower-cost centers reporting declines in general call volume, and one of each type of center reporting decreases in exposure calls as a percentage of all calls, drug information calls, and other types of calls (e.g., animal exposures). The decreases in information and other calls are largely a function of the centers' termination of these services.

Interorganizational Relationships

Five respondents indicated that their centers have joined some type of coalition or collaboration with other centers in the past 4 years. These formal arrangements include state poison center networks and regional consortia. Four respondents mentioned informal collaborative arrangements—such as data sharing or call coverage—that their centers have with another center or centers. Six centers have entered into partnership or joint venture arrangements with other organizations. These vary, and include partnerships to provide education and outreach services, multicenter research projects, and joint programs with other hospital or university departments. None of the centers indicated that they have been involved in a merger. (However, at least three of the centers mention in other parts of the survey that they have expanded their service areas in the past 10 years because other centers have closed.)

Of the five centers joining some sort of coalition or other formal collaboration with other centers in the past 4 years, four were centers serving larger populations. Of the four centers mentioning informal collaborative arrangements, two were centers serving larger and two smaller populations. Only one center serving a smaller population has entered into a partnership or other joint venture arrangement with another organization, compared with five of the large-population centers. Two higher-cost centers have entered into a partnership or other joint venture arrangement with another organization compared with four of the lower-cost centers.

Most centers provide services to other providers, the majority of which are related to professional education. A few centers provide data collection services for other providers. One regional center provides bilingual services to the rest of its state and to another state. Both large- and small-population centers are equally likely to provide services to other providers. Centers serving smaller populations have more shared staff and shared databases than centers serving larger populations. Both higher- and lower-cost centers are equally likely to provide services to other providers. Centers with higher costs have more shared staff and shared information technology (including databases) than centers with low costs, but neither type of center reported extensive sharing of administrative support or other services.

Few centers share administrative support services, information technology, or staffing with another center or organization; one center shares these with another service of the hospital in which it was based, while another two share education personnel with other hospitals. One center shares its bilingual staff, as described above. Two centers share information technology, some staff, and call volume with other poison control centers in their state. Two other centers share only databases for joint research projects, and three others mention providing occasional coverage for another center or handing off patients to a nearby center.

Referrals in and out of the centers vary considerably. While one center indicated it receives "minimal" referrals from other providers, another said 58 percent of its clients are so referred. Most centers indicate a percentage closer to 15 to 20 percent. Similarly, most centers refer about 15 to 20 percent of their clients to outside providers, usually hospitals, for evaluation, treatment, and monitoring; this percentage ranges from 7.5 percent to 30 to 40 percent. There seems to be no clear difference in the estimated number of referrals between types of centers in either referrals to the poison control center or referrals from the center to other providers. Centers serving smaller populations tend to fall at the higher end of the distribution in terms of referrals out to other health care providers, but there are some centers serving larger populations with similar figures. However, centers with lower costs tend to have a somewhat lower percentage of referrals to other health care providers (7.5 percent to 19 percent of calls, versus 11 percent to 35 percent of calls for higher-cost centers).

Quality Improvement and Assurance

All programs surveyed use some kind of written procedures for handling telephone calls. These procedures are of two types. Most of the centers have written policies or administrative procedures for handling calls that address topics such as how to answer the phone, what information to collect from callers, and other general triage guidelines. Most of the centers also have guidelines addressing evaluation and treatment of specific exposures. One center has a library of about 200 management guidelines on hand, not only for its own use, but to fax to treatment facilities.

All but one center practice case management as defined in the survey. This includes conducting a comprehensive assessment of clients' needs at intake, making referrals for services, following up on referrals to make sure that clients received services, and contacting clients periodically to check on their progress. Most centers indicate that they do this kind of case management at minimum for anyone they refer to a hospital for treatment; however, some also follow up with less serious exposures.

All centers have a formal written quality assurance plan, with two indicating they are in the process of revision. Quality assurance activities focus on two areas, customer service and appropriateness of treatment. Many centers assess this through review of telephone calls, either recordings or transcripts. Some do this for a random selection of all calls, while a few also target specific categories of calls for review (e.g., those clients referred to hospitals, deaths). Some centers also have daily or twice-daily reviews of currently active cases. In most instances, these reviews are done by senior personnel, and in particular the medical director; however, a few centers also involve other staff in quality reviews, including specialists in poison information, as a part of their ongoing training. Two centers have mission and vision statements related to quality improvement.

Finally, staff in all centers undergo at least in-house, in-service training related to their jobs, with some centers conducting such training on an ongoing basis. In terms of more formal training, 25 to 100 percent of centers' staffs have participated in formal continuing education or in-service training during the past year.

There are no discernable differences between high- and low-population areas and higher- and lower-cost centers regarding quality assurance activities.

Research and Training

FTEs devoted to research vary from zero to six, with higher numbers devoted to research at sites that have fellowship programs. FTEs devoted to staff training are difficult for respondents to estimate, given that none of the centers have dedicated personnel for this task. Instead, responsibility for training tends to be shared among senior staff and, sometimes, more experienced SPIs. Estimates range from 0.1 to 1 FTE. The estimated length of SPI training varies from 8 weeks to 12 months, with the most common length of approximately 3 months. Respondents note that it can take from 6 months to 2 years for new hires to come fully up to speed.

Few of the sites have funding for poison control center research, and those that do have grant funding. Not surprisingly, research funding is generally considered insufficient. Similarly, there is little dedicated funding for training, except for some fellowships (and these funds do not necessarily come through the center). A few centers have limited travel funds or small grant funds for training, but none consider them sufficient.

Staff-training FTEs also appear to be roughly equal, again with those centers having fellowship programs reporting additional FTEs. Almost none of the sites have funding for poison control center research or staff training, so this does not vary by type of center. The exception is three centers serving high populations that have fellowship funding; two of these centers are higher cost and one is lower cost. Neither research nor staff training funding was considered sufficient by any of the centers.

Future Organizational Challenges

General Three centers have a formal written strategic plan and one is in the process of developing such a plan. Four centers are included under the strategic plan of the organization (usually a hospital) of which they are a part. Two of these either also have their own plan, or are included under the plan of the state poison control network. Two centers do not have a strategic plan of their own, but have written objectives or a mission and vision statement specific to the center.

The most often mentioned organizational challenge the poison control centers face is staff recruitment and retention, particularly for SPIs. Respondents complain of difficulty in finding qualified staff to hire, not only because of competition from better paying jobs, but also because of problems finding people with the right mix of skills.

Some respondents also describe organizational challenges arising from the complex, multidisciplinary nature and structure of some of these centers. For example, a center may be part of a hospital affiliated with a university whose staff and fellows have appointments in a number of different departments and schools, and whose funding comes from multiple sources (e.g., the hospital, the university, the state department of health, and grants). In circumstances such as these, it can be difficult for the center to function as an autonomous, "cohesive" organizational entity. One respondent noted that the result can be delays in addressing important issues.

Related to this, some respondents point to the difficult balance between "core" functions (e.g., answering calls) and other activities that are important, but perhaps not considered central to the poison control center mission (e.g., research). In most cases, these are viewed as tasks that centers are well placed and well suited to do, such as bioterrorism and emergency preparedness. Some respondents express the opinion that poison control centers have been overlooked and should be more involved in these "noncore" issues. However, they also recognize the difficulties of coordinating multiple missions, given the realities of multiple, separate funding streams and already fragmented organizational structures.

Several respondents note problems related to HIPAA and the difficulty convincing provider organizations that they can share routine patient-level follow-up data with the center. This has hampered toxicosurveillance efforts and research efforts.

Finally, one respondent notes the language and cultural barriers that need to be addressed as the number of linguistic minorities in the United States continues to grow. These groups' utilization of poison control center services is low; yet, they are at perhaps a higher risk than the English-speaking population. This is because of the younger average age of some of these populations and the fact that they may have difficulty reading packaging in English.

Larger versus smaller populations More centers serving larger populations have a formal written strategic plan, are in the process of developing one, or have written objectives or a mission and vision statement specific to the center (five centers). Only one center serving a smaller population indicated having a formal strategic plan specific to its center. Two of each type of center are included under the strategic plan of the larger organization of which they are a part.

Staff recruitment and retention is a pressing organizational challenge for many centers serving larger populations, but it is a challenge for *all* centers serving smaller populations. Interestingly, low-population centers are as likely as high-population centers to describe organizational challenges arising from the complex, multidisciplinary nature and structure of their organizations. They are also as likely to report difficulties in balancing "core" poison control functions and other functions, such as research and bioterrorism response and preparedness. Overall, the challenges poison control centers face seem to be similar, no matter the size of the population they serve.

Lower- versus higher-cost centers More lower-cost centers have a formal written strategic plan, are in the process of developing one, or have written objectives or a mission and vision statement specific to their center (five centers). Only one high-cost center indicated having a formal strategic plan specific to its center. Three high-cost centers are included under the strategic plan of the larger organization of which they are a part, as is one lower-cost center.

Staff recruitment and retention is a pressing organizational challenge for most centers, and this does not differ by type of center. Higher-cost centers are much more likely than lower-cost centers to describe organizational challenges arising from the complex, multidisciplinary nature and structure of their organizations. They are also more likely to experience difficulties in balancing "core" poison control functions and other functions, such as research and bioterrorism response and preparedness.

SUMMARY

Many of the studies on examining the costs and effectiveness of poison control centers lack the necessary or appropriate data needed to reach strong conclusions, particularly regarding effectiveness. Nonetheless, taken as a whole, this literature makes a convincing case that, at least in terms of treatment management guidance for the public, poison control centers save the health care system economic resources and save members of the public time, lost wages, and anxiety. These studies, however, do not examine the cost-effectiveness among poison control centers, but rather compare them with other health care providers such as emergency departments.

There is wide variation among poison control centers on a number of operational characteristics, including total costs, personnel costs, and non-personnel costs per (1) population served, (2) total calls, and (3) human exposure calls. There is little conclusive evidence that economies of scale operate with respect to size of population served and poison control center costs, particularly for centers serving populations of 2 million or more. Costs are best predicted by variables related to staffing patterns and wage rates rather than hardware expenses or funding source.

Regarding staff time, higher-cost centers estimate fewer hours spent weekly on all poison control center activities, whether direct client response or not. However, the two types of centers do not differ greatly on total number of FTEs employed. Centers with higher costs have more shared staff and shared information technology (including databases) than centers with low costs, but neither type of center engages in extensive sharing of administrative support or other services. Centers serving smaller populations employ, on average, fewer professional education and public health education FTEs. They also have more shared staff and shared databases than centers serving larger populations.

In the area of quality assurance and planning, there was little variation in quality assurance activities among centers, with virtually all engaging in them. The staff of all centers undergo at least in-house, in-service training related to their job. Only two centers (both high cost) have no strategic plan; the eight others either have their own formal written strategic plan or objectives, or are included under the strategic plan of the organization of which they are a part.

Regarding affiliation and interorganizational relationships, four out of five higher-cost centers described themselves as public not-for-profits, while four out of five lower-cost centers are private not-for-profits. Centers serving larger populations and lower-cost centers described themselves as more likely to have joined a coalition, other formal collaboration, partnership, or other joint venture.

Centers serving smaller populations all experienced declines in service related to the inability to fund activities. Declines in service are somewhat different between higher- and lower-cost centers. Three higher-cost centers reported declines in professional education or fellowship training, while one lower-cost center experienced a decrease in resident training. Both types of centers, however, mentioned declines in call volumes.

The most often-mentioned organizational challenge the poison control centers face is staff recruitment and retention, particularly for SPIs. Lower-cost centers, all of which serve larger populations, are much less likely to mention low pay as contributing to turnover rates. Higher-cost centers are much more likely than lower-cost centers to describe organizational challenges arising from the complex reporting and conflicting accountabilities of their organizations. Higher-cost centers are more likely to experience difficulties in balancing "core" poison control functions and other functions, such as research and bioterrorism response and preparedness.

Appendix 6-A

Poison Control and Prevention Organizational Interview Questions

GENERAL

In what year was your Center established?

Please describe the current organizational structure of your poison control center, including: size, functional divisions/units, administrative support. Do have an organizational chart that you would be willing to send us?

Describe the governance of your center. Does it have a formally designated board (is it a separate 501(c)(3)? A steering committee? Who sits on the board and how is board composition/membership determined?

Describe how your center has changed organizationally over the past 10 years and what has precipitated those changes.

How has your center changed functionally (e.g., services provided) over the past 10 years and what has precipitated those changes?

STAFFING

Did your poison control center receive support services from volunteers during the most recent complete fiscal year?

Describe the extent of turnover in center staff over the past 5 years. What categories of personnel are most likely to experience turnover and why?

How many consultants or independent contractors, either part-time or full-time, are used to provide poison control and prevention services at your center?

During a typical week, what is the *total number of hours* that your staff, including consultants, independent contractors, and administrative staff, work at all activities for your poison control center?

What is the total number of hours per week spent by these staff members conducting services other than direct client response, but *relating to poison control and prevention*? These services may include, for example, outreach activities, community collaboratives, or prevention workshops.

INTERORGANIZATIONAL RELATIONS

In the last 4 years has your center done any of the following: Joined a coalition or association with other centers? Formed a partnership or entered into a joint venture with other groups or organizations? Merged with another poison control center or organization?

What types of services does your *poison control center provide to other* provider organizations: data analysis, education, consulting, client referrals, other?

Does your poison control center share any services with another poison control center or organization: administrative support services (clerical), information technology, staffing, other?

In the most recent complete fiscal year, about what percentage of your poison control center's clients were referred by other provider organizations or individual providers?

Of your poison control center's clients, what percentage are referred to outside providers for additional treatment or services? What are the major services/treatments for which your clients are referred?

SERVICES AND ACTIVITIES

What specific services does your center provide and to what markets or groups (e.g., handling exposure calls, handling nonexposure calls, research, education, professional training, residency training, contract data collection, outreach activities, or prevention workshops)?

How many staff FTEs (or staff hours per week) were assigned to each of these activities/services during the most recent fiscal year?

Which of the activities/services you listed have experienced the most growth in the past 3 years (top 2)?

Which of the activities you listed have experienced the most declines in the past 3 years (top 2)?

AFFILIATION

Is your poison control center private for-profit, private not-for-profit, or public not-for-profit?

Is your poison control center part of a larger organization (through ownership or affiliation)?

Which of the following describe the organization or type of facility that your poison control center is part of? Is it a hospital, an emergency department, a unit of county government, a network or consortium of providers, center, or some other kind of organization?

Does the organization that your poison control center is part of have to approve major management decisions such as those involving programs, services, staffing, or finances?

Describe the reporting/accountability relationship between your center and the entity with which your center is affiliated.

Of all the hours worked by all of your poison control center's employees each week (including consultants and independent contractors), what percent of total *staff hours are typically devoted to nontreatment activities* such as administrative work, supervision, hiring or program development, or clerical and support services, such as accounting, billing, and other recordkeeping activities?

QUALITY IMPROVEMENT AND ASSURANCE

Does your poison control center use standard, written procedures that indicate steps to follow in dealing with clients—these may be termed "practice guidelines," "protocols," or "critical pathways"?

Does your poison control center use case management?

Case management can consist of several different activities. Considering current poison control center clients who receive case management, for about what percent does the case manager:

b1. . . . conduct a comprehensive assessment of clients' needs at intake?

b2. . . . make referrals for clients to receive various services?

b3. . . . follow up referrals to make sure clients receive services?

b4. . . . contact clients periodically to check on their progress?

Does your poison control center have a formal written quality assurance plan?

During the most recent fiscal year, what percent of your center's staff underwent in-service training or took continuing education related to their jobs in the center?

RESEARCH AND TRAINING

How many staff FTEs are assigned to research activities that focus on (a) poison control center service delivery or (2) the use poison control center data (e.g., poison control center case data or comparisons of treatments conducted through the poison control center)?

How many staff FTEs are assigned to staff training (SPI and PIP) and how long does it take to complete the training? How many staff FTEs are assigned to medical toxicology fellowship training?

Is there funding specifically for poison control center research and is it sufficient? Is there funding specifically for staff training and toxicology fellowship training and is it sufficient?

FUTURE

Does your poison control center have a formal written strategic plan, or is the center a part of a larger unit/organization with such a strategic plan?

Other than sustainable funding, what are the most pressing organizational challenges faced by your center currently? What are those challenges likely to be in the future?

7

Data and Surveillance

THE NEED FOR DATA AND SURVEILLANCE SYSTEMS

This chapter examines the available data and assessment methodology as it relates to poison control and prevention. We discuss the need for data and surveillance systems, describe the currently available systems, and outline an approach for evaluating the data and surveillance systems.

Information on the epidemiology, treatment, and outcomes of poisonings can help inform regulatory decisions and compliance, public policy initiatives, and the development and assessment of clinical management guidelines. Data from various sources are used by federal, state, and local health agencies and others for surveillance of poisonings and their sequelae. Surveillance generally consists of the systematic and ongoing collection, analysis, and interpretation of health data for use to prevent and control disease (Thacker and Berkelman, 1988). A systematic assessment approach has been proposed by the Centers for Disease Control and Prevention (CDC) (2001a). The approach to evaluating data systems performance recommended by the CDC Guidelines Working Group consists of an assessment of usefulness and a description of system attributes (German, 2001). The attributes are simplicity, flexibility, data quality, acceptability, positive predictive value, representativeness, timeliness, and stability.

Although surveillance systems were originally developed to control communicable diseases, they now play a role in addressing other important problems, including chronic diseases and environmental issues. In

TABLE 7-1 Roles for Data and Surveillance as Applied to Poison Prevention and Control

Role	Uses	Example
Outbreak/cluster identification	Public health agencies can assess, then respond with investigation	Arsenic poisoning in Maine
Implementing and evaluating prevention and control measures	Examining temporal association of changes in exposures in relation to programs	Poison Prevention Packaging Act implementation and assessment
Planning and managing resources	Providing adequate levels of poison prevention service	Tracking volume of contacts by time and day
Epidemiology, including identification of trends	Determining prevalence and detecting increases in types of poisonings	Annual reports of the AAPCC (TESS)
Identification of emerging problems	Multiple, including environmental and occupational	Pesticide-related illness and injury
Research	Assessment of hazard to focus primary and secondary prevention	Acetaminophen overdose for Nonprescription Drug Advisory Board (September 2002)

public health, surveillance data can be useful for multiple purposes: (1) identifying and investigating outbreaks or clusters of diseases; (2) implementing and evaluating prevention and control measures; (3) planning and managing resources and establishing priorities; (4) identifying trends in occurrences of interest; and (5) identifying emerging problems or new populations at risk of disease (adapted from Calvert et al., 2001). Each of these types of needs is briefly described in Table 7-1 with examples relevant to poisoning prevention.

CURRENTLY AVAILABLE SYSTEMS

The following section describes the characteristics and the strengths and weaknesses of current data systems, beginning with the Toxic Exposure Surveillance System (TESS) and other poison-specific data sources, and followed by data sources derived from health records and health care datasets, other exposure-related data sources, and survey data sources. Table 7-2 provides a tabular description of these datasets. This review focuses on existing data resources, including national surveys, that have been designed at least in part for epidemiological tracking purposes or

TABLE 7-2 Surveillance Data Sources Relevant to Poisoning and Overdose in the United States

TABLE 7-2A Poison Control-Specific Data Sources

Data Source	Basis of Data Collected	Data Strengths	Data Limitations
American Association of Poison Control Centers Toxic Exposure Surveillance System (TESS)	Cases reported to the poison control centers by public and by health care providers; systematically collected and categorically defined variables.	Wide range of exposures and case severity; standardized format for data collection; immediate case detection and follow-up possible at local and national levels; access open to public directly.	Variable geographic coverage nationally; variable penetration within established centers; likely reporting biases in types of exposures and case severity; access to data for independent analysis costly.
Individual poison control center case detection	Source of case data supplied to TESS, but can also include supplemental information.	Can be prioritized for local or regional surveillance questions; supplemental data collection can address targeted management strategies; structured case follow-up studies possible.	Case numbers may be small with limited power; local factors may introduce case reporting biases; resources may be too limited to carry out surveillance activities.

TABLE 7-2B Data Sources Derived from Health Records or Health Care Datasets

Data Source	Basis of Data Collected	Data Strengths	Data Limitations
National Vital Statistics System (National Center for Health Statistics [NCHS])	National data derived from death certificates providing principal and contributing causes of death.	Comprehensive national collection of fatalities; no cost access to data; submission and review of data requests are required.	Broad categorization of cause by ICD codes; selection biases may impact which specific causes are attributed.
Medical examiner case series	Cases referred to medical examiners at the county level; data often include toxicological testing.	Medical examiners series comprise a subset of deaths with greater detail and possible anatomic/toxicologic confirmation. Research access variable.	Variable referral patterns and use of toxicological testing; relatively small subset biased to prehospital deaths.
Hospital discharge data at the state level	Data capturing all hospital discharges within a state.	Comprehensive data where collected; hospital admission less likely to be "discretionary."	Not all states collect and make available; limitations of ICD coding magnified by variable E-code assignment.
Hospital-based emergency department (ED) data at the state level (or participating region)	Data capturing all emergency department-treated cases within a state (or region).	Comprehensive data, where collected; excludes other sites of urgent care.	A minority of states have such data systems; same ICD coding issues as above.
National Hospital Discharge Survey (NCHS)	Data on hospital discharges from a national sample of hospitals. Selected representative hospitals are chosen and sampling weights assigned.	Allows national incidence estimates by extrapolation; multiyear pooling and time trend analyses possible; dataset access at no cost.	Sample may not generate reliable estimates for uncommon events. As above, ICD coding may not allow detailed analysis at desired level of exposure.

Continued

TABLE 7-2B Continued

Data Source	Basis of Data Collected	Data Strengths	Data Limitations
National Hospital Ambulatory Medical Care Survey (NHAMC)	Sampling data derived from hospital EDs and associated clinics. Selected, stratified facilities; 4-week sampling period each year.	Similar data strengths to the National Hospital Discharge Survey. Valuable source for health care cost data. Survey numbers balanced between ED and clinics.	Similar data limitations to National Hospital Discharge Data. Also has non-ICD global diagnosis codes (e.g., "poisoning" that may capture additional cases).
National Ambulatory Medical Care Survey (NAMCS)	Sampling data derived from outpatient practice care settings. Selected outpatient clinics in selected states. Sampling of cases over 1 week; physicians comprise the sampling unit (approximate n = 2,400).	Similar data strengths to the National Hospital Discharge Survey and NHAMC.	Similar coding limitations to NAMCS; sampling is more limited, with fewer observations than the NHAMC.
Healthcare Cost and Utilization Program National Inpatient Sample	Survey managed by the Agency for Healthcare Research and Quality. Stratified sample of hospitals that varies year by year.	Similar data strengths to the National Hospital Discharge Survey. Health care cost focus.	Dataset can be queried but not downloaded for free; ICD and sample limitations apply; not all states participate; sample size varies.

TABLE 7-2C Other Exposure-Related Data Sources

Data Source	Basis of Data Collected	Data Strengths	Data Limitations
National Electronic Injury Surveillance System (NEISS) All Injury Program	Survey sample of EDs for nonfatal external injuries and poisoning managed by Consumer Product Safety Commission working with the Centers for Disease Control and Prevention (CDC).	Sampling design allows national incidence estimates (state-level estimates are not generated); independent coding system focusing on injury provides rich information source.	Coding may not allow direct comparison to ICD-9 data. Adverse drug effects excluded. Sampled sites are limited in number (n = 64); an additional sample (n = 36) collects more limited NEISS data.
Drug Abuse Warning Network (Substance Abuse and Mental Health Services Administration [SAMHSA])	Data extraction for cases of medication toxicity (not limited to drugs of abuse) treated in urban hospital EDs.	Widely used surveillance source for drug overdose, especially for street drugs.	Urban settings only in selected EDs. Does not include non-medication-related toxic syndromes.
MedWatch (Food and Drug Administration)	Consumer or health care provider-initiated reports of adverse drug effects. Pharmaceutical companies are required to report all new or unexpected adverse events.	Identifies postmarketing effects with mechanism for timely analysis of emerging problems if reported.	Relies on health care provider initiation of report; obscure or chronic associations less likely to be identified. Dataset can be queried, but not downloaded.
Hazardous Substances Events Surveillance System (Agency for Toxic Substances and Disease Registry)	State-based surveillance system for hazardous substance release events with a public health impact.	Uses reporting from a variety of field sources at the state level. Focuses on exposures to the public.	Only a limited number of states participate; mainly useful for inhalation exposures.
Census of Fatal Occupational Injuries (Bureau of Labor Statistics)	Multisource census cross-referencing vital statistics, insurance reporting, and inspection data.	Unique multisource integrated surveillance scheme.	Focused on source of injury; toxic exposures comprise a small subset of the data.

Continued

TABLE 7-2D Survey Data Sources

Data Source	Basis of Data Collected	Data Strengths	Data Limitations
National Health Interview Survey	National annual health survey. Items on self-reported conditions include poisoning and drug overdose.	National yearly stratified sample; allows national prevalence estimates; trends over time captured; free downloading of data.	Cause-specific data for subsets of poisoning limited; local-level data cannot be derived; fatalities not captured.
National Health and Nutrition Examination Survey	Periodic (not annual) health survey, including physical examination and biologic testing.	Biological sampling has been a major source of data on lead; expanded analysis for other xenobiotics being developed.	Stratified sampling of home health-based surveys; not relevant for acute poisoning injury surveillance.
Behavioral Risk Factor Surveillance System (CDC)	Annual survey focusing on risk factors, with optional added state-level modules.	Widely used for national estimates of behavior-related risk. Relevant to drugs of abuse.	Has been only minimally exploited to assess issues related to poisoning injury beyond drugs of abuse.
National Household Survey on Drug Abuse (SAMHSA)	National survey with extensive batteries related to substances with abuse potential.	Detailed data with national estimates for the prevalence of use and abuse of selected substances; includes data on overdose events.	Excludes persons under 12 years of age; does not collect data relevant to poisoning from causes other than substances of abuse potential.

Selected websites:
Toxic Exposure Surveillance System (TESS): http://www.aapcc.org/poison1.htm
National Vital Statistics System: http://www.cdc.gov/nchs/nvss.htm
National Hospital Discharge Survey (NCHS): http://www.cdc.gov/nchs/about/major/hdasd/nhdsdes.htm
National Hospital Ambulatory Medical Care Survey (NHAMCS): http://www.cdc.gov/nchs/about/major/ahcd/nhamcsds.htm
National Ambulatory Medical Care Survey (NAMCS): http://www.cdc.gov/nchs/about/major/ahcd/namcsdes.htm
Healthcare Cost and Utilization Project (AHRQ): http://www.ahcpr.gov/data/hcup/hcup-pkt.htm

National Electronic Injury Surveillance System (NEISS): http://www.cpsc.gov/cpscpub/pubs/3002.html
Drug Abuse Warning Network (DAWN): http://dawninfo.samhsa.gov/about/howdawn.asp
MedWatch (FDA): http://www.fda.gov/medwatch/what.htm
Hazardous Substances Emergency Events Surveillance (HSEES) of the Agency for Toxic Substances and Disease Registry (ATSDR): http://www.atsdr.cdc.gov/HS/HSEES/
Census of Fatal Occupational Injuries (Bureau of Labor Statistics): http://www.bls.gov/iif/oshcfoi1.htm
National Health Interview Survey (NHIS): http://www.cdc.gov/nchs/about/major/nhis/hisdesc.htm
National Health and Nutrition Examination Survey (NHANES): http://www.cdc.gov/nchs/about/major/nhanes/history.htm http://www.cdc.gov/nchs/nhanes.htm
Behavioral Risk Factor Surveillance System (CDC): http://www.cdc.gov/brfss/about.htm
National Household Survey on Drug Abuse (SAMHSA): http://www.samhsa.gov/oas/nhsda/methods.cfm

can be readily exploited for such purposes. It should be acknowledged, however, that electronic medical records systems and automated pharmacy systems not specifically addressed in this chapter also might be used to identify poisoning and drug overdose events, an approach that has been applied to adverse drug events, particularly those occurring during hospitalization (Bates et al., 2003; Honigman et al., 2001; Thurman, 2003). Although these systems may hold promise for future approaches that might augment existing surveillance resources, a review of them—which would require analysis of potential limitations related to sensitivity and specificity, data access and coding, the scope of such systems, and data requirements as they pertain to surveillance for poisonings as opposed to therapeutic misadventure—is beyond the scope of this chapter.

Data Sources Specific to Poison Control

Toxic Exposure Surveillance System

General features When the Surgeon General created the National Clearinghouse for Poison Control Centers (NCHPCC) in 1957, the mandate included coordination of a data collection system, or the flow of information from the poison control centers to NCHPCC. In the absence of funding or regulatory power, NCHPCC hand tabulated the voluntary reports with an annual report and provided product ingredient and treatment summaries on a 5" by 8"card file to the centers, with a lag time of 18 to 24 months. About 150,000 case reports of exposures were reported annually by 150 to 400 centers using another set of cards (3" by 5") with an attached lightweight carbon copy to be retained by the reporting center. Regionalization of poison control center services in the 1970s brought a need for a more detailed, higher-quality data collection system.

In 1983, the American Association of Poison Control Centers (AAPCC) assumed responsibility for the data collection system. It developed a minimum set of data elements and made complete reporting a condition of certification as a regional poison control center. The first annual report from this system included 250,000 exposures and was published in 1984 (Veltri and Litovitz, 1984).

AAPCC has subsequently published exposure data in the *American Journal of Emergency Medicine* in a standardized (evolving) form each year. With data on approximately 2.4 million exposures from 64 poison control centers in the 2002 report, TESS has accumulated 33.8 million cases (Watson et al., 2003). Although most poison control center exposure calls, by their very nature, are not supplemented by direct review of medical records or specific, direct confirmatory toxicological testing, they include systematically gathered data delineating clinical effects, case severity, and estimates of exposure levels relevant to dose response. Figure 7-1 shows the variability and repeatability of exposure calls for the past 3 years. TESS data have been used to perform the following functions:

- Identify exposure hazards
- Focus prevention efforts
- Conduct clinical research
- Direct training programs
- Prompt and support regulatory actions (reformulations, repackaging, recalls, and bans) (Litovitz, 1998)

Specific uses of TESS in product safety assessment have included:

- Postmarketing surveillance of newly marketed drugs
- Routine reviewing to prevent poisoning and limit morbidity and mortality
- Comparing brands with product categories

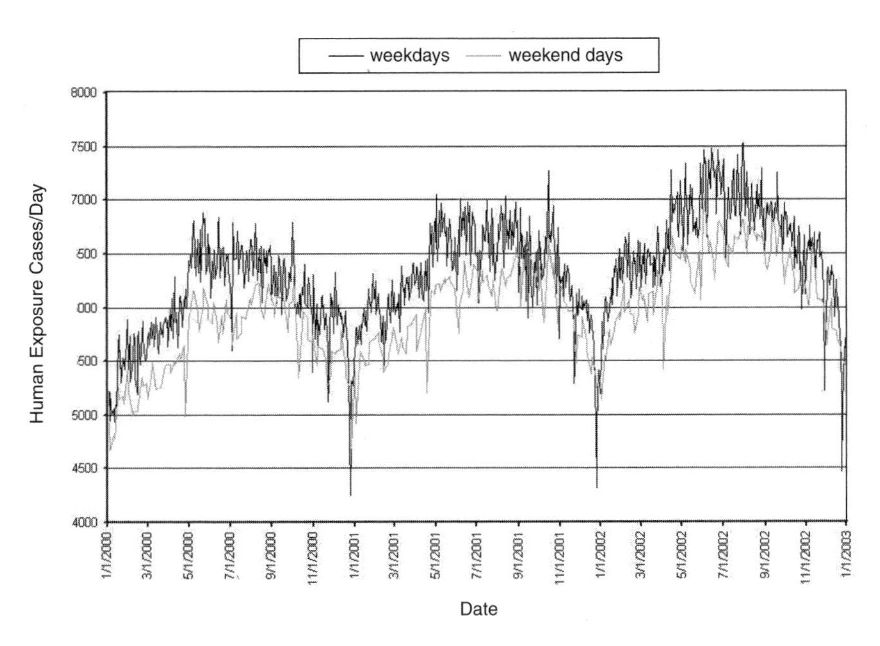

FIGURE 7-1 Frequency of human exposures reported to U.S. poison control centers.
SOURCE: Watson et al. (2003).

• Demonstrating product safety to regulatory agencies or consumer groups or fulfilling regulatory requirements
• Limiting animal testing (Litovitz, 1998)

A major benefit in the change from the 3" by 8.5" case report form (CRF) to the current TESS CRF was greater detail in reporting. The extensive coding that evolved over the 19 years of use enables efficient and powerful analyses. No patient identifiers are provided to TESS, thus assuring patient confidentiality. Among the recent changes to the TESS program are:

• In addition to the encoded fields submitted to TESS, poison control centers must also document each case through a narrative medical record. Guidelines for the narrative portion of the medical record are developed by each poison control center.
• As of January 1, 2002, all centers were required to submit reports of information (nonexposure) calls to TESS (most centers were already submitting them). Historically, information calls (calls not involving an exposure) did not "count" toward the poison control center's rate of contacts per population base served (sometimes referred to as "penetrance") and were not reported by all centers to AAPCC. A concordance observed between information calls and seizures of the same prescription drug (diversion) in the Cincinnati region (Krummen et al., 1999) suggested the use of information calls in detecting and monitoring drug abuse. This observation plus the prospect that information calls might provide an early signal for bioterrorism or other evolving public health events serve as arguments in favor of reporting information as well as exposure calls. Figure 7-2 summarizes data on the frequency of information calls reported to AAPCC from January 2000 through January 2003.
• By September 30, 2002, all centers were required to submit data to TESS automatically every few minutes (Auto-Upload). This real-time data collection is intended to enhance the value of TESS data for chemical/bioterrorism detection and for prompt identification of emerging drug and product hazards. The potential value of the real-time reporting is described in the toxicosurveillance discussion that follows.

Most fatality narratives collected in TESS are indirect (poison control center receives data from other health care professionals rather than from direct patient care) and lack external validation. The fatality narratives are used by some pharmaceutical companies to meet regulatory reporting requirements. Because many serious cases involving prescription medications reported in TESS are also reported in MedWatch and other systems, care must be taken to avoid duplication when utilizing multiple datasets.

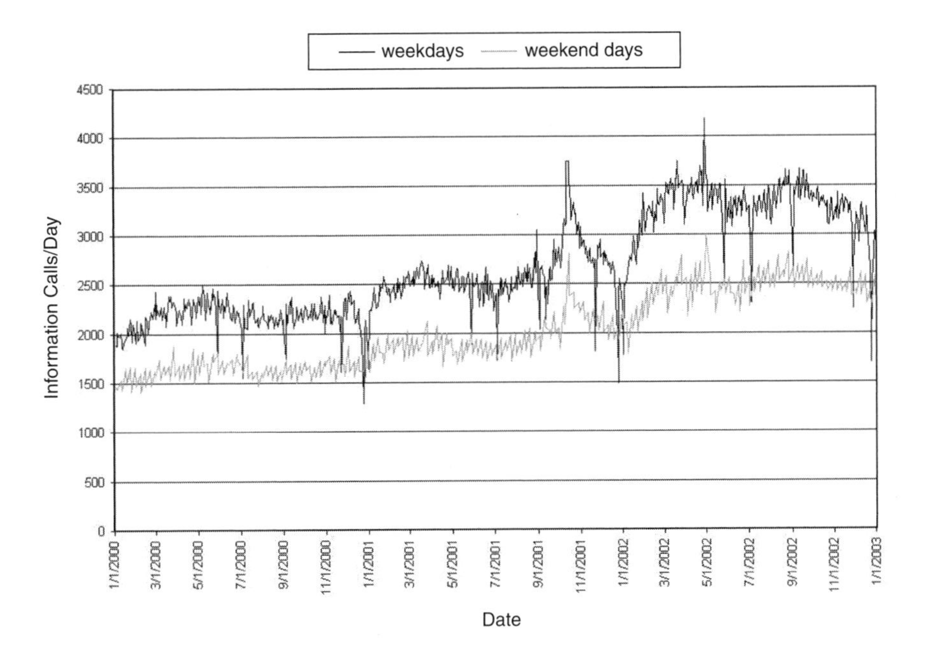

FIGURE 7-2 Frequency of information calls to U.S. poison control centers.
SOURCE: Watson et al. (2003).

The AAPCC "clinical guidelines group" and some AAPCC member centers use the narratives for research purposes. Older abstracts are being added to the fatality database. The AAPCC states that it is planning to support a "text mining system" that could enhance the analysis of clinical effects and therapies embedded in these narrative reports.

Table 7-3 lists the four commercial firms that provide the data capture and data submission software for TESS. Data from 62 of 63 participating centers are submitted to TESS in a prescribed format and are rapidly available for analyses.

Among the essential features of a high-quality data system are definitions of the data elements, setting of standards, and monitoring and reporting of data quality. The 128-page TESS instruction manual (American Association of Poison Control Centers, 2001) contains details of these definitions and standards. The following excerpt is the beginning of the Quality Assurance section in the TESS manual (p. 8):

> Maximum acceptable error rates have been set by the AAPCC Board of Directors for most TESS fields (See QF report, Appendix 2). For each of these fields the maximum acceptable rates of invalid, missing and un-

TABLE 7-3 Commercial Providers of Data Capture and Data Submission Software (listed in order of frequency of use by poison control centers)

Toxicall®
Computer Automation Systems, Inc., 6718 South Richfield Street, Aurora, CO 80016
http://www.cas-co.com/

DotLab®
1122 East Quincy Avenue, Fresno, CA 93720
http://www.wbmsoft.com

PathTech Software Solutions, Inc.
6601 Southpoint Drive North, Suite 200, Jacksonville, FL 32216
http://www.pathtech.com

CasePro
St. Anthony Main, 219 Southeast Main Street, Suite 306, Minneapolis, MN 55414
http://www.damarco.com/casepro.htm

known data have been specified. Each participating center receives a quality report at least annually, with a summary quality factor. High error rates lead to a reduction in a center's quality factor. Submission of data containing <75 percent product specific data likewise leads to a reduction of this factor. High quality factors are obtained by meticulous history-taking and coding and by editing and recoding rejected cases.

Such attention to data quality and routine monitoring contribute substantially to the quality of the TESS data. The training and commitment of the specialists in poisoning information and poison information providers who manage the calls and enter the data represent a major strength of TESS.

Each product reported in TESS is linked to a product-specific code contained in the Poisindex® system (Thomson MICROMEDEX, Greenwood Village, Colorado). Poisindex maintains (and updates) the product codes and provides them at no cost to the poison control centers and AAPCC. In practice, it appears that all U.S. poison control centers subscribe to Poisindex, but generic codes are available, so that theoretically the centers are not *required* to subscribe to Poisindex to submit data to TESS.

Toxicosurveillance Toxicosurveillance involves the identification of sentinel events that may represent emergency nonexposures such as intentional bioterrorist events or other toxin or chemical exposures. The Auto-Upload feature described earlier allows each poison control center to provide case data to TESS every 4 to 10 minutes. To assist in improving public health surveillance, CDC's National Center for Environmental Public Health, the Agency for Toxic Substances and Disease Registry (ATSDR), and AAPCC are working to utilize this new feature to convert TESS into a real-time public health surveillance system. This conversion has the potential to generate more immediate and appropriate responses to public health threats that may be related to toxins or chemicals in the environment. On a pilot basis at present, these federal agencies are using TESS for continuous national toxicosurveillance of poison control center data to detect new hazards. Figure 7-3 shows an example, using cases of calls about contaminated water, of how center data may be analyzed and displayed to detect exposure excesses (Watson et al., 2003). It illustrates the detection prospects for the real-time data system even without the application of signal detection algorithms.

In addition to the national toxicosurveillance effort performed in conjunction with AAPCC, some poison control centers participate in local syndromic surveillance and report notifiable conditions affecting multiple individuals to local and/or state health officials. For example, the Minnesota Poison Control System, in cooperation with the Minnesota Department of Health, provides active surveillance to detect infectious

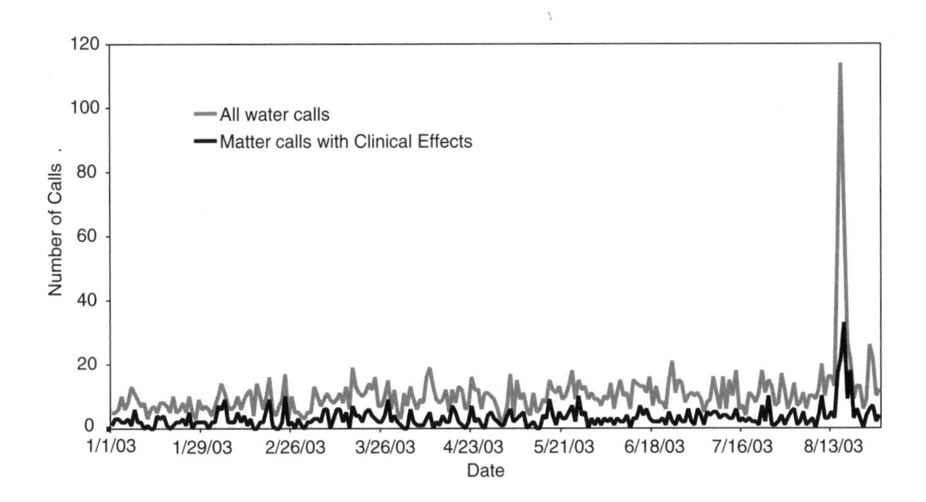

FIGURE 7-3 TESS—water and contaminated water—total and cases with any clinical effects.
SOURCE: American Association of Poison Control Centers TESS program and participating poison control centers.

disease outbreaks and chemical events (http://www.mnpoison.org/index.asp?pageID=194). Poison control centers can conduct active and passive toxicosurveillance and identify sentinel events. To be effective, center staff must be knowledgeable about biological and chemical agents, participate in standardized staff education programs, and utilize protocols, as appropriate. Funk et al. (2003) have described some approaches to using TESS data in the detection of chemical terrorism events. Information calls, for example, may provide a sensitive early indicator and exposure calls reporting symptoms may provide more specific indicators of a bioterrorist event. Some combined analyses of exposure and information calls will probably be developed for each toxicosurveillance application.

Surveillance at the Individual Poison Control Center Level

Beyond AAPCC's national system, TESS data collected at the level of individual poison control centers have independent relevance for surveillance. Based on analysis of published studies originating from poison control centers, the surveillance uses of local or regional center data generally fall into one of three categories:

- Review of the data collected through routine case consultations to the poison control centers. These data include part or all of the structured variables comprising the TESS database, but may also include additional information solicited from the case contact during initial and follow-up calls (i.e., data that are unique to the regional poison control center).
- Review of the same retrospectively identified data, supplemented with additional clinical information usually garnered through medical record review or, in certain cases, through special toxicological testing facilitated by the poison control centers as part of a study. It should be noted that TESS uses a similar approach at the national level in its supplemental surveillance summaries for fatal poisoning cases.
- Use of poison control center reporting for initial, specific case detection, with a structured approach to obtain further data (often prospectively), either from the exposed person or from health care providers.

Activities falling into the first category of poison control center surveillance are documented by a number of publications focusing on specific exposures or poisoning syndromes. Some of the exposures have had particular regional prominence, consistent with a focus of surveillance specific to a single center or cluster of centers. Examples include:

- Illness among tobacco harvesters related to transdermal nicotine absorption (McKnight et al., 1994);

- Envenomations (Brubacher et al., 1996);
- Food-related toxins, geographically concentrated (Barton et al., 1995; Pond et al., 1986); and
- Impact of a local natural disaster on poison control center operations (Nathan et al., 1992).

Although such case series and surveillance studies have been predominantly descriptive in nature and are based on local or regional data, they can focus on topics of general applicability, such as occupational illness or geriatric health (Blanc and Olson, 1986; Kroner et al., 1993). In addition to a regional perspective, this category of surveillance has the advantages of being relatively low cost (the internal data are freely available to the poison control centers) and supportive of collaboration with local and state public health authorities. A major drawback to this data source is the possibility of small sample size, case selection biases (often magnified by local factors), and the limitations of passively collected data. These limitations can be overcome, in part, through multicenter collaborations (McIntrie et al., 1984; Seifert et al., 2003; Spiller and Krenzelok, 1997).

Poison control center surveillance activity in the second category (supplementation of core TESS-formatted data with additional clinical case records or targeted toxicological data) is most frequently available through peer-reviewed publication of case reports or small case series of poisonings. Because the TESS-formatted dataset may not be as complete as that of a full medical record, such case reports and series typically reflect the poison control center consultation supplemented with additional information extracted from inpatient or outpatient charts. Often, these reports represent collaborations between center staff and health care providers outside the center. They can be valuable insofar as they serve to document and disseminate information on novel exposures or unusual manifestations of established toxins. Examples of such surveillance have included emerging issues such as gamma-hydroxybutyrate abuse (Dyer et al., 2001), metabolites in herbal toxicity (Anderson et al., 1996), and metabolic complications of acetaminophen (Roth et al., 1999).

In addition to case reports and series, this category of surveillance has also included the tracking and evaluation of clinical management interventions, such as a study of the predictors of use of head CT scanning in overdosed patients and its impact on management (Patel et al., 2002). This type of surveillance activity can also have elements of a noncontrolled clinical trial, such as a center-based report on clinical outcomes following introduction of a protocol for nebulized bicarbonate to treat chlorine inhalation cases (Bosse, 1994). The value of these surveillance activities includes the detection and dissemination of data on emerging hazards in a fairly rapid time frame; the tendency to strengthen collaboration between center-based personnel and community-based providers (through coauthored reports); and the initial evaluation of novel approaches to clinical management. Limitations include the potential reporting biases inherent in case reports, the observational noncontrolled nature of the clinical management studies (often invoking historical controls for comparison), and the likely chilling effect that the Health Insurance Portability and Accountability Act (HIPAA) restrictions will have on merging poison control center data with supplemental information derived from medical records.

The third category of poison control center-based surveillance, using case reporting as a starting point for subject recruitment for additional direct data collection or follow-up, is less common. This approach frequently collects survey data beyond the TESS format. Examples include center-based studies of occupational poisoning (Blanc et al., 1989, 1990), inhalation injuries (Blanc et al., 1991, 1993a), ocular exposure outcomes (Saunders et al., 1996), herbal supplement effects (Palmer et al., 2003), and outcomes of snake envenomation (Spiller and Bosse, 2003). This type of surveillance has the advantage of prospective detailed data collection targeted to a specific set of study questions and therefore has some aspects of an active rather than a passive surveillance approach. Limitations include the need for substantial supplemental personnel effort with associated costs, often requiring supplemental funding through extramurally funded research support. Because many poison control centers do not have a research-oriented infrastructure, this category of surveillance is typically beyond their capabilities.

Data Sources Derived from Health Records and Health Care Datasets

As shown in Table 7-2, a number of different datasets with vital statistics or health care information derived from medical records can be used for poisoning and drug overdose surveillance purposes.

National Vital Statistics System

National mortality data have provided a particular focus for surveillance analyses relevant to poisoning and drug overdose (Chyka and Somes, 2001; Cobb and Etzel, 1991; Fingerhut and Cox, 1998; Hoppe-Roberts et al., 2000; Klein-Schwartz and Smith, 1997; Shepherd and Klein-Schwartz, 1998). One major source for such data is the National Vital Statistics System of CDC's National Center for Health Statistics (NCHS). These data are derived from death certificates and include listed primary and contributing causes of death designated by ICD-9 codes (including relevant external cause of injury codes) through 1998 and ICD-10 codes thereafter. A number of peer-reviewed scientific publications have analyzed national mortality data from a poisoning surveillance perspective.

The advantages of this system are that it is national, comprehensive, and accessible (free online from 1981 onward, although requests for data analysis may need to be submitted in advance; nominal charges for earlier data tapes). In comparison with TESS surveillance:

- Death certificate-derived data capture far greater numbers of fatal poisonings and drug overdoses, in part because prehospital fatal poisoning events are unlikely to lead to poison control center consultations (Hoppe-Roberts et al., 2000).
- Mortality related to drug abuse is also underreported through TESS relative to death certificate-based data.
- Causal attribution among medical examiners has variable precision depending on practices in toxicological assaying, thus limiting the value of vital statistics mortality data.
- Identification of toxins for targeted surveillance purposes may be limited because the cause of death is available only by ICD-9- (or ICD-10-) defined categories (e.g., determining the species of mushroom or venomous snake involved; differentiation between methyl bromide and phosphine in fatal fumigant injuries; identifying hypochlorite bleach-acid mixing misadventures).
- Product identification beyond general classes, even through supplemental E-code inferences, is usually not possible using death certificate data.
- Correct attribution of intention using death certificate data is also fraught with difficulty.

Medical Examiner Case Series

Death certificate data can also be analyzed for poisoning surveillance at the local and state levels (Blanc et al., 1993c, 1995; Cone et al., 2003; Davidson et al., 2003; Landen et al., 2003; Linakis and Frederick, 1993; Soslow and Wolf, 1992). At these levels it may also be possible in some areas to exploit medical examiner-reviewed cases for surveillance purposes. Medical examiners' series are likely to be enriched for cases confirmed by toxicological testing. Narrative case files can provide a rich source of detailed information that is not available through death certificate data. Even summary data in annual medical examiners' reports are likely to provide a rich source for surveillance purposes. Many poisoning fatalities are not captured by medical examiner datasets, however, with a likely bias of underreporting for deaths that occur in an in-hospital setting

(Landen et al., 2003; Linakis and Frederick, 1993; Soslow and Wolf, 1992). Thus medical examiner data are not entirely captured in death certificates, while death certificate data include a far greater number of cases than those in medical examiner series.

Hospital Discharge Data Systems

Hospital discharge data systems (HDDS) at the local, regional, or state level represent a major source of health care data that potentially can be used for poisoning and drug overdose surveillance purposes (Agran et al., 2003; Hoyt et al., 1999; King, 1991; Smith et al., 1985, 1991; Sumner and Langley, 2000). The assignment of ICD-9 codes (routine and E-codes) draws on the direct medical evaluation documented in the hospital record. Hospital admission for poisoning or drug overdose can be presumed to be reasonably complete because a patient with a life-threatening poisoning or drug overdose is likely to be admitted to a hospital and included in this dataset regardless of health insurance status. Another strength of HDDS data is that, like death certificate vital statistics, the data are comprehensive for the states in which they are collected. Nonetheless, HDDS is not national and universal. As of 1998, 42 states collected such data. However, the consistency of E-coding was more variable, with only 36 states collecting some ICD-9 E-code data as part of their HDDS; less than half the states mandated such coding (American Public Health Association, 1998). To the extent that poisoning is a comorbid condition displaced within a longer list of diagnoses, it may not be captured in summary data. Iatrogenic causes of medication toxicity may be preferentially downgraded or obscured in such records.

The HDDS data exclude cases treated in an emergency department and discharged. Hospital-based emergency department data systems (HEDDS) exist, but they are limited. As of 1998, only 12 states had these systems in place. However, additional states reported plans to add similar systems. Once again, E-coding in these systems is variable. The data collection systems do not include freestanding urgent care centers or other outpatient treatment centers. As of 1998, only four states had a non-emergency department, statewide outpatient data system. The extent to which HDDS and HEDDS data are publicly available with minimal or nominal charge and the lag time between data collection and public data access vary by state. The potential limitations of ICD-9 nosology apply to HDDS and HEDDS data and are likely to be magnified by E-code deficiencies. Both the HDDS and HEDDS datasets are comparable to vital statistics death certificate data in that they are meant to capture all of the eligible events within the geographic areas they cover.

Sample-Based Health Care Data

Some datasets provide health care data derived from sampling methods. These datasets are not universal; rather, they are based on selected numbers of events. Depending on the sampling approach used, total prevalence or incidence estimates can be generated.

NCHS oversees three national sampling surveys of health care utilization that contain data relevant to poisoning injuries. These are:

- National Hospital Discharge Survey (a national sample of hospital data);
- National Hospital Ambulatory Medical Care Survey (a national sample of hospital-based emergency departments and ambulatory care centers); and
- National Ambulatory Medical Care Survey (a national sample of outpatient visits).

In addition to NCHS's National Hospital Discharge Survey, the Agency for Healthcare Research and Quality (AHRQ) oversees its own hospitalization survey, the Health Care Cost and Utilization Program National Inpatient Sample. Because these are all designed as representative, weighted samples, each survey can yield national estimates of health care utilization. The datasets include ICD-9 condition codes and are available electronically for downloading free of charge (free query of the dataset without downloading in the case of AHRQ). Because these survey data are collected annually with consistent sampling methods, they allow data merging across years as well as surveillance tracking of trends over time. Several of these datasets are particularly relevant to developing estimates of direct health care cost. Despite their potential as a rich surveillance data source, relatively few peer-reviewed research publications have exploited these surveys for poisoning and drug overdose surveillance purposes (Klein-Schwartz and Smith, 1997; McCaig and Burt, 1999; Powell and Tanz, 2002; Rodriguez and Sattin, 1987).

The datasets face the same ICD-9 coding limitations discussed in relation to death certificate national vital statistics data. Moreover, because they are based on samples, uncommon events may be undetected or have few sampled observations with a wide margin of statistical error. Combining survey years can sometimes, but not always, address this shortcoming. Because of the sampling design, estimates for discrete geographic areas (e.g., at the state level) usually cannot be generated from these surveys.

Other Exposure-Related Data Sources

The previously described datasets capture poisoning injury as a small proportion of their overall surveys. In addition to these resources, there is an ongoing survey of health care utilization specific to injury, the National Electronic Injury Surveillance System (NEISS). This system was initiated by the Consumer Product Safety Commission (CPSC) to capture consumer product-related injuries (including poisonings). NEISS relies on data collected from a stratified national probability sample of 100 hospital emergency departments, with datasets available for electronic query at no cost dating from 1991.

In 2000, the NEISS program was expanded (in collaboration with CDC's National Center for Injury Prevention) to include nonfatal injuries from all external causes (Centers for Disease Control and Prevention, 2001a, 2003). This expanded All Injury Program (NEISS-AIP) is based on data from 66 of the 100 NEISS hospitals. As with the parent survey, NEISS-AIP also yields nationally extrapolated incidence data. Although NEISS-AIP captures a wider range of cases than the parent survey, it excludes cases of adverse effects from therapeutic drugs or medical care. Thus it would not include data on safety packaging for pharmaceuticals, for example, even though CPSC has jurisdiction in this area. Although the collection system does not exclude fatal injuries, these cases are excluded from directly available summary data and most published analyses of NEISS.

NEISS has been used to a limited extent for surveillance purposes specific to poisoning injuries (Henneberger et al., 2002; Woolf and Shaw, 1998). Because NEISS is injury focused, it has employed its own non-ICD-driven coding scheme, with more targeted and detailed information on cause and intentionality than can be derived from the other datasets described previously. Although such detail may be useful, it can also be a limitation because it complicates direct comparison with data derived from these other datasets. Multiyear data are available for NEISS; NEISS-AIP is relatively new.

Several databases derived from focused data collection activities are also relevant to specific aspects of poisoning injury surveillance. The Drug Abuse Warning Network (DAWN) is a surveillance program overseen by the Substance Abuse and Mental Health Services Administration (SAMHSA). DAWN has two components. Reporting through one DAWN system draws on emergency department chart extraction from a national sample of hospitals with oversampling in selected metropolitan areas (U.S. Department of Health and Human Services, 2003). Case eligibility is not limited to drugs of abuse; it also includes cases treated for adverse effects attributed to prescription and over-the-counter medications. Nonetheless, its primary surveillance use has been applied to issues of illicit drug effects. Sampling is meant to capture any medical complication related to acute or chronic abuse (e.g., infection) and not simply direct drug toxicity consistent with poisoning ICD classifications. Children under age 6 are excluded. This DAWN program can provide national estimates.

The second DAWN program consists of collection of information based on medical examiners' or coroners' case data from 128 jurisdictions in 42 metropolitan areas (U.S. Department of Health and Human Services, 2002). Case eligibility is limited to illicit drugs or drugs used for nonmedicinal purposes (although this can include suicide as such a purpose) as a principal or a contributing cause of death. This DAWN system does not include a national probability sample and thus cannot yield national estimates. The metropolitan base may be a source of other biases (e.g., cases of amphetamine overdose treated in nonurban settings would not be captured by this surveillance system).

MedWatch, a program maintained by the Food and Drug Administration (FDA), is another example of a specialized surveillance activity that can be highly relevant to a subset of poisoning injuries (Chyka, 2000). Through MedWatch, the FDA monitors medical products, including prescription and over-the-counter drugs, biologics, medical and radiation-emitting devices, special nutritional products (e.g., medical foods, dietary supplements, infant formulas), and medication errors. Reporting to MedWatch's Adverse Events Reporting System is mandatory for manufacturers and voluntary for health care providers and the public. Reports can be submitted via mail, telephone, fax, or the Internet. These reports are then available to staff in the appropriate FDA center for evaluation. MedWatch focuses on unexpected and serious adverse events (i.e., death, life threatening, requiring or prolonging hospitalization, resulting in disability, congenital anomaly, or requiring therapeutic intervention). It is particularly valuable for new drugs because adverse event information for these agents is limited to the patients exposed during clinical trials. The FDA uses this information for signal (hypothesis) generation, subject to mechanisms to verify potential exposure-disease associations. MedWatch is likely to underrepresent delayed adverse effects. Health care provider reporting may be influenced by concerns over liability related to adverse events. Illicit drug use and nondietary substances are not well represented in MedWatch. MedWatch has been relatively underexploited as a surveillance tool in published research reports (Bennett et al., 1998; Chyka, 2000).

MEDMARX is a voluntary, Internet-accessible, anonymous medication error-reporting program that allows a selected number of subscribing facilities to access and share information. The United States Pharmacopoeia (USP) operates MEDMARX, which contains more than

580,000 released records (United States Pharmacopoeia, 2002). This program's relevance to poisoning and drug overdose surveillance systems is limited given that it is a voluntary, subscription-based method for reporting (no public access for analysis), weighted heavily to therapeutic misadventures.

The Hazardous Substances Emergency Events Surveillance system is a state-based program managed by the Agency for Toxic Substances and Disease Registry. As of 2002, the ATSDR compiled data detailing hazardous substance emergencies derived from 15 states. Although the primary unit of analysis in this system is the release event, details on the number of persons affected and associated health care utilization are obtained. The data in this system are most relevant to airborne releases, particularly of irritant gases (Berkowitz et al., 2003; Horton et al., 2002; Orr et al., 2001; Weisskopf et al., 2003).

A specialized surveillance dataset of potential interest is the Census of Fatal Occupational Injuries, maintained by the federal Bureau of Labor Statistics (BLS) (Valent et al., 2002). The goal of this system is to capture all occupationally related fatalities in the United States, including cases resulting from toxic exposures. This dataset is notable in that it includes data from all 50 states and attempts to integrate and cross-check data from multiple sources, including death certificate data, workers' compensation insurance claims, Occupational Safety and Health Administration (OSHA) fatality reports, and media reports. BLS also oversees two national surveys for nonfatal occupational injuries; one survey is based on a sample of workers' compensation claims and the other on OSHA-mandated injury forms. Neither system is effective in detecting toxin-related events; thus, each is marginal in relation to poisoning surveillance generally.

Survey Data Relevant to Poisoning and Drug Overdose Surveillance

In addition to data derived from poisoning and drug exposure sources and from general vital statistics and medical encounters, national survey data are also relevant to poisoning surveillance. Of these, the most comprehensive is the annual National Health Interview Survey (NHIS). This survey, based on an extensive national stratified random sample, generates data on poisoning through a series of injury-related items asked of all survey recipients. Beginning in 2000, coverage was expanded to better capture poisoning events. Items address the general nature of the exposure as well as some details of circumstances involved, including whether a poison control center was contacted. Only limited analyses of poisoning data derived from NHIS have been published by independent investigators (Fleming et al., 2003; Polivka et al., 2002).

The National Health and Nutrition Examination Survey (NHANES) is a periodic, but not an annual, national survey that includes a direct physical examination and biological testing, in addition to interviewing. Although not relevant to acute poisoning, this survey has provided pivotal data for national lead exposure paint prevalence and time trends. More recently, NHANES biological sampling data have been used to assess exposure to a variety of xenobiotic chemicals.

CDC also oversees several other surveys that may have tangential relevance. The Behavior Risk Factor Surveillance System provides a platform for added modules of survey items that can be used by the states. In 1992, some states employed a module asking whether respondents have the telephone number for the poison control center in their area and whether they keep syrup of ipecac in their home. Two Injury Control and Risk Surveys (ICARIS) have been carried out: ICARIS I was conducted in 1994 and ICARIS II in 2003. The ICARIS II survey included a single poisoning-related preventive care survey item ascertaining whether a child's treating physician had provided the family with poison control center contact information. The Youth Behavior Risk Factor Surveillance System survey has ascertained drug abuse, but has not elicited general information on poisoning.

Outside of NCHS and other CDC branches, the public interview survey most relevant to poisoning and drug overdose is SAMHSA's annual National Household Survey on Drug Abuse. In addition to illicit drugs, this survey addresses a variety of prescription and nonprescription medications with abuse potential, but does not collect data on poisoning or adverse drug effects generally (Kozel, 1990; Rouse, 1996). The survey is not limited to adults, but does exclude persons under 12 years of age.

COMPARISONS AMONG SURVEILLANCE SYSTEMS

There is no published systematic review of surveillance data sources for poisoning and drug overdose in the United States, although one relevant abstract was recently presented on this topic (Gotsch and Thomas, 2002).

A number of pairwise cross-comparisons have been made, particularly in relation to standard poison control center case detection through TESS. The most frequent comparison has been between poisoning deaths detected through vital records or medical examiner surveillance and deaths recorded through TESS-derived data (Blanc et al., 1995; Hoppe-Roberts et al., 2000; Linakis and Frederick, 1993; Soslow and Wolf, 1992). These analyses have consistently observed that TESS surveillance detects only a fraction of the fatal cases—approximately 1 of 20. Although the fatal cases reported in TESS have been reviewed by the medical director of the reporting poison control center, inadequate data are provided to demonstrate cause and effect. This is because the case information is secondary from hospital or other sources and the center rarely has access to complete case data. TESS data are therefore most useful as a signal along with other data sources, but cannot be utilized for policy making. This problem is exacerbated under HIPAA regulations, which decrease information flow to the poison control center once the patient has been admitted to a health care facility. In addition, the characteristics of the fatal cases differ proportionally for variables such as intent, type of poisoning, and demographics. Although one source of underreporting by TESS is attributable to out-of-hospital deaths, there is also substantial underreporting for fatal hospitalized cases. Despite the smaller numbers, there are also fatal cases detected through TESS surveillance that are not detected by death records. Comparison between death certificate data and medical examiner data has shown that these sources do not wholly overlap (Landen et al., 2003; Linakis and Frederick, 1993; Soslow and Wolf, 1992). A single study of fatal cases in the MedWatch system also found poor overlap with death certificate data (Chyka, 2000).

Direct review of hospital charts has demonstrated that only about 20 to 30 percent of poisoning cases managed in the emergency department are reported to poison control centers (Blanc et al., 1993b; Harchelroad et al., 1990; Hoyt et al., 1999). Successful case detection of medically treated cases by the DAWN system appears to be in a similar range (Roberts, 1996). In contrast with these patterns, surveillance based on the National Health Interview Survey yielded lower population estimates for poisoning incidence than those derived from TESS data (Polivka et al., 2002). This analysis was limited to pediatric cases ages 5 and younger.

Linkages among the various datasets are limited. The United States does not have a universal identification number that is used in medical records and surveys allowing for interlinking of disparate datasets. Although such linkages are desirable, no such identification system is likely to be developed and applied in the foreseeable future.

The available data suggest that no single surveillance source can provide a universal data source from which to draw a complete picture of all aspects of poisoning and drug overdose morbidity and mortality. The strengths and limitations of each source should be taken into account in interpreting surveillance data.

8

Prevention and Public Education

A useful framework that helps organize thinking about poison prevention is the Haddon Matrix (Table 8-1), named for William Haddon, an early leader in the injury prevention field, and widely used in injury control to guide the development of strategies for prevention and treatment.

The Haddon Matrix is organized along two dimensions. The first is a categorization of the timing of injury into three levels: "preexposure" factors that influence the likelihood that a poisoning will occur; "exposure" conditions that influence the exposure itself; and "postexposure" conditions that influence the consequences of the exposure once it has taken place. The other dimension is organized according to the classification of host, agent, and environmental factors that influence a poisoning occurrence. Finally, the cells identify risk factors and potential interventions for all three temporal periods. For example, removing the availability of pills from the environment of a young toddler through appropriate storage is an example of a preexposure environmental intervention. Alternatively, immediate action to contact a poison control center and administer the recommended treatment for ingestion is an "exposure"-level intervention strategy based on human host action. Finally, initiating new regulations of a hazardous product that has been newly identified as causing poisoning, after the fact, is an example of a policy-level environmental response at "postexposure" level.

Public education and community outreach can contribute at all temporal levels—it can be used to teach safe product storage practices at the

TABLE 8-1 Haddon Matrix for Poison Prevention and Control

Timing of Injury	Human Host	Agent (vector of metabolic injury)
Preexposure (preexisting)	Developmental aspects: —children: explore —teens: experiment —elderly: mistakes; drug interactions —addicts: overdose —workers: job related Unnecessary use of hazardous agents	Toxicity of chemical Availability of agent Unanticipated exposure Eliminate production Eliminate chemicals as weapons
Exposure	Knowledge of what to do Knowledge of poison control center toll-free number Emergency actions: —antidotes —emetics —correct treatment Access to expert advice	Dose of exposure Route of exposure Body's reaction to agent Immediate use of antidote/ treatment
Postexposure	Rehabilitation: —lungs —esophagus —neurological Educate public based on cases, experience Postexposure follow-up and monitoring of victims Knowledge about correct care of poisoned host	Track exposure to agents Modify hazardous agents based on information Repackaging agents

NOTE: Two principles of the Haddon matrix: (1) It is not when you do something, such as eliminating production, but when the action is relevant to the time frame of the injury occurrence or mitigation. (2) An action that prevents an injury is "preevent" even if it is based on after-the-fact knowledge.

Physical Environment	Social-Economic Environment
Agents available: —pills —consumer products —illegal substances —terrorism Product labeling Poor storage Too many pills Safety packaging; blister packaging Protective gear for workers Workplace safety equipment	Poor storage practices Poor supervision of young and old Drug-taking society Substance abuse Poor children at greater risk Substitute safer products Design homes with childproof storage areas
Antidote information available Availability of emergency medical services and emergency departments Health system functioning Available antidotes	Availability of telephone and transport Availability of poison control center information
Regulate products Invent new products that are less toxic to replace those more toxic Repackage agents Protect workers	Political and public support to regulate chemicals and change manufacturing Replace hazards Workers actually use protective gear Public knowledge and perception of poison control system

preexposure level, provide awareness of poison center services and contact information at the exposure level, and provide input to developing a positive public perception of poison control centers and poison control at the postexposure level.

An important message that can be drawn from experience with the Haddon Matrix is that injury prevention can only be achieved through a multifaceted approach. Integral components of effective programs incorporate the following elements, known as the "E's" of injury prevention and control:

- Education
- Environmental/Engineering modifications
- Enactment/Enforcement
- Economic incentives
- Empowerment
- Evaluation

Education includes any efforts to reach children, parents, caregivers, the public, practitioners, the media, policy makers, and other target groups to change knowledge, attitudes, and behavior (e.g., National Poison Prevention Week, Spike's Poison Prevention Adventure). In addition to changes in the physical environment, engineering and environmental modifications include the design, development, and manufacture of safe products (e.g., child-resistant packaging of prescription medications). Enactment and enforcement include the passage, strengthening, and enforcement of laws; the issuance and enforcement of regulations; and the development of voluntary standards and guidelines (e.g., safety cap regulations, packaging and labeling of baby aspirin and medications containing iron). As noted earlier in this report, regulations can be effective in reducing the number of human exposures (a complete listing of regulations is provided in Chapter 4). For example, an estimated 460 deaths among children ages 4 and under were prevented from 1974 through 1992 through the use of child-resistant packaging of prescription medications, a 45 percent reduction in the mortality rate from levels predicted without such packaging (Rodgers, 1996). In particular, the use of child-resistant packaging was associated with a 34 percent reduction in the aspirin-related child death rate (Rodgers, 2002). Several other studies providing evidence for the effectiveness of the Poison Prevention Packaging Act are summarized by the Harborview Injury Prevention and Research Center (http://depts.washington.edu/hiprc/childinjury/topic/poisoning/pcc.htm).

Economic incentives influence the socioeconomic environment of communities through the distribution of safety products at no cost or low cost to families in need and working with manufacturers to improve safety devices without increasing their costs (e.g., distribution of cabinet latches to prevent access to medications and household cleaners). Empowerment includes activism at the grassroots level as well as the formation of federal and private-sector advisory panels and injury prevention coalitions or partnerships at the national, state, and local levels (e.g., Health Resources and Services Administration Stakeholders Group; Poison Prevention Week Council). Evaluation includes research, data collection, and surveillance, as well as evaluation of program and product effectiveness (e.g., the Toxic Exposure Surveillance System, or TESS).

This chapter focuses on public education efforts in poison prevention. These efforts have the potential to influence the health behavior of individuals in positive ways, yet as described above, they constitute only one of many factors to be considered. Public education efforts should be considered a necessary but not sufficient component in preventing and mitigating poisonings.

Poison control centers have two relatively distinct education activities—primary and secondary prevention. The goal of primary prevention is to avoid the occurrence of a poisoning exposure. Examples include advising parents to lock up medication and household cleaners to keep them out of reach of young children; requiring employees who could be exposed to hazardous chemicals to wear safety equipment such as gloves, goggles, and protective clothing; and recommending storage techniques to older adults to avoid medication mishaps. Secondary prevention strives to reduce the effect of a poisoning exposure through improved access to poison control services (e.g., raising awareness of the poison control center telephone number as well as urging adults to keep activated charcoal available for use as an antidote on the explicit advice of a physician or the poison control center staff). Educators in centers offer both types of prevention by distributing brochures, refrigerator magnets, and checklists; airing videos and public service announcements; and making presentations in classrooms, at senior centers, and at health and safety fairs. Much of the educational material distributed to the public contains both primary and secondary prevention messages. An example is the widely used Mr. Yuk campaign material, which contains warning labels for poisonous substances and stickers with the poison control center telephone number for rapid access to advice and needed treatment.

Although the size, content, and reach of the public education programs vary among the poison control centers, implementation generally includes a variety of cooperative arrangements involving health departments, health care facilities, pediatricians and family care practitioners, pharmacies, local retail outlets, and various private and public organizations (e.g., National SAFE KIDS Campaign, Red Cross, National Fire Prevention Association's Risk Watch Program). In 2001, 71.3 full-time equivalent poison control center personnel across all centers were devoted to public education. During that same year, approximately 16.8 million materials were distributed at a cost of $2.4 million, and more than 287,000 individuals attended poison education sessions organized and delivered by poison control center staff or by individuals trained by them to be trainers in their service area (American Association of Poison Control Centers, 2001, 2002a). Additional information on prevention and poison recognition is provided to the public through follow-up to exposure calls and by answering information calls.

PROGRAM DEVELOPMENT AND EVALUATION: THEORIES AND MODELS

As noted earlier, the purpose of public education is to change health behavior related to poisonings; that is, to influence people to (1) take the recommended actions to avoid poisonings; and (2) contact a poison control center should a poisoning occur. There is evidence from surveys and focus groups suggesting that, although a large percentage of people say they are knowledgeable about poisons, most do not take appropriate actions. According to a survey conducted by the Home Safety Council (2002), more than half of homes with children ages 6 and younger have household chemicals (e.g., cleaners, bleach, kerosene) stored in unlocked locations. The question remains: How do we provide people with the necessary knowledge and also influence them to take positive actions based on that information?

According to theories of communication and behavior change, a number of variables influence the intention of an individual to perform a behavior (Institute of Medicine, 2002a). The most central of these include attitudes (how favorable a person is toward the behavior), perceived norms (the degree to which a person perceives that a given behavior is viewed as appropriate or inappropriate by members of the individual's social network), and personal agency (the belief that one has the necessary skills and abilities to perform the behavior). In addition, there are positive and negative influences associated with the individual's environment (e.g., lack of a telephone). Figure 8-1 provides a general model of the determinants of behavior change. The variables shown are also central in the Theory of Reasoned Action (Fishbein et al., 1991), Social Cognitive Theory (Bandura, 1977, 1986, 1991, 1994), the Theory of Planned Behavior (Ajzen, 1985, 1991; Ajzen and Madden, 1986), and the Health Belief Model (Becker, 1974; Rosenstock et al., 1994).

The PRECEDE model, developed by Green in 1968 and later elaborated on by Green and Krueter in the late 1980s, provides a framework for the systematic development and evaluation of health education programs

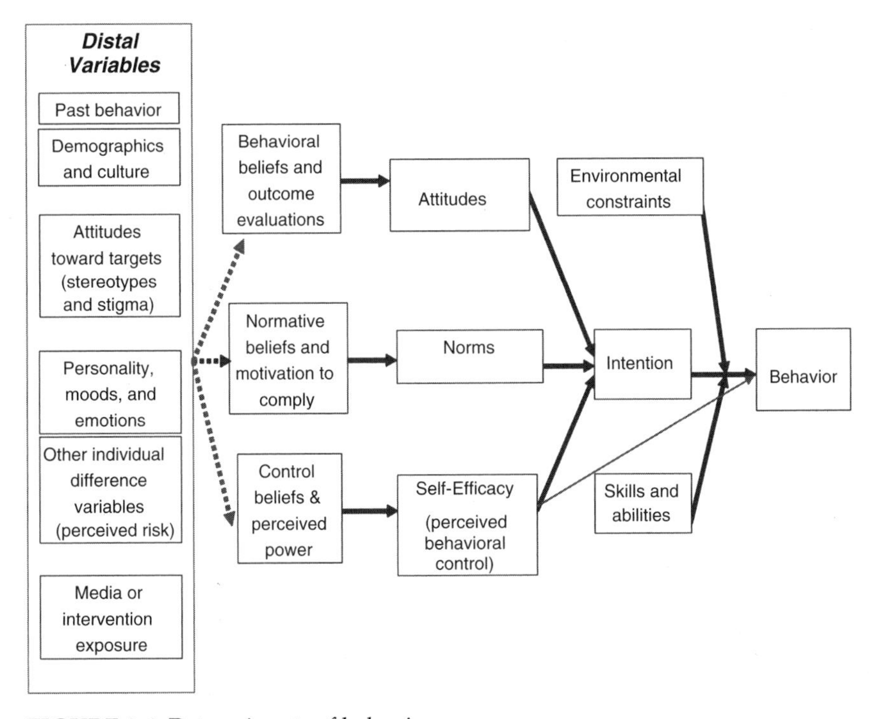

FIGURE 8-1 Determinants of behavior.
SOURCE: Institute of Medicine (2002b).

that takes into account the variables discussed previously (Green and Krueter, 1991). The elaborated model, PRECEDE/PROCEED, relies on the premise that effective health education is dependent on the voluntary participation of the client in both identifying current behavioral practices and changing those practices. Furthermore, it is based on the notion that the degree of change in knowledge and practice is directly related to the degree of active participation by the client. The model specifies nine phases. The first five phases focus on program development, and the last four phases involve implementation and evaluation activities. This model has been used for a variety of applications, including developing collaborative partnerships in community health (Fawcett, 1995), promoting breast cancer screening (Taylor, 1994), and assisting in the development of public health campaigns by the Centers for Disease Control and Prevention (CDC) (Donovan, 1995).

Another model used in health education-related program development and evaluation is the Program Evaluation Logic Model. It is organized in terms of program inputs (e.g., the problem being addressed, characteristics of the client's circumstances, resources needed, and content of the program); the activities or outputs of the program; and the outcome objectives (including increases in knowledge and skills, modified behavior). The evaluation portion of this model focuses on establishing quantitative performance standards (e.g., how many individuals will change their behavior in the required direction and over what time period). This model is currently being used in Arizona to develop, implement, and evaluate the "Tell a Friend Campaign." This campaign, discussed later in the chapter, is designed to increase poison control center access across the state through community partnerships and the distribution of multilingual educational materials.

A third communication model entitled "A Su Salud" aims to change behavior by using positive role models and volunteers from the community to provide positive social support. This model has been implemented by the Texas Department of Health to encourage the use of screening for breast and cervical cancer in diverse communities (Suarez et al., 1993) and is currently being employed by the South Texas Poison Center in an attempt to increase awareness of poison center services in the Hispanic community.

THE CHALLENGE OF ETHNIC DIVERSITY

Current demographic trends show that the U.S. population is growing larger, older, and more ethnically diverse. Hispanics are now the largest minority and are projected to grow to 23 percent of the population by 2045. In some states, such as California, many cultures and languages are represented in the population. What are the challenges these trends pose for poison control center educators?

Penetrance data available from the American Association of Poison Control Centers (AAPCC) show different levels of poison control center use in ethnically diverse and low-income communities (American Association of Poison Control Centers, 2003b). One study that illustrates different poison control center use patterns by different population segments was conducted at a Texas medical center by Kelly et al. (1997). The purpose of the study was to characterize and compare caretakers of children who failed to contact the poison control center about unintentional poisonings prior to visiting an emergency department with those who used the center first and were then referred to the emergency department. Comparisons were made by age, gender, relationship to the child, ethnicity, language preference, education, and marital status. The results showed that whites were most likely to call a poison control center, followed by Hispanics and then blacks. Also, caretakers schooled in Mexico were significantly less likely to call a poison control center than those schooled in the United States. Although knowledge about poisonings and the availability of a poison control center differed between those who called and those who did not, 68 percent of those who did not call indicated that they knew about the center. Thus, having knowledge of the existence of the poison control center does not guarantee that an individual will call the center.

Kelly et al. (2003) found that lower utilization of poison control centers by Spanish-speaking parents was attributable to a lack of confidence in center staff, greater trust in their own family physicians, and a lack of knowledge about the severity of different poisons. In this study, videotapes were used successfully to both increase knowledge and encourage changes in behavior of women attending classes at a Special Supplemental Nutrition Program for Women, Infants, and Children clinic. Conclusions were based on pre-post questions administered to matched control and treatment samples. Significant differences between the groups were found in knowledge, attitudes, and behavioral intentions following the videotape intervention. It was suggested by the authors that this intervention might prove useful in other low-income, ethnically diverse areas.

CURRENT PRACTICES IN POISON PREVENTION EDUCATION

Although several populations are at disproportionate risk for poisoning (e.g., the elderly, alcohol and drug abusers, workers in certain high-risk occupations), the majority of public education materials focus on preventing unintentional childhood injuries. Although poison control center education efforts cover the United States, there is substantial variability among centers in the amount of material distributed (under 7,000 pieces to more than 2 million) and the number of offsite activities (6 to more than 1,500). Some centers rely heavily on materials produced by AAPCC, while others develop their own. The focus of these efforts has been on unintentional, minor poisonings in children under 6 years of age; little effort has been directed toward prevention activities for the broader population, for serious poisonings, or for individuals intentionally poisoning themselves.

A few programs have been developed based on a careful analysis of target audience characteristics such as age, ethnicity, and socioeconomic status, and have attempted to follow models in developing and evaluating their campaigns. The following sections briefly discuss the education activities of AAPCC; describe the education staff at the poison control centers; provide examples of model-based educational programs developed by centers; and describe collaborative efforts, including activities involved in National Poison Prevention Week.

Poison Prevention Education Activities of the American Association of Poison Control Centers

According to a cooperative agreement with CDC, APPCC is currently working on the following tasks in the public education arena:

- Maintain a national toll-free telephone number for poison control services.
- Develop and implement a public service media campaign to familiarize health care professionals, public health professionals, and the public with poison control services.
- Establish a media campaign stakeholder committee composed of poison control center health educators, state health department injury prevention professionals, and representatives from relevant national organizations to guide this effort.
- Promote the broad use of the toll-free number by poison control centers, professionals, and the public by using materials developed by AAPCC in 2002.
- Conduct an independent evaluation of materials developed in 2002, such as English- or Spanish-language promotional brochures or preschool educational materials. Use formative research methods to test the effectiveness in target audiences.

In positioning the national toll-free number, and in deciding on appropriate poison messages and logos, AAPCC worked with KRC Research and Consulting. The approach was to use focus groups representing senior citizens, parents (who had completed college or high school), teenagers, and preteens. These groups discussed attitudes and behaviors in poison situations, perceptions of poison control centers, messages regarding poison prevention and the use of poison control centers, various logos, and the desirability of a national toll-free number (KRC Research and Consulting, 2001). The results were used to design campaign materials in consultation with experts in public education and poisoning prevention. The primary public education messages from AAPCC are:

- Read the label each and every time you take a medicine or use a product or chemical.
- Prevent poisoning by using child-resistant packaging; locking cabinets containing medications, cleaning supplies, and other toxic substances; and putting potentially harmful products out of reach.
- Put your poison control center's emergency phone number on or near every phone: 1-800-222-1222.
- Call the poison control center immediately in case of possible poisoning.

- Call your poison control center for educational materials and for assistance with planning poison prevention education programs in your community.

The national toll-free telephone number has been in effect since January 2002, and all poison control centers have been provided with stickers, brochures, and posters advertising it. KRC Research conducted a tracking survey to evaluate the usage and effects of this number (KRC Research, 2003). The results are based on a comparison of responses from random samples drawn prior to the introduction of the toll-free number and then 14 months later. The samples consisted of adults 18 years and older living in private households and representing the demographic and geographic distributions present in the population. The results showed that over the 14 months of operation, the percentage of respondents who would call the poison control center first increased from 19 to 30 percent, while the percentage who would call 911 first decreased from 63 to 49 percent. Examining the results across age groups shows that adults aged 25 to 34 were most likely to call a poison control center (45 percent) while adults aged 65 and older were least likely (14 percent). With the new national toll-free telephone number, total call volume at the Pittsburgh Poison Center increased by 11.2 percent. Comparing 2001 and 2002 trend analysis data revealed a 10 percent increase in exposure volume. Information calls dramatically increased in volume as well, probably because of toll-free access (Krenzelok and Mvros, 2003).

Monthly data tracking of the number of calls to the national toll-free number shows an increase from approximately 40,000 per month in January 2002 to approximately 140,000 per month in April 2003. During the same time period, data from TESS show a fairly constant level of exposure plus information calls to poison control centers (250,000 to 300,000 per month).

The cooperative agreement with CDC has been used to fund the development of a prevention program for preschoolers and their parents. The package (containing a teacher's guide and video along with music and take-home materials) is free to schools and is distributed through the poison control centers. The Head Start program has ordered kits for its own internal distribution. This child-based program was developed though the use of focus groups to test key messages. A similar program for adults is under development. It is anticipated that the adult program will be provided in venues such as senior centers, parent/teacher association and parent/teacher organization meetings, and workplace safety meetings.

The AAPCC website offers information to various audience segments. For example, poisoning fact sheets are available for specific target groups:

parents and adult child care providers, children, and teens/babysitters. It also provides links to all poison control centers as well as to various teaching aids.

To date there has been no systematic evaluation on the part of AAPCC of the education efforts launched by its office or by the poison control centers. According to AAPCC, evaluations are planned for calendar year 2004.

Public Education Staff

Approximately 110 poison control center staff work as health educators, and of these, 40 percent perform this function full time. Most staff working as part-time educators have multiple responsibilities; more than half spend some of their time answering exposure and information calls and another 13 percent perform administrative tasks. The educational background of the staff includes nursing, pharmacy, medicine, education, and public health. There are two M.D.s, five Pharm.D.s, and eight M.P.H.s serving as health educators. Furthermore, there are seven certified health education specialists and 18 staff with either a bachelor's or master's degree in education. Thus, many who are working in this area do not have a formal background in educational theory, program design, and evaluation (American Association of Poison Control Centers, 2003b). In addition, there are no formal training programs for poison control center health educators and no criteria for specifying the capabilities necessary for professional performance. However, there is a health educators' track at the annual meeting that provides seminars on various aspects of education program development and marketing. These seminars are at a general level and are offered by individuals filling health educator positions in the poison control centers.

The public education function is represented at AAPCC by the Public Education Committee. The mission of this committee is to "assist poison control centers and poison center educators in their efforts to provide the population of the United States and Canada with poison prevention awareness programs, in an effort to reduce morbidity and mortality due to poisoning" (American Association of Poison Control Centers, 2003b, p. 1). The committee is composed of a chair, a co-chair, a secretary, and seven steering committee members. These members are selected by poison control center health educators; they are not required to have any special credentials in health education. The strategic goals of the committee for 2002–2004 are provided below: these goals relate primarily to increasing visibility and participation of the committee, and not to program planning, development, and evaluation.

- Promote awareness of the public education committee to educators through orientation, mentoring, networking, and provision of materials on health education activities.
- Promote awareness of the public education committee to the full AAPCC membership through liaison with the board of directors, by representation on standing committees, and by provision of materials on health education activities.
- Promote participation of education committee members in official activities of AAPCC, such as certification, scientific review, and development of valid outcome measures.
- Compile a comprehensive resource library to include a catalog of existing material and uniform curricula. An important and useful product is the *Poison Prevention Education Materials Resource Guide* (American Association of Poison Control Centers, 2003c), which describes resources and provides contact information. This allows staff to share presentations and materials among centers.

Other committee activities include planning the education track at the AAPCC annual meeting that includes sessions on theory and practice, facilitating regional educator meetings on particular topics such as serving the Hispanic population, and publishing a quarterly newsletter—*The Educators' Antidote*—covering outreach and marketing approaches.

Public Education Programs: Examples of Best Practices

Poison control centers engage in a wide range of public education activities: some develop programs using model-based systematic approaches, some create messages based on local poisoning sources, and others rely primarily on material produced at the national level. Typical prevention messages for unintentional poisoning focus on some of the following categories: medicines; household products; pesticides; environmental hazards; plants and mushrooms; and bites from snakes, spiders, and scorpions. These messages are printed in brochures and provided on posters, stickers, activity sheets, and websites—the most widely used are from the Mr. Yuk campaign. Outreach activities include visits to schools, hospitals, and doctors' offices, and local community venues such as safety fairs. Distribution of materials also occurs through community organizations, retail outlets, and special interest groups. All poison control centers have websites that contain information on prevention and provide links to other poison information websites. A detailed listing of the resources available at each poison center website is provided in Appendix 8-A.

Four programs will be discussed briefly as examples of various systematic approaches to program development or program implementation. They include: (1) Arizona's "Tell a Friend," featuring special outreach to Native American tribes and Hispanic communities; (2) South Texas's education program focusing on increasing usage of the poison control center by the Hispanic community by overcoming language and cultural barriers to center use; (3) Ohio's "Be Poison Smart," providing a common message in Ohio though networking and training stakeholders to carry the message forward in a standardized form; and (4) California's "Don't Guess, Be Sure," which was developed using rigorous market research methods leading to different message strategies for various audience segments. Unfortunately, all four programs are in the early stages of implementation and, as a result, little data are available for purposes of evaluation.

The Arizona Poison Center developed "Tell a Friend," an education program that focused on prevention and poison control center access and use for all residents of Arizona, with a particular emphasis on increasing center use by the Hispanic/Latino and Native American populations (Krueger, 2003). Other goals were to increase the use of the national toll-free number; the number of pediatricians providing poison prevention anticipatory guidance; the use of the Internet; and knowledge of partners and community collaborators regarding poison control centers' services.

The development of this program was guided by the Program Evaluation Logic Model. Focus groups in the Hispanic and Native American communities were used to determine levels of knowledge and barriers to poison control center use. Partners in the education effort included Native American communities, the Phoenix Indian Medical Center, county health departments, Head Start, the State Maternal and Child Health Bureau, and the statewide Medicaid programs. As a result, educational materials have been developed in English and Spanish, and Native American outreach programs have been conducted with 22 Arizona tribes. Early results showed that outreach programs led to an increase in penetrance of 200 calls per 1,000 population compared with a decrease of 160 calls per 100,000 population in the control group.

The South Texas Poison Center program on language barriers to poison control center use was designed to increase awareness and use of center services by the Hispanic community (Griffin et al., 2001). The initial study involved the collection of data from five Texas counties along the Mexican border to determine (1) the level of knowledge about poison control services and (2) the perceived barriers to using them. The survey was conducted though cooperation with health care facilities and community service organizations. The results showed that Hispanics who spoke only Spanish were less likely than whites to know how to contact the poison control center and more likely than whites to believe that the services would cost money. Bilingual-speaking individuals were most concerned about confidentiality. The resulting educational campaign was based on an A Su Salud methodology and involved collaboration with a community network that had applied this methodology successfully to encourage breast and cervical cancer screening. The materials were produced in Spanish and English and featured role models as communicators. The campaign included large media presentations, newsletters, and activities in schools, places of worship, colonial centers, migrant health programs, clinics, and businesses. Call volume from the targeted communities showed a slight increase from 2000 to 2002 (Watson and Villarreal, 2002).

The Ohio "Be Poison Smart Campaign" was designed to provide a consistent poison prevention and control message to all residents of the state. The three poison control centers— Cleveland, Columbus, and Cincinnati—formed the Ohio Poison Control Collaborative. The program involved network partnerships with private and public agencies. Individuals in the network were formally trained to implement the program throughout the state. The training included sessions on in-the-home interactions with individuals living in underserved communities. Protocols and materials were provided for all those who carried the education into the field. In 2002, 777 staff from 100 stakeholders were trained. Evaluation will involve pre-post testing of knowledge, attitudes, and behaviors (Krueger, 2003).

The message strategies and content used in California's program, "Don't Guess, Be Sure," were developed though gaining an in-depth understanding of the potential customers. The process began with audience analysis and segmentation based on demographic characteristics, lifestyle choices, and attitudes toward health care. Four segments were selected: mainstream suburban families; African Americans of low socioeconomic status; bilingual Latinos; and monolingual Latinos. Focus groups and individual face-to-face interviews were used to define the attitudes and behaviors of the various groups and develop appropriate messages and message-delivery strategies. One result was to refocus the core message from "call in case of a poison emergency" to "call if you're not sure." Focus group discussions indicated that the first message could be perceived as alarmist and confusing. Furthermore, monolingual Latinos and Latinos with limited English-speaking skills wanted more information than other groups regarding what happens when a call is made to a poison control center.

Poison Control Center Websites

In addition to providing education through the distribution of materials and presentations, each center maintains a website containing an array of information on poison prevention and education. Information on poison control center websites was collected during July 2003 by staff at the Institute of Medicine using Microsoft Internet Explorer. The table provided in Appendix 8-A describes each website in terms of the material provided and the links to other organizations (including government agencies, private organizations, and other poison control centers). All of these websites can be accessed through the AAPCC website using a directory list. The Internet links provided in the directory connect to a standardized cover page for each center, which includes the name, address, and other pertinent contact information for the center and, for most of the centers, a direct link to the center website. A wealth of important and potentially lifesaving information is provided by the websites. Most of this information is clearly marked and readily accessible; only infrequently is navigation of the website difficult—primarily when the poison control center links are at separate host hospital sites. In addition to a description of the center and the emergency contact information, numerous and varied documents (usually not standardized across sites nor found at multiple websites) with detailed information about poisonings in general and in relation to specific geographic locations are provided for the public, health educators, and health care workers. These documents can generally be downloaded free of charge. English is the primary language used on the websites and Spanish is common; however, Chinese, Vietnamese, Tagalog, Arabic, Laotian, Russian, Korean, Hmong, Tigrinya, and Cambodian are also found. A wide range of information is offered about topics such as household safety and lookalike household products, medications, poisonous and nonpoisonous plants, seasonal hazards, snakes and insects, inhalants, pesticides, possibly dangerous food, alcohol, hazardous materials, and biological and chemical agents that could possibly be used in terrorist threats.

Websites also provide materials for use by parents and classroom teachers to educate children, as well as contact information to arrange for health presentations and other community events. News and important updates about legislation, national and community events (including the annual National Poison Prevention Week), education and training programs, and other current issues related to poison are often available. Internet links to journals and databases that deal with relevant epidemiological topics are also provided for educators, health care professionals, and more informed members of the public.

It is important to mention that each center has designed its own website and selected its own content to be displayed—there is no sharing of design from one center to another and no common standard of quality for content coverage, content display, or website navigation.

Evaluation

Most evaluations of the programs developed to educate the public about primary and secondary poison prevention have relied on measuring the number of materials distributed or the number of presentations given. Other evaluations have focused on increases in the number of exposure or information calls to the poison control center. For example, according to Oderda and Klein-Schwartz (1985), Mr. Yuk stickers distributed by the Maryland Poison Center raised awareness of the center's existence and telephone number. Krenzelok and Dean (1988) found that Mr. Yuk raised awareness of poisoning and of the poison control center particularly among the college educated. In 1984, center staff from Virginia carried out a randomized trial of the deterrent value of the widely used Mr. Yuk stickers for children ages 12 to 30 months before and after an educational program. Toddlers were randomly assigned to a control or experimental group. During the first trial exposure (preeducation), toddlers in both groups showed no preference for containers labeled versus not labeled with Mr. Yuk stickers. In the second trial, children in the experimental group were provided with education. The results showed that children in the educated group were more likely to handle containers with stickers whereas those in the control group continued to show no preference, suggesting that *the stickers did not deter children* from manipulating the containers (Vernberg et al., 1984).

There are few published studies evaluating the effects of poison prevention education on behavior. One study used a pre-post test design with two intervention groups of children from 3 to 6 years of age to examine the effects of a 30-minute classroom presentation by a health educator (Brogan and Lobell, 1999). One group answered the posttest questions immediately following the presentation; the second group answered the posttest questions 7 months later. Both groups showed a significant increase in knowledge about poison prevention. In the pretest 36 percent of the children answered the questions correctly; the posttest results were 96 percent correct for the immediate group and 95 percent correct for the 7-month follow-up groups.

In another study, researchers examined the effects of poison prevention education on kindergarten and third-grade students and their parents (Liller et al., 1998). They used a posttest, control group design. Children in the intervention schools were given the *More Health* 40-minute interactive teaching lesson. This was supplemented by the *Always Ask First* video for kindergarten students, and reading and vocabulary lists for third graders as well as take-home materials for parents and caregivers. The results show that students in the intervention groups consistently answered more questions correctly than students in the control groups.

Also, the majority of parents/caregivers of intervention group students indicated, in a follow-up questionnaire, that their homes were made poison safe.

Emergency departments also have been used as a venue for educating the public about poison prevention. Wolf et al. (1987) examined the effects of medical counseling on poison prevention practices of inner-city families. Comparison of control and intervention groups indicated a positive effect of the counseling on ipecac storage and use. Reddy et al. (1999) conducted a prospective study of the effects on behavior of a videotape presentation supplement to written material handed out in an emergency department and as part of a general pediatric visit. The results indicated that caregivers who viewed the tape (intervention group) were three times more likely to read the material than caregivers who did not view the videotape (control group). Comparisons of pre- and posttest home safety scores indicated that individuals in the video group showed a higher incremental increase in knowledge than the control group.

Perhaps the most complete evaluation of poison prevention education was conducted between 1975 and 1979 in Monroe County, New York (Fisher et al., 1981). This project was designed to heighten public awareness and reduce risks and incidents of poisonings in the home. The results demonstrate that a comprehensive, multifaceted approach can have significant impact on prevention behavior. Interventions were conducted over a 3-year period and included community outreach seminars (to community workers in touch with families); school curriculum seminars and checklists; retail outreach efforts (sale of safety latches and reshelving of hazardous products); distribution of prevention materials to new mothers at the time of childbirth; and mass media campaigns. Follow-up surveys in the home showed significant reduction in accessibility to children of potentially hazardous products (36 percent fewer homes had accessible aspirins, 32 percent fewer had available drain cleaners, and 27 percent had fewer furniture polishes). Although all facets of the intervention were useful, the most frequently cited sources of information were (1) booklets provided in hospitals at the time of birth or other prevention material provided with a birth certificate and (2) mass media. Shelf warnings and materials that older children brought home from school were also useful sources. Data from surveys of retail stores showed marked improvement in stocking products with appropriate packaging, reshelving, posting warning signs, and stocking safety latches. In 1981 the Monroe County project was extended into five adjoining counties (Fisher et al., 1986). The results confirmed the findings of the original study showing an increase in knowledge about poisoning, an increase in calls to the poison control center, and a decrease in visits to emergency departments in the short term. No long-term follow-up of the effects of the interventions was conducted.

COLLABORATIONS AND CONTRIBUTIONS BY OTHER ORGANIZATIONS

National Poison Prevention Week

National Poison Prevention Week was established by President John F. Kennedy in 1961 in Pub. L. No. 87–319 (75 Stat. 681). It is held each year during the third full week of March (for details, see http://www.aapcc.org/poison2.htm; http://www.nsc.org/poison.htm). The intent of Congress in passing this law was to provide a way for local communities to make their citizens aware of the dangers of unintentional poisonings and to promote prevention measures (http://www.nsc.org/poison.htm).

Subsequent to the passage and signing of the law, the Poison Prevention Week Council, a coalition of national organizations, was established to coordinate the event. The 37-member council serves as a focal point for member activities (see Box 8-1 for list of council members). The council's basic theme is "Children Act Fast . . . So Do Poisons!"

The council joins with the American Association of Poison Control Centers and the U.S. Consumer Product Safety Commission (CPSC) to promote and sponsor a poster contest to mark the annual event. Furthermore, the council urges parents to (1) use products with child-resistant packaging; (2) keep medicines and chemicals locked away from children; and (3) use the toll-free number for poison control centers when needed. Video clips in both English and Spanish about poison prevention and poison control centers are linked to the CPSC website (http://www.cpsc.gov/cpscpub/prerel/prhtml03/03092.html).

National Poison Prevention Week also presents an opportunity for member organizations to separately educate the public and encourage vigilance about poisonings, particularly in households with children. Each member organization develops a program suited to its interests in poison prevention and promotes it directly through radio, television, or print media or indirectly through its own chapters and affiliates. For example, the National Safety Council and the U.S. Environmental Protection Agency (EPA) have a cooperative agreement to emphasize the need for consumers to read labels on household cleaners, pesticides, and insecticides.

Distribution of Poison Prevention Information by Other Organizations

Some organizations promote poison prevention and control throughout the year (Table 8-2 lists a number of organizations and the types of information they provide on their websites). Some of these organizations are associations of clinicians (e.g., American Academy of Pediatrics, Amer-

BOX 8-1
National Poison Prevention Week Council Members

American Academy of Clinical Toxicology
American Academy of Pediatrics
American Association of Poison Control Centers
American College of Emergency Physicians
American Dental Association
American Nurses Association
American Petroleum Institute
American Pharmaceutical Association
American Public Health Association
American Red Cross
American Society of Health-System Pharmacists
Art & Creative Materials Institute, Inc.
ASTM Committee on Packaging
Boy Scouts of America
Center for Proper Medication Use
Closure Manufacturers Association
Consumer Healthcare Products Association
Consumer Specialty Products Association
Cosmetic, Toiletry, and Fragrance Association
Council on Family Health
CropLife America
Food Marketing Institute
Healthcare Compliance Packaging Council
National Association of Broadcasters
National Association of Chain Drug Stores
National Association of Pediatric Nurse Practitioners
National Community Pharmacists Association
National SAFE KIDS Campaign
National Safety Council
Pharmaceutical Care Management Association
Soap and Detergent Association
U.S. Consumer Product Safety Commission
U.S. Department of Agriculture
U.S. Department of Health and Human Services
U.S. Environmental Protection Agency

ican Academy of Emergency Physicians, American Nurses Association); some are nonprofit organizations (e.g., American Red Cross, Alliance for Healthy Homes, Children's Safety Network, National SAFE KIDS Campaign); and some are federal agencies (CDC, CPSC, EPA). Many offer information in both English and Spanish. Pharmacists provide poison prevention and control information and other activities throughout the

TABLE 8-2 Education Materials from Outside Organizations

Name	Audience	Content
Clinical Membership Organizations		
American Academy of Pediatrics *Medium:* Downloadable Internet documents and publications and kits that can be purchased	Pediatricians/ parents/ general audience	• The Injury Prevention Program (TIPP) description and fact sheets. TIPP is an educational program for parents of children newborn through 12 years of age to help prevent common injuries from a number of incidents, including poisoning; TIPP materials are available for purchase • TIPP fact sheets • News releases/surveys about common injuries
American College of Emergency Physicians *Medium:* Downloadable Internet documents and publications that can be purchased	Emergency physicians/ parents	• Tips on how to protect your child from poison • Pediatric Emergency Guide: tips on what do when your child has an emergency • Poison information and treatment systems policy statement • News releases
American Medical Association *Medium:* Downloadable Internet documents	Physicians	• Technical articles about poisonings • Policy statements related to various poisonings, and the responsibility of physicians and public health departments in preventing them • Council on Scientific Affairs reports, including reports on medical preparedness for terrorism and other disasters
American Nurses Association *Medium:* Downloadable Internet documents	Nurses	• Technical articles about poisonings (Children's Health and the Environment articles)—including news and updated information • Articles related to poisonings from the *American Journal of Nursing*

Continued

TABLE 8-2 Continued

Name	Audience	Content
American Pharmaceutical Association *Medium:* Downloadable Internet documents	Pharmacists	• Information about National Poison Prevention Week • Poison Lookout Checklist (from U.S. Consumer Product Safety Commission) • Explanation of pharmacist classification system • Guidelines for pharmacists
American Trauma Society *Medium:* Searchable Internet website	Professional community	Tangentially related to poisoning events, the Trauma Information Exchange Program (TIEP) is a program of the American Trauma Society in collaboration with the Johns Hopkins Center for Injury Research and Policy and is funded by the Centers for Disease Control and Prevention; TIEP maintains an inventory of trauma centers in the United States; collects data and develops information related to the causes, treatment, and outcomes of injury; and facilitates the exchange of information among trauma care institutions, care providers, researchers, payers, and policy makers
Emergency Nurses Association *Medium:* Downloadable Internet documents	Nurses/other health care professionals	• E-code articles • Speaker/course presentations (e.g., 2002 Scientific Assembly) • Abstracts/research work (e.g., boating injuries/carbon monoxide poisoning) • Subjects index from *Journal of Emergency Nursing* • Weapons of mass destruction preparedness resources listing • Nursing news updates

TABLE 8-2 Continued

Name	Audience	Content
National Nonprofit or Voluntary Health Agencies		
Alliance for Healthy Homes (formerly Alliance to End Childhood Lead Poisoning) *Medium:* Downloadable Internet documents	General audience	*Fact sheets on:* • Housing/environmental exposures information (toxics, irritants, allergens, gases) • Risks to children (higher risk, disparities among ethnic and socioeconomic groups, safety and injury prevention, community-based solutions) • Impacts on communities • Health hazards and methods of prevention *Action agendas:* • Lessons learned about lead poisoning prevention • Policy documents *Web resources and publications:* • Holistic/Multi-Topic Web Resources • Listservs • List of Publications (Adobe Acrobat) • Resources listing in Spanish
AARP (American Association of Retired Persons) *Medium:* Downloadable Internet documents and publications (mostly unrelated to poisonings)	Retired adults	Advice about using medications wisely—asking questions about medications before using them and following the directions when taking prescription drugs to be safe
American Red Cross *Medium:* Downloadable Internet documents	Adults/general audience	• Health and safety tips and services • Prevention fact sheets • Emergency/disaster information • Safety inspection checklist • Family information cards that can be filled out and printed out • Description of contents of supply kits and essential supplies

Continued

TABLE 8-2 Continued

Name	Audience	Content
Children's Safety Network *Medium:* Downloadable Internet documents and online ordering of publications	Epidemiologists/ adolescent health coordinators/ state and territorial injury prevention directors/ injury researchers and educators/ emergency medical services for children (EMSC) staff/ national organizations devoted to injury prevention/ members of the insurance community/ other public health professionals	• Extensive list of documents and publications concerning external use of injury codes (e-codes), economic measures of unintentional injury and prevention, school environment, rural injury prevention, etc., from various sources, including journals, presentations, and fact sheets
Consumer Federation of America *Medium:* Downloadable Internet documents	Consumers (general audience)	• Fact sheets (e.g., indicating the success of the Poison Prevention Packaging Act) • Updates on legislation (e.g., providing regional grants to poison control centers and to educate the public about poisoning prevention)
Emergency Medical Services for Children/National EMSC Resource Center *Medium:* Downloadable Internet documents/ links to publications	Health care and injury professionals/ general audience	• Health Resources and Services Administration of the Maternal and Child Health Bureau awarded the new contract for the EMSC National Resource Center to Children's Hospital of Washington, DC (September 25, 2003) • Legislation/public policy announcements (e.g., Poison Control Center Enhancement and Awareness Act, announcement of national toll-free poison emergency hotline, updates on legislation in

TABLE 8-2 Continued

Name	Audience	Content
		emergency medical services for children in different states, Risk Watch Champion Aware Program announcements) • Resource listing, including links to other organization home pages, such as National SAFE KIDS Campaign and National Injury Data Technical Assistance Center • Specific poison control publications from the EMSC National Resource Center and links to resources available from other organizations
Home Safety Council (Lowe's) *Medium:* Downloadable Internet documents	Adults/parents	*Some of the information in the following pertains to poisonings:* • Surveys of Americans' attitudes, practice, and knowledge of home safety practices • Newsletters about home safety • Informative press releases, particularly about safety during holidays • Salute to Home Safety Excellence award presented annually to a home industry supplier who displays leadership in home safety by conducting an educational program targeted to either employees or consumers
National Lead Information Center	General audience/ health care professionals/ other professionals	• Contact information in both English and Spanish • General information packet that can be ordered • Basic information/fact sheet with links to additional resources • Current news and related links • Rules and regulations and related links

Continued

TABLE 8-2 Continued

Name	Audience	Content
		• Education and outreach materials (downloadable brochures and posters, program activities documents, links to current grants, the center's link and contact information) • Technical studies (downloadable or able to be ordered) in various media formats • Other lead links and short descriptions of their functions • Environmental Protection Agency regions map and contact information
National SAFE KIDS Campaign *Medium:* Downloadable Internet documents, electronic newsletter distribution, publication downloads, information about speakers	Adults/parents/ teachers	• Fact sheets/safety tips on preventing unintentional child poisonings, relating to the responsibility of both parents and the wider community *Some of the information in the following pertains to poisonings:* • Global network focus, local organizational chapters • SAFE KIDS worldwide study, *Childhood Unintentional Injury Worldwide: Meeting the Challenge* • Monthly newsletters • Brochures • Children's activities to promote safety • Links to many organizations dedicated to injury prevention and safety • Academic references
National Safety Council (including the Environmental Health Center) *Medium:* Downloadable Internet documents featuring publications (with web links) that can be purchased by sections	Adults/parents/ professionals	• Injury Facts DataBytes, one section of which contains graphs of deaths and death rates for six leading causes of unintentional-injury death, including poisoning • How to Prevent Poisonings in Your Home fact sheet with link to the American Association of Poison Control Centers and the National Poison Prevention Week websites

TABLE 8-2 Continued

Name	Audience	Content
NOTE: National Lead Information Hotline and Clearinghouse no longer operational—see the National Lead Information Center (Environmental Protection Agency)		• Lead Poisoning Prevention Outreach Program (operated by the council's Environmental Health Center)—fact sheet about lead poisoning, e-mail link to community outreach activities, and Internet link to the Environmental Protection Agency's National Lead Information Center
Parents Anonymous *Medium:* Downloadable Internet documents and order forms for publications	Parents/ professionals	• Information in both English and Spanish • Not specifically relevant to poisonings; rather, aimed at helping parents with their children's health and well-being
Poison Prevention Week Council *Medium:* Downloadable Internet information and links	General audience/ professionals	• Information about the annual National Poison Prevention Week • U.S. Consumer Product Safety Commission link (see below for extensive information the commission provides)
Federal Agencies		
Centers for Disease Control and Prevention (CDC) (National Center for Injury Prevention and Control) *Medium:* Downloadable Internet documents	Public health professionals/ professionals in agriculture/ general audience/ parents	*Extensive list of "hits" (about 700) related to poisonings, including:* • Articles on specific poisoning issues from *Morbidity and Mortality Weekly Reports,* with references, geared more toward health care professionals • Fact sheets on various poisons, including coverage of food safety issues (i.e., National Ag Safety Database) • Poisoning prevention and other safety tips for families (national toll-free poison emergency hotline recommended) from the National Center for Injury Prevention and Control

Continued

TABLE 8-2 Continued

Name	Audience	Content
		• National Center for Injury Prevention and Control "Poisoning Prevention" provision of contact information for its Poisoning Prevention Partner Organizations, including the American Association of Poison Control Centers, National SAFE KIDS Campaign, Poison Prevention Week Council, American Academy of Pediatrics, National Center for Environmental Health, National Lead Information Center, and U.S. Consumer Product Safety Commission
CDC National Center for Environmental Health *Medium:* Downloadable Internet documents, publication downloads, searchable journal index	General audience/ parents/public health practitioners/ other health care professionals	• Information in both English and Spanish • Carbon monoxide poisoning: Health tips (with a link to more detailed information from CDC); basic facts (checklist, questions and answers, information from the National Institute for Occupational Safety and Health, reports, news articles, testimonials); education brochures, photos, posters, and other tools; boat-related (link to *Morbidity and Mortality Weekly Reports*); policy statement link to the *Journal of the American Medical Association* • Lead poisoning: Information about CDC's Childhood Lead Poisoning Program, general information and publications, news links to *Morbidity and Mortality Weekly Reports*, policy statement from the American Academy of Pediatrics • Children's health information and links to other organizations with similar concerns about the built environment (Designing and Building Healthy Places website)

TABLE 8-2 Continued

Name	Audience	Content
Substance Abuse and Mental Health Services Administration *Medium:* Downloadable Internet information and links	Health care professionals/ general audience	*Documentation related to poisoning in several areas:* • Statistics • Workplace • Mental health • Substance abuse • Other
U.S. Consumer Product Safety Commission *Medium:* Downloadable Internet documents	Parents/ grandparents/ pharmacists/ physicians	• Information in both English and Spanish • Legislation from the *Federal Register* and the U.S. Consumer Product Safety Commission regarding poison prevention packaging and the Poison Prevention Packaging Act • Press releases about National Poison Prevention Week • Poison Prevention Packaging: A Text for Pharmacists and Physicians • Material related to child-resistant packaging • Documentation of deaths associated with the use of consumer products, including non-fire carbon monoxide • Fact sheets with tips on childproofing homes, locking up poisons from children, checklists for poisons in the home • Safety alerts for grandparents to prevent grandchildren from being poisoned • Safety alerts about possible poisonings from medications, carbon monoxide, inhalants, etc.
U.S. Department of Housing and Urban Development *Medium:* Downloadable Internet documents and links	General audience/ parents/ communities/ health care professionals	• Documentation in both English and Spanish • Office of Healthy Homes and Lead Hazard Control site with highlights, recent publications (and ordering information), and quick links to other relevant information and websites

Continued

TABLE 8-2 Continued

Name	Audience	Content
		• Fact sheet about potential for poisonings in the bathroom • Legislation as related to poisons (particularly lead) and housing issues • Community planning and development with reference to poisons (particularly lead) and citizen participation and consultation
U.S. Environmental Protection Agency *Medium:* Downloadable Internet documents; HTML format pages	General audience/ parents/ health care professionals/ industry professionals	• Information in both English and Spanish • Tips to protect children from pesticide and lead poisonings • Advertisement of national toll-free poison emergency hotline • Recognition of Management of Pesticide Poisonings report and updates, corrections, and ordering information for the report • Worker safety and training information • Pertinent government memoranda about poisoning incidents and poisoning incident data (pirimphos methyl, oxamyl, oxydemeton mehyl, terbufos, phorate, dimethoate, propetamphos, parathion)

year. Pharmacies are accessible and convenient and virtually every household in the United States is within 5 miles of a community or hospital/institutional pharmacy.

Some professional associations, including the American Pharmaceutical Association, American Medical Association, and American Nurses Association, focus on scientific and practice-related information while others provide general advice to the public and support the information needs of their membership. For example, the American Academy of Pediatrics promotes the national toll-free number on its TIPP® (The Injury

Prevention Program) sheet, "Protect Your Child . . . Prevent Poisoning," and offers an educational program for parents of children through 12 years of age (http://www.aap.org/family/poisonwk.htm). The academy is also studying the effectiveness of physicians offering poison prevention anticipatory guidance to parents of children less than 1 year of age as part of scheduled office visits. This guidance was recommended in *Healthy People 2000* (www.healthypeople.gov) as one of a number of health promotion and disease prevention strategies. According to the *Guide to Clinical Preventive Services* (http://www.ahrq.gov/clinic/cps3dix.htm), the actual proportion of children whose parents receive injury prevention counseling is 39 percent, and the implementation of such measures decreases with level of income and education. Even so, the U.S. Preventive Services Task Force supports counseling parents in strategies to prevent injury.

Websites of nonprofit organizations provide a variety of safety materials such as checklists, fact sheets, and news releases for the public (parents, caregivers, babysitters, senior citizens). The National SAFE KIDS Campaign, another nonprofit organization, provides one of the more active websites in poison prevention for children. It offers a wide range of information on the magnitude of poisonings and their sources; the effectiveness of various prevention programs, such as childproof caps; tips about prevention; the national toll-free number; and links to related organizations and websites. In addition, safety fact sheets on prevention are available for parents and other adults caring for children.

Federal government websites offer professional articles, information on legislation, and materials to assist the public in preventing poisoning of seniors, adults, and children. The most active agencies include the CDC, CPSC, EPA, and U.S. Department of Housing and Urban Development.

CONCLUSIONS

1. Public education efforts are necessary but not sufficient to accomplish primary and secondary prevention of poisonings. Education efforts must be integrated with other programs in the broader public health system at the state and federal levels. For example, many of the maternal and child health programs in the states have an injury prevention program that might serve as a focal point for coordinating poison prevention and education programs (see Chapter 9 for elaboration). In addition, prevention is best accomplished through a multifaceted approach combining education, engineering and environmental modifications, enactment and enforcement of regulations and legislation, economic incentives, involvement of local health care providers, community empowerment, and pro-

gram evaluation. Although poison control centers have been active in public education, the discussions in this chapter have shown that (1) these efforts are not well coordinated among centers, and (2) there is no convincing evidence of a positive impact on poisoning prevention.

2. Public education efforts should separate primary and secondary prevention messages. Most existing materials mix these messages. If education is effective for primary prevention, one would expect fewer poisonings and thus fewer calls to poison control centers. On the other hand, if the educational message is awareness and use of poison control services (secondary prevention), then one would expect more calls to them. With both messages in the same package, it is hard to measure the effectiveness of either. There is some indication that public education has raised awareness of poison control center services, but there is little evidence concerning the impact of these efforts on mortality and morbidity.

3. The focus of most education programs is prevention of unintentional poisoning of children less than 6 years of age. Little effort has been directed toward serious poisoning or toward other age groups, drug and alcohol abusers, and workers in certain high-risk occupations.

4. A repository of best practices in public education should be established. The description of each practice should include information on target audiences, literacy level, and how the program was developed, implemented, and evaluated.

Appendix 8-A

Education Materials from AAPCC and Poison Control Center Websites

TABLE 8-A-1 Education Materials from AAPCC and Poison Control Center Websites

Name	Accessible on Web Page	Website Links to Other Organizations
American Association of Poison Control Centers	• U.S. Poison Center Educator List (full contact information) • Preventing Poisonings in the Home (brochure) • *Poisoning Fact Sheets:* Parents/Childcare Providers; Children; Teens/Babysitters • Prevention Tips • Emergency Action for Poisoning Card • Teaching Aids list and links to specific examples (geared toward younger children, several marked as developed by the Oregon Poison Center): Ideas for Bookmarks; Which Poisons Are in Your House Puzzle; Draw a Line to Poisons; Color Items Safe to Taste; Color—Never Take Drugs That Belong to Someone Else; Poison Word Game; Are You a Poison Expert Quiz; Choose Items Safe to Taste; The Children's Corner; Bathroom Clean Up; Plants and Flowers; How Many Snakes Can You Find in This Picture?; Know Your Poisons; Food Safe to Eat; The Poison Jungle	• Link to each poison control center website at http://www.aapcc.org/director2.htm, each of which begins with a short standardized page with address and contact information for the center and a center's website link, if applicable
Alabama Poison Center (Tuscaloosa. AL)	• Emergency Action for Poisonings • Poison Prevention Guide (free upon request) • Poison Prevention Stickers (free upon request) • Checklist to Poison Proof Your Home • Poison Prevention Tips • Seasonal Fact Sheets (e.g., snakebites, West Nile virus, insect bites and stings, jellyfish) • Kid's Page —Preschool (Find Objects Safe to Play With) —Grades 1-3 (Help Officer Ugg Get Home Safely) —Grades 4-6 (Poison Word Find) —Middle School and Junior High (Word Find)	• Website: www.alapoisoncenter.org • Link to website of the American Association of Poison Control Centers

	• Educational Programs presented upon request to professional and nonprofessional groups throughout the state of Alabama on poison prevention, seasonal poisonings, "look alikes," biological and chemical warfare agents, and current developments in clinical toxicology • Train the Trainer Program (interested community Poison Prevention Providers receive training from a certified poison information specialist and are provided with educational materials to use in poison prevention presentations) • Staff listing with short description of related American Academy of Clinical Toxicology, American Board of Applied Toxicology, American College of Medical Toxicology, Certified Specialist in Poison Information • Alabama Poison Center's Historical Milestones	
Regional Poison Control Center (Birmingham, AL)	*Children's Health System Information* • General Safety (includes link to poisoning information) • Prevent Childhood Injuries • Poison Prevention (home dangers, prevention tips); gives http://www.poisonprevention.org website as a reference and a link to the poison control center, which shows the center's number and link to the latest Toxicology bulletin • Center website (cited above) links to Alabama SAFE KIDS Campaign (with links to its newsletter); information on lead poisoning, toxic or poisonous plants, and venomous snakes; and bulletins (carbon monoxide poison, Central Serotonin syndrome) • Telephone triage information	• Website: Children's Health System: http://www.chsys.org, with a link to website of the Southeast Child Safety Institute and the Poison Control Center

Continued

TABLE 8-A-1 Continued

Name	Accessible on Web Page	Website Links to Other Organizations
Alaska Poison Control System (Juneau)	• This center provides poison prevention education and does not manage exposure cases • Description of National Poison Control System • Poisonings and Poison Control in Alaska • Alaska Poison Control System (PowerPoint slide presentation), advertising national toll-free number and giving information about deaths from poisoning, substances most common in Alaska's poison exposures, medications, food poisoning, paralytic shellfish poisoning, poisonous mushrooms, carbon monoxide poisoning, and poisoning prevention and first aid • Poison Prevention for Elders: Home Safety Tips (brochure) • Substances Most Common in Alaska's Poison Exposures (brochure) • Be Prepared to Handle a Poisoning (brochure) • Why Should You Learn About Poison Prevention? (brochure) • First Aid and Treatment for Poisonings (brochure) • Know Your Plants (brochure)	• Website: http://www.chems.alaska.gov/ • Link to website of the American Association of Poison Control Centers • Link to website of the Oregon Poison Center
Arizona Poison and Drug Information Center (Tucson)	*Information Provided About:* • Venomous Critters: arthropods (bees, black widows, brown spiders, centipedes, conenose bugs, scorpions, tarantulas, velvet and other ants) and reptiles (gila monster and snakes) • Pregnancy Riskline: potential problems with poisonous substances • Plants • Home safety	• The University of Arizona's College of Pharmacy website: http://www.pharmacy.arizona.edu/centers/apdic/apdic.shtml

	• Poisoning and Pets • Hazardous Materials in the Household and Environment • Medicinal Herbs (including links) • Research Centers' Toxtrivia Question Sets 1and 2 • Medication for the Masses: Pharmacy's Role in Times of Crisis (brochure/advertisement)	
Arkansas Poison and Drug Information Center (Little Rock)	• No additional information found at the website	• No website provided; web search for the name of the center links to a page of contact information only provided by the University of Arkansas for Medical Sciences (Departmental Phone Listings)
ASPCA—Animal Poison Control Center (Urbana, IL)	• Read the Label First? Protect Your Pet • Events (such as Toxicology Short Course for Wildlife Profession) with links • Toxicology publications listing with links • ASPCA Animal Poison Control Center News Bulletins with links *Make Your Pet's Home Poison Safe*: What to Do for a Poisoned Animal; Poison-Proof Your Home; List of Toxic Plants; List of Non-Toxic Plants	• Website: http://www.apcc.aspca.org • Links to affiliations: University of Illinois College of Veterinary Medicine, Veterinary Support Personnel Network, Environmental Protection Agency, AAPCC, Veterinary Emergency and Critical Care Society, and Veterinary Information Network
Banner Poison Control Center (Phoenix, AZ)	• Documentation in both English and Spanish • Description of the center's services • *Arizona Critters:* black widow, gila monster, rattlesnakes, scorpion • *Babysitters' Guide:* Banner Poison Control Center information; What can be poisonous to children?; When do poisonings happen?; Where do poisonings happen?; What can a babysitter do?; What if a poisoning happens?	• Website: http://www.bannerhealthaz.com/services/ poison/poison.html • Links to websites of the Class and Physician Search, New Gilbert, AZ, Location, and Oncology Conference

Continued

TABLE 8-A-1 Continued

Name	Accessible on Web Page	Website Links to Other Organizations
	• *Current Topics Concerning the Banner Poison Control Center Community:* spiders and amphetamines • *Prevention Guide:* Prevention Guide; What can you do?; Hiking and camping?; Bitten or stung? • National Poison Prevention Week information	
Blue Ridge Poison Center (Charlottesville, VA)	• Information in English and Spanish • Link to education section provided at the website • *Education:* What Is a Poison?; Potential Poisons?; First Aid for Poisoning • *For Adults:* Parent Poison Fact Sheet; Alcohol; Asbestos; Autumn Hazards; Caterpillar Stings; Christmas Hazards; Copper in Water; Drug Disposal Guidelines; Farm Hazards; Flea Bombs; Inhalants; Lead Poisoning; Mushrooms; Plants; Pokeweed; Snakebites; Spanish Poison Information; Spider bites; Summer Hazards; Winter Hazards • *For Teens:* Teen Poison Fact Sheet; Alcohol; Caterpillar Stings; Drug Disposal Guidelines; Inhalants; Mushrooms; Plants; Pokeweed; Snakebites • *For Children:* Activity Sheet to help identify poisons; Child Poison Fact Sheet • News section (e.g., Moonshine Is Still a Problem)	• Website: http://www.healthsystem.virginia.edu/brpc/ • Links to websites of the American Association of Poison Control Centers, Center for Disease Control and Prevention, http://www.CharlottesvilleCERT.org, Consumer Product Safety Commission, Drug Help, Environmental Protection Agency, Food and Drug Administration, Food Safety and Inspection Service, Meat and Poultry Hotline, National Animal Poison Control Center, National Inhalant Abuse Coalition, National Institutes of Health, National Institute of Mental Health, National Pesticide Information Center, Occupational Safety and Health Administration, Outreach Virginia, Undersea and Hyperbaric Medical Society, U.S. Department of Agriculture, Virginia Department of Health
Bon Secours-Holy Family Hospital (Altoona, PA)	• This center provides poison prevention education and does not manage exposure cases • Indicates only that there are education programs and activities such as teaching, seminars, health fairs, and health screenings	• Website: http://www.hso.blairco.org/BONSECOU.HTM

<table>
<tr><td>California Poison Control Center (Fresno, Sacramento, San Diego, San Francisco)</td><td>• Information in English, Spanish, Chinese, Vietnamese, Tagalog, Arabic, Laotian, Russian, Korean, Hmong, Tigrinya, and Cambodian
• Information for Health Care Professionals: Health Professional Phone Stickers, "Because We Care" Tear Off Pad, Poisoning and Drug Overdose (3rd edition)
• Information for Health Educators: California Poison Control System Health Education Program: An Overview of Research, Development and Implementation, Understanding Bilingual and Monolingual Latino Consumers
• You and Your Family: Emergency Action for Poisoning, How You Can Prevent Poisonings, Protect Yourself from Breathing Household Poisons, For Grandparents, Babysitter's Guide to Poison Prevention, Inhalant Awareness Newsletter, National Poison Prevention Week poster
• Materials Catalog and Online Orders: pamphlets, stickers, books, videos, posters
• News and Information</td><td>• Website: http://www.calpoison.org
• Links to websites of the American Association of Poison Control Centers and University of California San Francisco School of Pharmacy</td></tr>
<tr><td>Carolinas Poison Center (Charlotte, NC)</td><td>• Quick Facts border with poisoning information on each webpage
• Community Outreach
• Emergency First Aid
• How Safe Is Your Home?
• Insect Bites and Stings
• Snakes
• Plants
• Poisons and Poison Prevention</td><td>• Website: http://www.carolinas.org/services/poison/</td></tr>
</table>

Continued

TABLE 8-A-1 Continued

Name	Accessible on Web Page	Website Links to Other Organizations
	• Stop Childhood Poisoning (including a list of other organizations working to produce a year-long program to educate North Carolinians about the potential hazards of pesticides, household chemicals, cosmetics, and other materials commonly found in homes)	
Central New York Poison Control Center (Syracuse)	• Pop-up poison emergency number advertisement • *Comprehensive Education:* General Information on Anthrax, General Information on Cyanide • *Educational Opportunities for the General Public/Prevention Tips:* Train the Trainer, Detailed Explanation of Look-Alike Products, The ABC's of Poison Prevention: A Teacher's Guide, Poison Prevention Newsletters, Poison Prevention Checklist, brochures, posters • *Poison Prevention Tips from Billie:* Fun Pages for Kids, Teachers' Guide, Billie's Poison Prevention Checklist, Poison Awareness (interactive CD-ROM for kids) • *Brochures:* Child Care Provider's Guide to Poison Prevention, Poison Prevention Information, Poisonous Plants • *Posters:* New National Telephone Number, Poison Emergency?, Is It Candy or Medicine, Poison Prevention Information	• Website: http://www.cnypoison.org/ • Links to websites of the Cornell Cooperative Extension, American Association of Poison Control Centers, National Inhalant Prevention Coalition, Cornell University Plant Home Page, National SAFE KIDS Coalition
Central Ohio Poison Center (Columbus)	• Poison Safety • Courses for Parents and Kids (Poison Prevention) • Courses and Conferences (Be Poison Smart!® Program) • Family Resources (brochures about poison)	• Website: http://www.bepoisonsmart.com

Central Texas Poison Center	• Poison Prevention Week news *Public Education Section* • Information in English and Spanish • Free emergency action cards, magnets, telephone stickers, and poison brochure • National Poison Prevention Week is March 16–22, 2003 • Lives We Touch (photo advertisements) • Kids Page • *Articles:* Bites and Stings (Critters), Food Poisoning, Pesticides, Seasonal (Summer, Winter, Spring, and Fall), Carbon Monoxide, Lead Poisoning, Know Your Plants, Home Safety Checklist • Mikey Learns About Poison Safety • Information About Anthrax • List of suggested antidotes for emergency departments	• Website: http://www.PoisonControl.org • Links to websites of the American Academy of Clinical Toxicology, American Association of Poison Control Centers, American Council for Drug Education, American College of Pediatrics, Central Texas Poison Center, North Texas Poison Center, More National Poison Information, Southeast Texas Poison Center, Texas Panhandle Poison Center, U.S. Environmental Protection Agency Kids Page, Virtual Field Trip Network, and West Texas Regional Poison Center
Children's Hospital of Michigan Regional Poison Control Center (Detroit)	• *Parents:* Keeping Our Kids Safe: Poison Control for Parents, Babysitters' Guide to Poison Control, Reduce Poison Risk at Home, Common Household Poisons, Lead Poisoning/Visit the Virtual Lead House, Virtual Safety House, Bioterrorism and Nuclear Threats, Poison Brochures, Emergencies, Silly String Fact Sheet, Summer Safety Quiz • *Kids:* What Is a Poison? • Help! The Power Is Out . . . tips about perishable foods • Poison Control Order Form for pamphlets and stickers: Plant Guide, Poison Guide, telephone stickers, magnets, Holiday Hazards, Ipecac Fact Sheet, subscription to newsletter • *Resources:* Terrorism: An Overview, Prepare Yourself for Bioterrorism, Poison Control Brochures (carbon monoxide, plants, other general information) • Make Your Pet's Home Poison Safe	• Website: http://www.mitoxic.org • Link to resources available on the Internet

Continued

TABLE 8-A-1 Continued

Name	Accessible on Web Page	Website Links to Other Organizations
Children's Hospital of Wisconsin Poison Center (Milwaukee)	• Poison Center and link to poison information • First-Aid for Poisonings • Emergency information for Wisconsin families • Protect Your Child from Poisons (on injury prevention webpage) • Take Steps to Prevent Outdoor Poisonings • Facts About Poisons • Syrup of Ipecac • Poison Center Materials (Poison Fact Sheet, Mr. Yuk Brochure, Carbon Monoxide Brochure, Inhalants Brochure, Medication Abuse Brochure, Pesticide Brochure, Toxic Plant List, Kitchen Poison Search, X Marks the Poison, Poison Word Search) • Emergency Contact Information • Beware Outside Poisons During Summer Fun • Sitemap • Keep Children Safe During the Holidays • Preventing Injuries—How You Can Help Your Child • Outpatient Clinics • Are Your Kids Getting High? • Programs and Clinics (link to Poison Center) • Childproof Your Home for Poisons • Pediatric Environmental Health (information about doctors) • Information of Biological Threats (anthrax)	• Website: http://www.chw.org/Emergency/emergency.htm, which is no longer active; information can be found at the Home Page selection at Page Not Found • Link to poison information on site • Link to website of the Centers for Disease Control and Prevention on biological threats site page
Cincinnati Drug and Poison Information Center Regional	• *Drug and Poison Information Center (DPIC):* Overview with links to Emergency Actions for Poisoning, Safety House: Guide to Preventing Household Injuries and Accidental Poisonings (created by Cincinnati Children's Hospital), and	• Website: http://www.cincinnatichildrens.org/dpic/ • Links to websites of the National SAFE KIDS Campaign®, Children's National Medical Center in

Poison Control System (OH)	Cincinnati SAFE KIDS Coalition; What Is a Poison Center?; 24-Hour Crisis/Hotline Services; Poisoning Statistics, Emergency and Information Hotline Data • *Plants:* house, poisonous, nonpoisonous • *Carbon Monoxide Poisoning:* Frequent Questions about the Colorless, Odorless Gas; Prevention; Symptoms and Treatment • Poison Prevention Tips with link to babysitter checklist • Syrup of Ipecac • Why and When You Should Call the Poison Center	Washington, DC, and Johnson & Johnson
Connecticut Poison Control Center (CPCC) (Farmington)	• Emergency information • Information about the center • Information to provide when calling the CPCC • About Poisons (links to documents about berries, carbon monoxide, lead, medications [acetaminophen, children and medicine, eye drops, iron poisoning, safety tips]) • On-line Resources (links to a number of organizations in the areas of Connecticut resources, toxicology, animal poison control, occupational and environmental, pharmaceutical, injury prevention, and substance abuse)	• Website: http://poisoncontrol.uchc.edu • Link to website of the American Association of Poison Control Centers
DeVos Children's Hospital Regional Poison Center (Grand Rapids, MI)	• Regional Poison Center information • Children at Risk for Lead Poisoning • Emergency Action for Poisoning • Poison Prevention Week 2002 • Poisons Overview • Poisoning • Chemical Poisoning and Syrup of Ipecac • Emergency Information Form • Mushroom Poisoning in Children • Preventing Unintentional Injuries Overview	• Website: http://www.spectrum-health.org • Link to website of DeVos Children's Hospital • Links to websites of the American Academy of Family Physicians, American Academy of Pediatrics, American College of Emergency Physicians, American Red Cross, Centers for Disease Control and Prevention, National Center for Infectious Diseases, National Eye Institute, National Institute of Allergy and Infectious Diseases, National Institutes of Health, National Safety Council

Continued

TABLE 8-A-1 Continued

Name	Accessible on Web Page	Website Links to Other Organizations
	• Household Safety and Household Safety Checklist • Online Resources links	
Finger Lakes Regional Poison and Drug Information Center (Rochester, NY)	• About Us Fact Sheet about services and counties serviced • National toll-free emergency number advertisement with alert to National Poison Prevention Week news updates • Poison Facts • What to Do if a Poisoning Occurs • Poison Prevention for Babysitters and Older Brothers and Sisters • Tips for Older Adults • Poison Prevention at Home • Prevent Plant Poisoning • 100 Poisonous Plants in Northeast United States and Canada • What You Need to Know About Trash and Recycling • Headlines (poison control news items as available)	• Website: http://www.stronghealth.com/ • Link to website of the University of Rochester Medical Center
Florida Poison Information Center Network	• Background about the Network (Jacksonville, Miami, Tampa centers) • Toxicology/Poison News Alert! • Florida Poison Network Topics/Links (numerous) • Bio Terrorism Links (numerous) • *e-Government Services:* State Statutes, Red Cross/Poison Center License Plates, Florida Census Quick-Facts, Health License Renewals, Florida Poison Center Network Statutes, Florida Health Statistics • Providing Patient Data to Poison Control Centers is HIPAA-compliant	• Website: http://www.fpicn.org • Links to websites of the Jacksonville, Miami, and Tampa centers • Link to website of the Florida Information Network Statewide Data Reports • Links to websites of Partners: Shands Jacksonville, University of Miami, Tampa General Hospital, University of South Florida, Jackson Memorial Hospital Jackson Health System, University of Florida

Florida Poison Information Center (Jacksonville)	• Poison Prevention for Seniors with link to Grandparent's Checklist • *Children:* Become a Poison Patrol Deputy!!, Word Search!!, Unscramble the Words to Find the Poisons, Sing Along jingle for Poison Hotline (words and audio), Coloring Fun!! • *Poison Information:* food, pets, snakes, marine, critters, prevention, lead, carbon monoxide, pesticides, plants, brochures, drug abuse • *What to Do:* When to Call, What to Expect, first aid, ipecac, FAQ, brochures • Poison Center/Toxicology Links (numerous) • Bio/Chemical Terrorism Links (numerous) • *RealPlayer* video links about poisonings	• Website: http://fpicjax.org/ • Links to websites of the Florida Poison Information Network, American Association of Poison Control Centers, University of Florida College of Medicine, Shands Jacksonville, The University of Florida Health Science Center/Jacksonville, Children's Medical Services/State of Florida Department of Health
Florida Poison Information Center (Miami)	• *Poisonous Animals and Critters of South Florida:* Animals: snakes; Critters: spiders, stinging insects, caterpillars, bees, wasps, and hornets • *Poisonous Plants of Florida:* Visual Guide to Poisonous Plants, Safety Tips for Parents, List of Poisonous and Non-Poisonous Plants • *Household Hazards:* Look-Alikes: Don't Be Fooled, Is Your Home Poison-Proof?	• Website: http://www.miami.edu/poison-center/ • Links to websites of the American Association of Poison Control Centers, University of California/Davis Poison Center, Maryland Poison Center, Arizona Poison and Drug Information Center, Finger Lakes Regional Poison Center, UC San Diego Regional Poison Center, Central Texas Poison Center, Southeast Texas Poison Center, Florida Poison Information Center/Jacksonville • Links to websites of the Toxicology Resource Sites: Summit Research Services, Clinical Pharmacy Drug Monograph Service, MedWeb: Toxicology, HyperTox, Hardin Meta Direct, Toxicology Internet URLs, Pesticide Poisoning Handbook, National Institute of Environment Health Sciences, Marine and Freshwater Biomedical Sciences Center—University of Miami

Continued

TABLE 8-A-1 Continued

Name	Accessible on Web Page	Website Links to Other Organizations
		• Links to the University of Miami, University of Florida, University of Miami School of Medicine, University of Florida Health Science Center Jacksonville, University of Miami School of Medicine Department of Pediatrics, Animal Poison Control Center
Florida Poison Information Center (Tampa)	• In both English and Spanish • What to Do in a Poisoning • Poisonings Can Be Prevented! • Herbal Products Plus Prescription Medications: Potentially Dangerous Combinations • Facts About Medication Poisonings in Senior Citizens • *Venomous Critters:* black widow spider, brown recluse spider, stinging caterpillars, bees and wasps, scorpion, snakes, Eastern coral snakes, jellyfish, coral, man-of-war, anemones, stingrays, catfish, urchins, stonefish, scorpion fish, lionfish • Ordering of *Materials for Teaching Poison Prevention:* Poison Facts for Health Educators, Interactive Exhibit Ideas, Guide for Teaching Poison Prevention to Children, Guide for Teaching Poison Prevention to Adults, Guide to Teaching Poison Prevention to Babysitters • Ordering of brochures and phone stickers	• Website: http://www.poisoncentertampa.org/ • Links to websites of the American Association of Poison Control Centers, Tampa General Hospital, Florida Poison Information Centers, Florida Department of Health, Health Resources and Services Administration
Georgia Poison Center (Atlanta)	• *Poison Overview:* Common Poisons, First-Aid, Frequently Asked Poisoning Related Questions • *Poisoning Topics:* carbon monoxide, food poisoning, household product safety, insects, spiders, ticks, snake bites,	• Website: http://www.georgiapoisoncenter.org • Links to websites of the American Association of Poison Control Centers, Agency for Toxic Substances and Disease Registry, American

	medicine safety, pesticide safety, pets and poisons, poisonous plants, rabies and animal safety, lead • *Educational Information:* Poison Prevention Instructor Training Program, Other Public Awareness Programs • *Educational Programs:* Georgia Poison Center Poison Prevention Instructor Training Program (ITP) • *Educational Materials:* Poison Prevention Literature, Ordering Information and related links, audiovisual materials, brochures, promotional materials • News and Events • Poisoning Statistics	Academy of Pediatrics, American Veterinary Medical Association, Centers for Disease Control and Prevention, Consumer Product Safety Commission, Directory of Poison Control Centers, Georgia Cooperative Extension Service, Georgia Department of Human Resources, Georgia Department of Agriculture, Grady Health System, GA Emergency Medical Services, Leave A Legacy Georgia! Campaign, National Center for Health Statistics, National Center for Injury Prevention and Control, National SAFE KIDS Campaign, National Lead Information Center
Greater Cleveland Poison Control Center (OH)	No information was found at the site—only the AAPCC generic cover page with no e-mail link; when searched otherwise, the default was to the same cover page	
Hawaii Poison Center (Honolulu)	• This center provides poison prevention education and does not manage exposure cases • General center background and contact information are provided, but no specific information except for detailed list of links on specific topics when searching by "poisons" or "poisonings"	• Website: http://www.kapiolani.org/facilities/programs-hpc.html
Hennepin Regional Poison Center (Minneapolis, MN)	• Information in both English and Spanish • What Is Poison Help? (information about the national toll-free emergency number) • Poison First Aid: What to Do if Poisoned • *Poison Information for the Home:* Your Home, Poison Prevention Information, Information for Older Adults, Specific	• Website: http://www.mnpoison.org • Links to websites of the American Association of Poison Control Centers, Minnesota Department of Health, Centers for Disease Control and Prevention, Hennepin County Medical Center, Risk Watch, Animal Poison Control Center

Continued

TABLE 8-A-1 Continued

Name	Accessible on Web Page	Website Links to Other Organizations
	Poison Information, Seasonal Hazards, Plants and Mushrooms, Pet Poisoning, Kids Corner, Poison Trivia, FAQ, Other Languages, HomePacket (ordering) • *Information for Educators:* Train the Trainer, Teaching Children (activity sheets and games), Teaching Seniors (adverse effects of medication and food poisoning), ordering bulk material • *Information for Health Care Professionals:* PDF articles on specific topics of medical interest, newsletters and reviews, ordering bulk material • Information on the Midwest Regional Poison Interest Group (MR PIG), a not-for-profit organization of poison control centers and others interested in the prevention and treatment of poisoning in the states of Iowa, Minnesota, North Dakota, South Dakota, and Wisconsin—one of the objectives is to enhance education in all of the states involved	
Hudson Valley Poison Education Center (Sleepy Hollow, NY)	• This center provides poison prevention education and does not manage exposure cases • Poisons Information • *Educational Programs:* General Poison Prevention, Poisons of Terrorism, basic programmatic information • Bioterrorism Information • Monthly newsletters (The Antidote)	• Website: http://www.PoisonEducation.org • Links to websites of the American Association of Poison Control Centers, Central New York Poison Control Center, Long Island Poison Control Center, Centers for Disease Control and Prevention, Phelps Memorial Hospital Center—Emergency Life Support Programs, Phelps Memorial Hospital Center—Hospital Homepage
Illinois Poison Center (Chicago)	• Information in both English and Spanish • Complimentary information packet (two telephone stickers, one magnet, Your Guide to Poison Prevention brochure, toxic	• Website: http://www.mchc.org/ipc *Affiliated with these poison prevention education centers that do not manage exposure cases and provide reference*

	plant list, home safety checklist) • What Is a Poison? • Household Products • Medicine Safety Tips • *Plants:* Your Guide to Plants (house, garden, wild plants/weeds, trees/shrubs), How to Prevent a Plant Poisoning, First Aid for Plant Exposures • Carbon Monoxide Facts • Protect Your Children from Lead Poisoning • Seasonal Poisoning Hazards • First Aid and Treatment Recommendations • Emergency Kit Essentials • *Home:* Poison Prevention Starts at Home, 10 Steps to a Safe Home • *Information for Health Care Professionals:* Bioterrorism Treatment Guidelines, Upcoming Education Courses, Helpful Websites Links • *Information for Educators and Volunteers:* Become an IPC Volunteer Educator, IPC Calendar of Events, Register and Event, Pearls of Poison Prevention Newsletter, Activity Sheets and Handouts, IPC's Presentation Handbook, contact information, helpful websites • *Children:* Tic-Tac-Toe Poison Game, Poison Prevention Word Search, Spot the Poison, Poison Prevention Pro Certificate	*to the same website:* • Advocate Illinois Masonic Medical Center—Education Center (Alton, IL) • Carle Foundation Hospital Education Center (Urbana, IL) • Freeport Health Network Education Center (Freeport, IL) • Loyola University Health System Education Center (Maywood, IL) • Memorial Hospital of Carbondale Education Center (Carbondale, IL) • Mount Sinai Hospital Poison Prevention Education Center (Chicago, IL) • St. John's Hospital Education Center (Springfield, IL) *Other Links:* American Association of Poison Control Centers, Agency for Toxic Substances and Disease Registry, American Academy of Pediatrics, Centers for Disease Control and Prevention, Consumer Product Safety Commission, Illinois Department of Public Health, National Center for Health Statistics, National Center for Injury Prevention and Control, National Library of Medicine—Tox Town, National SAFE KIDS Campaign, National Lead Information Center
Indiana Poison Center (Indianapolis)	• *Dealing with Emergencies:* Poisoning Overview, Prevention, Home Treatment, Related Information (carbon monoxide poisoning, lead poisoning, nausea and vomiting [age 4 and older], nonprescription medications and products, Your Home Health Center)	• Clarian Health/Methodist Indiana University Website http://www.clarion.org/clinical/poisoncontrol • Select "poison control" at the website

Continued

TABLE 8-A-1 Continued

Name	Accessible on Web Page	Website Links to Other Organizations
	• Food Poisoning • Lead Poisoning • Poison Ivy, Oak, Sumac • What to Do About Poisoning	
Iowa Statewide Poison Control Center (Sioux City)	• *Hot Topics:* Anthrax and Bioterrorism, Childhood Lead Poisoning • What to Do if Someone Is Poisoned • Poison Prevention • Poison Information • Food Poisoning • Carbon Monoxide: A Silent Killer • Plants (with link to "mushrooms") • Herbal and Homeopathic Remedies • *Creepy Things That Bite and Sting:* chiggers, mosquitoes, bees and wasps, ticks, spiders, snakes • *Holiday Hazards:* Easter, Spring/Summer, Halloween, Christmas, Food • Drugs of Abuse • Lead Poisoning: A Childhood Disease • Preventing Pet Poisonings • Common Non-Toxic Ingestions • *Kids Corner:* Coloring Book, and coloring contests and more fun games to be added to website • Professional Information and links • Poison Information for Health Care Professionals Iowa Statewide Poison Control Center (ISPCC)	• Website: http://www.iowapoison.org • Link to website of the Food Safety and Inspection Service (food poisoning) • Links to websites of the Iowa Department of Natural Resources and Iowa Department of Natural Resources—Wildlife (creepy things) • Link to website of the Drug Enforcement Agency (drugs) • Link to website of the Animal Poison Control Center (pets) • Links to websites of the American Association of Poison Control Centers, St. Luke's Regional Medical Center, University of Iowa Hospitals and Clinics, Centers for Disease Control and Prevention, Food and Drug Administration, National Library of Medicine, Iowa's 24 Hour Substance Abuse Helpline, Emergency Medical Services, Herbal Information, Dr. Koop, Animal Poison Control Center

Kentucky Regional Poison Center (Louisville)	• *Recommendations for Parents:* Poisoning Proof Your Home Using Our Room by Room Checklist; Keep Activated Charcoal in the Home; Keep the Poison Center Phone Number Available; Top 10 Substances Involved in Childhood Poisoning • *Articles of Interest:* What Can I Do About Lead Poisoning?; Bees, Wasps, and Other Stings; Ticks; Do You Have a Carbon Monoxide Detector? • Teacher Curriculum (booklet of lesson plans and student exercises free of charge to all teachers)	• Website: http://www.krpc.com • Link to website of the Kosair Children's Hospital • Links to websites of the American Association of Poison Control Centers, Virtual Tour of a House, Consumer Product Safety Commission, Florida Poison Information Center/Miami, Georgia Poison Center, Indiana Poison Center, Maryland Poison Center, Massachusetts Poison Center, National Capital Poison Center, Rocky Mountain Poison and Drug Center, California Poison Control System, Toxikon—University of Illinois, Cornell University Poisonous Plants Page, Indiana Plants Poisonous to Animals and Livestock, Canadian Department of Agriculture Site on Poisonous Plants
Long Island Regional Poison and Drug Information Center (Mineola, NY)	• This is "a service to inform and educate the public and professionals on drugs and substances related to poisonings" and refers emergency cases to the national toll-free emergency number • *Seasonal Hazards:* Allergy Tips, Winter Hazards, March Is Poison Prevention Month (with links to fact sheets, including one in Spanish), Easter Egg Hunt for Salmonella, Poisonous Mushrooms, Summer Hazards, Sunscreens and Lotions, Halloween Safety Tips, Thanksgiving Turkey Tips, The Dangers of Holiday Decorations and Plants • *Poison Prevention Tips for Parents:* Poison Guidelines for Babysitters, Grandparents Alert, Hand Sanitizers, How to Poison Proof Your Home, Infant and Child Choking Hazards • *Poison Prevention for Everyone:* Carbon Monoxide: The Silent Killer, What You Should Know About Herbal Products,	• Website: http://www.lirpdic.org • Links to websites of the American Association of Poison Control Centers, American Academy of Pediatrics, American College of Emergency Physicians, American Public Health Association, Centers for Disease Control and Prevention—*Morbidity and Mortality Weekly Report*, Consumer Product Safety Commission Recalls, Environmental Protection Agency, Food and Drug Administration News and Press Releases, Food and Drug Administration Recalls and Enforcement Reports, *Journal of Toxicology-Clinical Toxicology*, *Lancet*, Nassau County Department of Health, National Center for Complementary and Alternative Medicine, National Institutes of Health, *New England*

Continued

TABLE 8-A-1 Continued

Name	Accessible on Web Page	Website Links to Other Organizations
	Mercury Poisoning, Pesticides in Your Environment, Let's Talk About Substance Abuse, West Nile Virus • *Facts About Biological and Chemical Agents:* Chemical Agents, Anthrax, Smallpox, Pneumonic Plague, Botulism, Tularemia • Seasonal Newsletters • Professional Education: advertisement of 2003 consultant meetings on specific topics, etc.	*Journal of Medicine*, New York State Department of Health, Partnership for a Drug-Free America, Safety Alerts, Suffolk County Department of Health, U.S. Department of Agriculture, Winthrop-University Hospital, Winthrop-University Hospital Community Training Center
Louisiana Drug and Poison Information Center (Monroe)	• National Poison Prevention Week • *Poison Prevention:* Children Act Fast—So Do Poisons!, Educational Material Order Form, Poison Lookout checklist, Louisiana Poison Control Center Recommends New Treatment for Poisoning • *Mosquito Information:* Recommendations for Using DEET Containing Insecticides, CDC West Nile Virus, LA Department of Health and Hospitals • Current Issue of The Poison Times • Current Issue of Toxicum • Ordering free materials • Poison Pearls on a number of specific topics for medical professional information and emergency management • Health Care Journals with Free Full-Text Access	• Website: http://www.lapcc.org • Links to websites of the University of Louisiana/Monroe School of Pharmacy and Health Technology and the American Association of Poison Control Centers
Maryland Poison Center (Baltimore)	• Order Form for Educational Materials (and reference to product listings and catalog) • Education Section (still under construction) • Babysitter Cards (10 copies free) • *Clin Tox FAQ's® patient management guidelines (selected*	• Website: http://www.pharmacy.umaryland.edu/~mpc/ • Links to websites of the California Poison System/San Diego Division, Arizona Poison Center, Finger Lakes Poison Center, North Texas Poison Center,

	topics): acetaminophen, iron, salicylates, tricyclic antidepressants (TCAs), digoxin, theophylline • Training Topics in Toxicology • Tox Alert downloadable documents	Florida Poison Center/Jacksonville, National Capital Poison Center, New Jersey Poison Center, Georgia Poison Center, Southeast Texas Poison Center, Samaritan Poison Center/Arizona, Western New York Poison Center, Kentucky Poison Center • Links to the websites of the American Association of Poison Control Centers and other useful links
Mid-American Poison Center (Kansas City, KS)		• No website could be located, except a page from the Kansas University Medical College Department of Pharmacy saying that it is also the site for the Mid-America Poison Control Center; no further information resulted from a search at the university website
Middle Tennessee Poison Center (Nashville)	• Outreach and Education Program: Free Poison Education Programs for Schools! • The Poison Center Outreach Program Around the Town past and current activities • *Interactive Poison Education Activities:* Poison House and Look-a-Likes • Household Dangers • Are You a Poison Expert? Can You Answer These Questions? • Ask Auntie Dote: Send Your Poison Questions to Auntie Dote • Poison Center News	• Website: http://www.poisonlifeline.org • Link to website of the Vanderbilt Center in Molecular Toxicology • Links to websites of numerous poison control centers, education and outreach documents, and governmental agencies, information, and databases

Continued

TABLE 8-A-1 Continued

Name	Accessible on Web Page	Website Links to Other Organizations
Mississippi Regional Poison Control Center (Jackson)		• No website for the center • Use website of the University of Mississippi Medical Center: http://pharmacology.umc.edu/main.html; however, the poison center is only mentioned briefly
Missouri Regional Poison Control Center (St. Louis)	• *Helpful Parent Sites:* ChiBro—Parenting Pages, Kid's Domain—Growing Place, Kid's Source Online Parenting, Parent's Guide to the Internet, SAFE KIDS Child Safety, Reality Check—Keeping Youths Drug Free http://www.cardinalglennon.com/internet/home/net10hom.nsf/SearchDocuments?searchview = 1&query = poison • Cardinal Glennon Library links to numerous sites, including journals • Community Outreach	• No website for the center • Use the SSM Cardinal Glennon Children's Hospital website: http://www.cardinalglennon.com/internet/home/net10hom.nsf/SearchDocuments?searchview=1&query=poison
National Capital Poison Center (Washington, DC)	• Questions About Calling the Poison Center • Poison Prevention Tips • Mr. Yuk (under construction) • Even Plants Can Be Dangerous (with list of poisonous and nonpoisonous plants) • Act Fast if Poisoning Happens • The Yukkiest Poisons • *Previous Issues of Poison Pearls and Perils:* Pearls from Recent Literature (July 1996), Calcium Channel Blockers, Isoniazid, Organophosphates, Carbon Monoxide, Salicylates, Pearls from Recent Literature (July 1995) • Toxicology Tidbits • Remember . . . (pets)	• Website: http://www.gwemed.edu/ncpc/Index.htm • Links to websites of the American Association of Poison Control Centers, Maryland Poison Center, UC Davis Poison Center, Minnesota Regional Poison Center, San Diego Regional Poison Center, Central Texas Poison Center, Louisiana Drug and Poison Information Center, Texas Poison Control Network at Galveston/Southeast Texas Poison Center

New Hampshire Poison Information Center (Lebanon)	• About NHPIC • Poisons in Your Home • *Plant Poisonings:* common plants and shrubs, Toxic Plants, Nontoxic Plants • Pet Poisonings • Bites and Stings • Insect Bites and Stings and Spider Bites • Poisoned? What to Do if a Poisoning Occurs • Inhalation Abuse Newsletter • Poison Prevention Week 2003	• Website: http://www.hitchcock.org • Links to numerous public and professional sites
New Jersey Poison Information and Education System (NJPIES) (Newark)	• NJPIES Newsletters • *Informational Brochures* (in black and white and in color; also in Spanish): general information, child care, food, pets, plants, medicine, pesticides, seasonal • Advertisement of New Jersey Observance of National Poison Prevention Week • *Stop & Think Sam Materials:* Teacher's Guide, Student Activity Book, About Sam, Poison Pointers, Quick Tips, Tips Press Release, Sam Tour Press Release • Holiday Reminders • Poison News • Website News • *Fun Stuff:* Carbon Monoxide, Lead Poisoning • *Just the Facts:* Carbon Monoxide, Lead Poisoning	• Website: http://www.njpies.org • Links to websites of Lycos, Family Friendly Site™, Family + Fun, SafeSurf, Achoo.com Healthcare, InfoGrove, Garden State Environet, KidsSafe, New Jersey Department of Health, The Healthcare Foundation of New Jersey

Continued

TABLE 8-A-1 Continued

Name	Accessible on Web Page	Website Links to Other Organizations
New Mexico Poison and Drug Information Center (Albuquerque)	• *Fall Season Poison Prevention Tips:* Peeling Green Chile, Carbon Monoxide, Halloween Safety, Thanksgiving • First Aid for Poisoning • Treatment Products • What Is a Poison? • Children and Poison, Poison Prevention and Children, Poison Prevention for Children Brochure • Poisonous Plants in New Mexico, Plant Poison Prevention Tips • Seniors and Poison, Poison Prevention Tips for Seniors • Venomous and Non-Venomous Snakes in New Mexico, Preventing Snake Bites, Snake First Aid Tips • *Poison Prevention Tips:* For Children, For Seniors, Plant and Mushroom Poison Prevention Tips, Snake Bite • National Poison Prevention Week • *Public Education:* New Mexico Poison Center Brochure, Poison Prevention for Children Brochure, Poison Prevention for Seniors Brochure, Poisonous Plants in New Mexico Brochure, Poison Help Logo, Frequently Asked Questions About Illicit Drug Use • *Professional Education:* Professional Outreach Program, Viewable Examples, Recent Continuing Medical Education Lectures, New Mexico Poison Center Snakebite Flow Sheet, Antidote Charts • Student education opportunities	• Website: http://hsc.unm.edu/pharmacy/poison • Links to websites of the American Association of Poison Control Centers, American Pharmaceutical Association, American Academy of Pediatrics, Learn About Chemicals Around Your House, National Animal Poison Control Center, National Poison Prevention Week, National SAFE KIDS Campaign, University of New Mexico College of Pharmacy, Centers for Disease Control and Prevention, U.S. Department of Agriculture, Environmental Protection Agency, Food and Drug Administration, Consumer Product Safety Commission, New Mexico Department of Health

New York City Poison Control Center	• NYC Poison Control Center Order Form for Materials/Brochures • National Poison Prevention Week • What if a Poisoning Occurs? • *Basic Information:* Understanding the Basics, Plants, You Can Prevent Carbon Monoxide Poisoning, National Toll-Free Number for Poison Control Centers • *Tips for Seasonal Safety:* Winter Poisons, Summer Poisons • *For Parents and Caregivers:* Poison Lookout Checklist, Poison Proof Your Home • *For Kids:* Put an X Through the Poisons, Emergency Telephone List, Poison Prevention Word Find • *For Teachers:* K–6 Curriculum Poison Prevention Education • *For Seniors:* Medicines: Use Them Safely • Health Education Literacy Program (HELP): brochures and instructor's guide (free of charge)	• Website: http://www.nyc.gov/html/doh/html/poison/poison.html • Link to website of the New York City Department of Health and Mental Hygiene
North Texas Poison Center (Dallas)	• Information in both English and Spanish • *Community Activities:* Poison Jungle Safari, Poison Prevention Poster Contest, Community Presentations and Special Events • National Poison Prevention Week and Poison Prevention Poster Contest • Virtual Field Trip • Emergency Action for Poisoning • Potentially Poisonous Products in Your Home • Poison Prevention Materials (ordering) • *Bites and Stings:* fire ants, tarantulas, black widow, scorpions, centipede/millipede, rattlesnakes, bees and wasps, brown recluse spider, sun spider/wind scorpion, coral snakes, nino de la tierra, first aid for bites and stings	• Website: http://www.poisoncenter.org • Links to websites of the American Association of Poison Control Centers, Thomason Hospital, Texas Poison Center Network, Southwest Center for Pediatric Environmental Health, Animal Poison Control Center, Consumer Product Safety Commission

Continued

TABLE 8-A-1 Continued

Name	Accessible on Web Page	Website Links to Other Organizations
	• *Seasonal Information:* Test Your Poison Summer Safety IQ; Winter, Spring, and Fall sites under construction	
Northern New England Poison Center (Portland, ME)	• General Information	• Website: http://www.fahc.org/ER/Services/Sub/vtpoison.html
Oklahoma Poison Control Center (Oklahoma City)	• *Poison Prevention Information:* Poison Prevention Week, Prevention Tips, Home Check List, What if a Poisoning Happens to You?, Is Your Purse or Diaper Bag a Deadly Weapon?, Arsine Fact Sheet, Carbon Monoxide Fact Sheet, Food Poisoning Fact Sheet, Head Lice Fact Sheet, Look-a-Like Products, Nicotine Fact Sheet, Information to Leave with the Babysitter, Poison Control Center Audio Jingle • National Poison Prevention Week • First Aid • *Education Resources:* Teacher's Guides/Handouts, Games, National Poison Prevention Week Activities • *Free Resources:* stickers and magnets, brochures, fact sheets, teaching resources for elementary school levels • *Poison-Related Links:* poisonous plants, insects, mushrooms, pesticides • News Archives • Medical Professional Development: 2002 Antidote Survey, Antidote Chart	• Website: http://www.Oklahomapoison.org • Links to websites of the American Association of Poison Control Centers (http://www.1-800-222-1222.org) and software (MDS Index, MSDS Search, TOXNET Web Search) • Useful links to websites of the American Academy of Pediatrics, American Association of Poison Control Centers, Animal Poison Control Center, Federal Emergency Management Agency, Hazardous Materials/Waste Disposal, National Library of Medicine Toxicology and Environmental Health and Safety, National Inhalant Prevention Coalition, N.Y.-Arts, Crafts and Theater Safety, Occupational Safety and Health Administration, SAFE KIDS Coalition, Southwest Center for Pediatric Environmental Health

Oregon Poison Center (Portland)		• Website: http://www.oregonpoison.org is no longer active
Palmetto Poison Center (Columbia, SC)	• Be Prepared • Epidemiology of Poisonings • Home Poison Safety Check List • First Aid Measures • An Ounce of Prevention • Order Form (not free; mostly small charge)	• Website: http://www.pharm.sc.edu/pps/pps.htm • Numerous links to best sites, medical textbooks online, medical dictionary, journals, MEDLINE® access, medical search engines, and links to medical sites, searches, and links
Pittsburgh Poison Center (PA)	• Mr. Yuk (with audio song) • Ordering poison prevention materials • Fighting the Poisons of Terror: Understanding the Threat of Biological and Chemical Weapons • Poison Education Information with link to poison prevention materials	• Website: http://www.chp.edu/clinical/03a_poison.php • Link to website of Children's Community Pediatrics (http://www.cc-peds.net/main/index.shtm)
The Poison Center (Omaha, NE)	• *Antidote Poster:* Emergency Antidotal Management of Poisonings • *2003 Poison Prevention Week:* National Poison Prevention Week • *First Aid for Poisoning:* What to Do if . . . • *Poison Prevention:* Steps to Prevent Poisonings, Safety Checklist • *Know Your Plants:* Poisonous Plants, Non-Poisonous Plants • *Brochures:* Steps to Prevention Poisonings, Holiday Hazards • *Coloring Book:* Poison Proof with Pinky • *Seasonal Hazards:* Spring and Summer Hazards, Autumn Hazards, Hints for Halloween, Winter Hazards	• Website: http://www.Poison-Center.org • Links to websites of the American Association of Poison Control Centers, Centers for Disease Control and Prevention, Consumer Product Safety Commission, Food and Drug Administration, National Animal Poison Control Center, National Capital Poison Center, National Clearinghouse for Alcohol and Drug Information, Safety and Health Council

Continued

TABLE 8-A-1 Continued

Name	Accessible on Web Page	Website Links to Other Organizations
	• *Teacher's Packet:* Parents Letter, Primary Program, Intermediate Program, Program and Materials Evaluation, Inhalant Fact Sheet, coloring pages • *Speakers Packet:* Speaker's Manual • *Videos:* Primary Poison Video, Intermediate Poison Video, Adult Poison Video	
PROSAR International Poison Control Center (St. Paul, MN)	• Health and Safety Call Center: "PROSAR manages medical and veterinary inquiries regarding product related adverse incidents from your consumers or end users 25 hours per day, 365 days a year." • Animal Poison Hotline • Consumer products, industrial and institutional, agricultural	• Website: http://www.prosarcorp.com • Link to website of Animal Poison Hotline
The Poison Control Center (Philadelphia, PA)	• Children's Education Program • *Poison Prevention Tips (listing from A–Z):* animal poisonings, berries and seeds, calling the poison control center, caustics, Children Act Fast . . . So Do Poisons, cough and cold medicines, the elderly, eye exposures, food safety, foreign bodies, glow jewelry, holiday hazards, home safety, hydrocarbons, inhalant exposure, insect repellents, mushrooms, personal care products, plants, poison facts and fiction, poison ivy, poison oak, poison sumac, rodenticides (rat poison), safer alternatives, spider bites, sting things, summertime skin irritants, syrup of ipecac, vitamins and pills • *Publications:* brochures, manuals, books, poster, poison pen notes • Resources for professionals	• Website: http://poisoncontrol.chop.edu/ • Link to the website of The Children's Hospital of Philadelphia (Emergency Medicine, Toxicology) • Links to website of the American Association of Poison Control Centers, National SAFE KIDS Campaign, SAFE KIDS Coalition of Southeastern Pennsylvania

Regional Center for Poison Control and Prevention Serving Massachusetts and Rhode Island	• The Education Program description • Prevention Tips • Basic Information About Poisons • *Health and Beauty Products:* perfumes and colognes, nail products, toothpaste and mouthwash, thermometers • Over-the-Counter and Prescription Medicines • *Plants:* Safe Plants, Poisonous Plants, How to Prevent Plant Poisonings, What to Do if You Suspect Someone Has Swallowed a Plant • Carbon Monoxide • Pesticides • Clinical Toxicology Review (alphabetical list and by month) • Annual Statistics	• Website: http://www.maripoisoncenter.com • Numerous links to websites: toxicology resources, environmental health resources, pharmaceutical information resources, injury prevention resources, prevention resources in Massachusetts
Rocky Mountain Poison and Drug Center (RMPDC) (Denver, CO)	• Information in both English and Spanish • Educational Materials and Presentations (to be added as available) • Developments in Medical Toxicology and RMPDC News • Environmental and Public Health Projects • Poison Prevention Information *News and Topics of Interest* • Web-Based BT (bioterrorism) Training Available • Para Halloween (Spanish) and Halloween Poison Prevention Safety Tips (English) • What You Need to Know About Carbon Monoxide As Colder Weather Approaches (English–Spanish) • Guía de Plantas and Plant Guide (Spanish and English) • Holiday Poison Prevention Safety Tips (English and Spanish) • Information on Severe Acute Respiratory Syndrome (SARS) (multiple languages at linked website) • Emergency Action Cards—For Parents (English and Spanish)	• Website: http://www.RMPDC.org • Link to website of the Metropolitan Medical Response System, Rocky Mountain Arsenal Medical Monitoring Program, Denver Center for Public Health Preparedness • Link to Colorado Department of Public Health and Environment website for SARS information • Links to websites: Proterics and Acetaminophen • Links to websites of the American Association of Poison Control Centers, Health Alert Network (HAN) Training Modules, U.S. Department of Health and Human Services, ASPCA Animal Poison Control Center

Continued

TABLE 8-A-1 Continued

Name	Accessible on Web Page	Website Links to Other Organizations
	• Poison Prevention Tips—Pets (English) • Plants and Children—How to Choose Plants for Your Home • Poison Safety Tips	
San Jorge Children's Hospital Poison Center (Santurce, Puerto Rico)	• Information in both Spanish and English • Prevention	• Website: http://www.poisoncenter.net • Links to websites of the American Association of Poison Control Centers, American Academy of Clinical Toxicology, American College of Clinical Toxicology, Texas Poison Control Network at Galveston Center, North Texas Poison Center, Arizona Poison and Drug Information Center, California Poison Control System, Central Texas Poison Center, Central Pennsylvania Poison Center, Fingerlakes Regional Poison Center, Florida Poison Information Center/Jacksonville, Kentucky Regional Poison Center, Maryland Poison Center, New Jersey Poison Information and Education System
Sioux Valley Poison Control Education Center (Sioux Falls, SD)	• This center provides poison prevention education and does not manage exposure cases	• Website: http://www.siouxvalley.org; however, "page cannot be displayed" for the poison control center webpage on the main hospital site
South Texas Poison Center (San Antonio)	• Short summary page with information about poisons and prevention	• Website: http://www.uthscsa.edu/surgery/poisoncenter • Website: http:////sthrc.uthscsa.edu/STIPRC/poison.htm

Southeast Texas Poison Center (Galveston)	• *Poison Primer:* Is Your Home Poison Proof?, Syrup of Ipecac, Mr. Yuk and Children, Things to Do to Prevent Accidental Poisonings	• Website: http://www.utmb.edu/setpc • Link to website of the University of Texas Medical Branch Campus in Galveston • Links to websites of the American Association of Poison Control Centers, Central Texas Poison Center, South Texas Poison Center, Texas Poison Control Network at Amarillo, West Texas Regional Poison Center
Southern Poison Center (Memphis, TN)		• Website is under construction
Texas Department of Health (Austin)	• This department administers the funding to the six poison control centers of the Texas Poison Control Network and conducts epidemiological analyses of poison data; it does not manage exposure cases • Links can provide educational information	• Website: http://www.tdh.state.tx.us/epidemiology • Link to Texas Poison Center Network • Links to websites of the Centers for Disease Control and Prevention, National Center for Injury Prevention and Control, National Institutes of Health, Texas Department of Health, State of Texas Public Health Information, World Health Organization
Texas Panhandle Poison Center (Amarillo)		• Website: http://www.panhandlepoison.org

Continued

TABLE 8-A-1 Continued

Name	Accessible on Web Page	Website Links to Other Organizations
University of Wisconsin Hospital and Clinics (Madison)	• This center provides poison prevention education and does not manage exposure cases • Poison Prevention in the Home • Poison Ivy, Poison Oak, Poison Sumac • Lead Poisoning • Heavy Metal Poisoning • Metal Toxicity: Your Brain Under Siege • *Resource Materials:* Pediatric Poison Prevention, Poisonous Plants Booklet, Poison Prevention for Seniors, Poison Through the Seasons, Home Poison Checklist • Poison Prevention Tips	• Website: http://www.uwhospital.org • Search results for poisoning—links to related topics provided • Links to websites of the American Association of Poison Control Centers, Environmental Protection Agency National Lead Information Center and Office of Pollution Prevention and Toxics, Centers for Disease Control and Prevention, American Academy of Dermatology, National Center for Environmental Health (CDC), Occupational Safety and Health Administration, Consumer Product Safety Commission, CDC Childhood Lead Poisoning Prevention Program, Agency for Toxic Substances and Disease Registry, National SAFE KIDS Campaign
Utah Poison Control Center (UPCC) (Salt Lake City)	• Information in both English and Spanish; also Vietnamese and Russian • Prevention Tips • *Education Materials:* Order Form (materials free of charge to Utah residents), Babysitters Brochure, Poisonous Plants, Emergency Action for Poisoning, Emergency Action—Vietnamese, Emergency Action—Russian, Poison Smart Utah —A Newsletter of UPCC • *More for Educators:* Train-the-Trainer Program, Preventing Poisonings: Lesson Plan for Adults, Poison Safety: Lesson Plan for Elementary School Children • *More for Parents and Kids:* Poisonous Plants, Test Your	• Website: http://uuhsc.utah.edu/poison • Links to websites of University of Utah College of Pharmacy, University of Utah Health Sciences Center, Utah Department of Health • Links to websites of the American Association of Poison Control Centers, Environmental Protection Agency, American Academy of Clinical Toxicology, American College of Medical Toxicology, Centers for Disease Control and Prevention, National Animal Poison Control Center, National Poison Prevention Week Council, Occupational Safety and Health Administration, Pregnancy Riskline,

	Poison Summer Safety IQ, activity sheets, Virtual Interactive Tour, Games for Kids • Health Information and Resources: search adult, pediatric; also in Spanish • *Fact Sheets:* Parents/Health Care Providers, Children, Teens/Babysitters • Health Information and Resources (searchable database of health information including, but not limited to, "poison" topics) • UPCC in the News	Consumer Product Safety Commission, Utah Safety Council, Utah Department of Health
Virginia Poison Center (Richmond)	• Educational programs on a variety of poison prevention topics can be requested by businesses, schools, child care centers, and community groups • The Virginia Poison Center at MCV Hospitals: An Introduction for Health Care Professionals • First Aid for Poisoning • Poisoning: Don't Let It Happen to Your Child • Poison Prevention Fact Sheet for Teen Babysitters • The Poisons of Summer • Plants: Many Are Pretty, But Some Are Poisonous • Educational Materials Order Form • Poison Prevention Videotapes	• Website: http://www.vcu.edu/mcved/; links to the center
Washington State Poison Center Information (Seattle)	• Mr. Yuk stickers • Bioterrorism/Disaster Information	• Website: http://www.doh.wa.gov.hsqa/emtp/poison.htm • Link to website of the Injury Control Resource Information Network

Continued

TABLE 8-A-1 Continued

Name	Accessible on Web Page	Website Links to Other Organizations
West Texas Regional Poison Center (El Paso)	• Information in both English and Spanish • *Community Activities:* Poison Jungle Safari, Poison Prevention Poster Contest, Community Presentations and Special Events • National Poison Prevention Week and Poison Prevention Poster Contest • Virtual Field Trip • Emergency Action for Poisoning • Potentially Poisonous Products in Your Home • Poison Prevention Materials (ordering) • *Bites and Stings:* fire ants, tarantulas, black widow, scorpions, centipede/millipede, rattlesnakes, bees and wasps, brown recluse spider, sun spider/wind scorpion, coral snakes, nino de la tierra, first aid for bites and stings • *Seasonal Information:* Test Your Poison Summer Safety IQ; other sites	• Website: http://www.poison.org • Links to websites of the American Association of Poison Control Centers, Thomason Hospital, Texas Poison Center Network, Southwest Center for Pediatric Environmental Health, Animal Poison Control Center, Consumer Product Safety Commission
West Virginia Poison Center (Charleston)	• *Hotline* Newsletters • Electronic School Newsletter • Order forms • Stickers • Pamphlets • Audio-Visual Materials List • Information about speakers and health fairs • *Plants:* Toxic Plants, Non-Toxic Plants, West Virginia Summer Wildflowers, Poisonous Plants of the Southern United States, Living With Holiday Plants	• Website: http://www.hsc.wvu.edu/charleston/wvpc/ • Link to West Virginia Chemical Emergency Procedures website • Links for kids to Sparky the Fire Dog, Agency for Toxic Substances and Disease Registry, Child Health Programs, Carbon Monoxide Information, EPA Explorers Club, EPA Student Center, Kidd Safety, Internet Public Library—Poison Prevention Site • Links for teachers to Join Hands Educational

- Hazardous Materials
- Inhalant Abuse
- *Kids Corner:* Product Safety, Career Rap: Toxicologist, Rap Sheet: How to Use Consumer Products Safely, How Many?, Riddles, Word Search, Where Does Mr. Yuk Go?, Poison Prevention Game
- "Prevention Central": Your Community Bulletin Board (listing of events, photographs)
- *Media Topics:* Party Mix—Alcohol, Holiday Safety, Jimson Weed, What's Your Resolution, Hobby Safety, Halloween, Poison Ivy, Poison Oak, Poison Sumac, Poisoning Myths, National Childhood Lead Prevention Week, Poison Prevention Week, New Nationwide Telephone Number
- *Pesticides:* Understanding a Pesticide Label, Pesticide Regulations
- *Your Home:* Prevention Tips, Unintentional Poisoning of Children: The Statistics, Protect Children from Iron Poisoning, Senseless Poisoning from Carbon Monoxide
- *Bites and Stings:* bees and wasps, spiders, caterpillars, snakes
- Lead Poisoning
- Poisoning and Pets
- National Poison Prevention Week

Foundation publications page and National SAFE KIDS Teachers' Desk

- Media links to local television stations and newspapers
- Link to websites of the American Association of Poison Control Centers, National Pesticide Information Center, Food and Drug Administration, National Lead Information Center, National Animal Poison Control Center, American Veterinary Medical Association, West Virginia Animal Shelter, Cornell University—Poisonous Plants Webpages, Dr. C. Everett Koop, Healthtouch Online, KidsHealth.org, National Safety Council, National Fire Protection Association, Partners in Health Network, West Virginia Medical Institute, West Virginia State Fire Marshal, RxFactStat, Marshall University's West Virginia Websites Index, West Virginia' Public Employees Insurance Agency "Pathways to Wellness" Program, State of West Virginia, University of Charleston, West Liberty State College, West Virginia Department of Health and Human Resources, West Virginia Kids Count Fund, West Virginia University, West Virginia Division of Natural Resources, Agency for Toxic Substances and Disease Registry, American Red Cross, Chemical Emergency Preparedness and Prevention Office, National Safety Council: Guides to Chemical Risk Management, Kanawha Putnam Emergency Planning Committee, South Charleston West Virginia Community Advisory Panel

Continued

TABLE 8-A-1 Continued

Name	Accessible on Web Page	Website Links to Other Organizations
Western New York Poison Center (Buffalo)	• What is a Poison? • What Substances Are Most Commonly Involved in Poisoning? • How Can Poisonings be Prevented? • What is carbon monoxide poisoning? • Information About Lead Poisoning • Information for Childcare Providers, Teachers, and Educators—presentations can be requested and teaching packets with guidelines, statistical data, examples for displays, activity sheets, and handouts for parents are available on request as well; videos that teach children about household dangers are available on loan	• Website: http://www.chob.edu/poison

9

A Public Health System for Poison Prevention and Control

The mission of public health is to "fulfill society's interest in assuring conditions in which people can be healthy" (Institute of Medicine, 1988, p. 17). Public health entities at the federal, state, and local levels of government are in place to assist with prevention of disease and promotion of health. The recent Committee on Assuring the Health of the Public in the 21st Century (Institute of Medicine, 2002a) focused attention on the collaborative efforts among potential system partners (e.g., private health care, academia, business) needed to achieve the vision of "healthy people in healthy communities." A strong public health system needs to be in place to support the goal of a consistent, comprehensive, and community-based Poison Prevention and Control System.

An approach to addressing the health care needs of the population is to set goals and objectives for the nation. One set of goals for the Poison Prevention and Control System is from *Healthy People 2010*, which set two target objectives for the field of poisoning (http://www.healthypeople.gov):

> Objective 15-7: Reduce nonfatal poisonings to no more than 292 per 100,000 population (based on emergency department visits) (baseline was 349 in 1997)
> (http://www.healthypeople.gov/document/html/objectives/15-07.htm)
>
> Objective 15-8: Reduce deaths caused by poisonings to 1.5 per 100,000 population (baseline was 6.8 per 100,000 in 1997)
> (http://www.healthypeople.gov/document/html/objectives/15-08.htm)

As noted in Chapter 3, these national objectives are very ambitious and may even be based on poor estimates of true incidence. Clearly, the United States has a longer way to go in reaching its 2010 objectives than originally anticipated. The Committee's estimate of 8.5 fatal poisonings per 100,000 population (Chapter 3) is far above the national 2010 objective of 1.5, and even higher than the 1997 estimate of 6.8 used as a baseline. Furthermore, our estimate of nonfatal poisonings of 530 per 100,000 population in 2001 (Chapter 3) is nearly twice the national 2010 objective of 292 and again even higher than the 1997 baseline estimate of 349 nonfatal poisonings per 100,000. These findings suggest that national efforts to reduce poisonings and fatalities must be more strongly linked to the nation's overall agenda for health promotion and disease prevention. In this chapter we develop the argument for how incorporating the Poison Prevention and Control System into the broader public health system will accomplish this health improvement.

CORE PUBLIC HEALTH FUNCTIONS

Since the publication of the 1988 Institute of Medicine (IOM) report, *The Future of Public Health*, the public health system has addressed the charge of "disarray" in the field by focusing scientific and technical knowledge into three core functions needed to improve the health of the public: assessment (e.g., monitoring health status of a population, surveillance to detect disease outbreaks), policy development (e.g., development of partners, implementation of legislation), and assurance (e.g., education of the public and providers, standards and regulations to promote quality services, provision of direct health care services). These three core functions are helpful in describing the components of an integrated Poison Prevention and Control System at the federal, state, and local levels in the United States as they relate to the public health system.

Assessment

Data collection and analysis on a populationwide basis serves to monitor health status in order to identify and plan solutions for community health problems; characterize and investigate health problems and health hazards in the community; evaluate effectiveness, accessibility, and quality of personal and population-based health services; and carry out research for new insights and innovative solutions to health problems and conditions. According to the Centers for Disease Control and Prevention's (CDC's) *Updated Guidelines for Evaluating Public Health Surveillance Systems* (2001b, p. 2), "public health surveillance is the ongoing, systematic collection, analysis, interpretation, and dissemination of data regarding a health-related event for use in public health action to reduce morbidity and mortality and to improve health. Data disseminated by a public health surveillance system can be used for immediate public health action, program planning and evaluation, and formulating research hypotheses." The importance of poisoning data tracking and surveillance is discussed in detail in Chapter 7.

Policy Development

The development and implementation of policies and plans that support individual and community health efforts are essential components of public health practice. Sound health policy development requires a combination of scientific guidance and analyses of existing policies, resources, research, and evaluation. Policy development and implementation may be expressed as legislation, regulation, executive orders, or policy. Processes for improving health in the community, including the use of performance measures to track progress on solving the health problem, are outlined in the IOM report, *Improving Health in the Community* (Durch et al., 1997).

Policy development is an important component of assuring a comprehensive Poison Prevention and Control System in every community in the United States. Data quantifying the incidence and prevalence of poisonings, along with evaluation and research findings about prevention strategies, are useful at all levels of government public health to implement policies to protect the public. Development of evidence-based policies best occurs through an informed process that includes input from a broad-based spectrum of disciplines, professional backgrounds, interest groups, community stakeholders, consumers, and others. A commitment by state and federal public health agencies to provide resources to assure a comprehensive poison control system of equal quality and accessibility in every jurisdiction is an example of a policy designed to decrease poisonings and improve outcomes of those poisoned.

Assurance

It is important to assure the public that services necessary to achieve the best health outcomes and quality of life are provided, either by encouraging action by other private or public entities, by requiring such action through regulation or legislation, or by public health agencies providing the services directly either with staff or through contracts with providers (Institute of Medicine, 1988).

Strategies related to assurance within a poisoning prevention and control system may include (1) education of the public about poisonings, including how to prevent them as well as what to do if one occurs; (2) education of health providers, including first responders, about poisonings; (3) training of health professionals with expertise in toxicology; (4) development and implementation of clinical standards and protocols for responding to individuals who are poisoned; (5) implementation of standards for poison control centers to assure consistent seamless coverage and responses in all communities; and (6) provision of resources for all components of the Poison Prevention and Control System. It is important that data systems be established to track the delivery of services and implementation of standards and policies so that quality improvement mechanisms are in place at all levels of performance within the system.

Using the model outlined in the IOM report (Durch et al., 1997) for improving health in the community, a set of performance indicators for poison prevention and control might be developed and implemented in every community in the country. This "shared accountability model" for the health of the community would designate who, either alone or together with another stakeholder (e.g., public health department, poison control center, law enforcement, hospital, provider), is responsible for various outcomes within the system. For example, state health departments have an infrastructure and experience with primary prevention and health education activities across the lifespan; they also have one or more individuals with training in health education and community prevention. These activities are funded with federal dollars through programs such as the Maternal and Child Health Block Grant, Substance Abuse Block Grant, and Women, Infant, and Children's Nutrition Service. Thus, the state is a good candidate for taking the lead in prevention and health education regarding poisonings.

Essential Services for Public Health Related to Poisoning Prevention and Control

The three core public health functions previously mentioned have been expanded into 10 essential services by the major stakeholder groups in public health. The document, *The Essential Services of Public Health* (http://www.apha.org/ppp/science/10ES.htm), outlines the practices needed to discharge these obligations for the entire population in a community. These roles are generally much broader than the provision of direct clinical services and they are used frequently today to assess the capability of local and state public health agencies. For example, they are currently used to assess local and state health capacity to be prepared to respond to "all disasters," including bioterrorism and chemical terrorism. Table 9-1 shows the relationship between the essential services and the core public health functions and provides a brief description of each service as it relates to poison prevention and control.

To best achieve the goals and function as a system, federal, state, and local agencies ideally would perform in a unified fashion so that all of the essential services related to poison prevention and control are performed or assured for every community across the nation. Sharing and linking of relevant data addressing poisoning exposures, therapeutic interventions, and outcomes among providers, institutions, and poison control centers will be essential to advance these goals. The remainder of this chapter will review what capacity and activities currently exist within federal, state, and local agencies related to the system functions needed to meet the *Healthy People 2010* goals for poisoning.

POISON CONTROL SERVICES IN THE CURRENT PUBLIC HEALTH INFRASTRUCTURE

Turnock has described public health infrastructure as the "nerve center of public health" (Centers for Disease Control and Prevention, 1998). *Healthy People 2010* includes objectives for the nation regarding public health infrastructure in several areas: data and information systems, skilled workforce, effective public health organizations, resources, and prevention research. Because of the importance of public health's role in national security and preparedness for all hazards, there is increased interest and attention on building an effective and sustainable public health infrastructure (Centers for Disease Control and Prevention, 2001c).

At present, the accountability for the establishment and maintenance of a population-based Poison Prevention and Control System is diffuse at all levels of governmental public health. Although there are a variety of interested programs and components of public health agencies involved in various aspects of poison prevention and control, some of which interact with the poison control centers, there is insufficient clarity in the roles of each entity in the maintenance of a system across any level of geography. For example, at a local and/or state public health level, it is not unusual for the following programs within a health department to have an interface with the poison control centers and larger Poison Prevention and Control System: emergency medical services, the injury prevention and control program, maternal and child health program, occupational health program, substance abuse program, health statistics, epidemiology, environmental health, emergency preparedness, and others. In addition, there are activities related to poisonings in agencies outside of the usual public health authority; these include the Environmental Protection Agency, the medical examiner's office, the mental health agency, and the Board of Pharmacy.

Each of these programs at the local and/or state level(s) has an associated link(s) to a federal agency that mirrors the state. The complexity

TABLE 9-1 Core Functions and Essential Services of Public Health as Applied to Poison Prevention and Control Services

Core Functions	10 Essential Services
Assessment Collection, assembly, analysis, and distribution of information on the community's health	1. Monitor health status to identify community problems. 2. Diagnose and investigate health problems and the health hazards in the community. 3. Evaluate the effectiveness, accessibility, and quality of personal and population-based health services.
Policy development Development of comprehensive policies based on scientific knowledge and decision making	4. Inform, educate, and empower people about health issues. 5. Mobilize community partnerships to identify and solve health problems. 6. Develop policies and plans that support individual and community health efforts.
Assurance Determination of needed personal and communitywide health services, and provision of these services by encouraging action by others, by requiring action by others, or by direct provision	7. Assure a competent public health and personal health care workforce. 8. Enforce laws and regulations that protect health and ensure safety. 9. Link people to needed personal health services and assure the provision of health care when otherwise unavailable.
Assessment, policy development, and assurance	10. Research for new insights and innovative solutions to health problems

SOURCE: Adapted from the IOM report, *The Future of Public Health* (1988).

and variety of agencies involved with activities related to poisoning is greater at the federal level. Currently there is no single state or federal plan or authority (e.g., legislative, regulatory) that gives one entity accountability for all poison prevention and control activities.

Local and State Health Department Involvement with Poison Prevention and Control Activities

There is no single point of accountability for poison prevention and control activities at most local or state health departments. The few tar-

Examples as Applied to Poison Prevention and Control Services
1. Monitor population frequency of poisonings across the lifespan. Assess outcomes. 2. Assess factors contributing to poisonings. Develop policies and services for primary and secondary prevention. 3. Evaluate public education activities related to poisonings. Continuously review and evaluate poison control center functions and their efficiency and effectiveness. Ensure the availability and accessibility of poison control information to the entire public.
4. Assess and enhance the public's knowledge about poison impact, prevention, and control. 5. Establish effective communication with community members regarding poisonings. 6. Apply population-based data to policy development for poison prevention and control.
7. Create and maintain a workforce that is competent in poison prevention and control. Educate health professionals on subjects related to poisonings. 8. Develop laws, statutes, and regulations that provide for optimal use of poison control centers and protect individuals in the workplace. 9. Create provisions for high-quality, culturally competent poison control center services. Ensure linkages among all parts of the public health and medical systems with poison control centers.
10. Identify best practices for poison control centers. Contribute to the evidence base for poison prevention and control through the funding and generation of new knowledge.

geted poison prevention activities that exist are located in the injury prevention and control program and/or the maternal and child health program. Data and surveillance are usually located in the health statistics and information unit with links to the state epidemiologist, the medical examiner, and the agency that collects hospital discharge data. Other programs within public health departments that may be involved with poison prevention and control activities are emergency medical services, maternal and child health, health education and promotion, emergency preparedness, substance abuse services, environmental health, occupational health, and the medical director. The following is a more in-depth discussion of specific roles of several state health department programs in relation to poison prevention and control, including links to the state poison control center(s).

Maternal and Child Health Linkages

To understand the current linkages between poison control centers and various programs within state health departments, an informal survey of activities or linkages with the state Title V agencies was conducted in 2003 by the Association of Maternal and Child Health Programs. Twenty-nine states and territories responded. Eleven additional states included information on poison prevention and control in their applications for 2003 or 2004 funding. All but 4 of the 40 indicated some degree of involvement with poison control centers, such as participation in advisory groups and partnering on health education and data reviews.

Since the early 1990s, states have provided the federal Maternal and Child Health Bureau (MCHB) with up to 10 state-selected performance measures, often related to *Healthy People 2000* or *Healthy People 2010*, established every 5 years as part of a mandated needs assessment and updated or revised as part of their annual plans. None of the reporting states and territories described a poison-specific performance measure, although many noted that poisonings were included in other measures such as hospitalizations or child deaths. Several noted that home visiting projects for new parents included information on poisoning prevention in early childhood.

Based on information from questionnaires and annual applications, three-fourths (30 of 40) of the states have working relationships between Title V and poison control centers. Examples include Healthy Child Care America initiatives, serving on injury prevention advisory groups such as SAFE KIDS, special projects including mercury thermometer removal from home or schools, and lead poisoning prevention. A few states reported specific agreements to share data or data analyses, participate on each other's advisory committees, collaborate on trainings, and so forth.

Nearly half (19 of 40) reported funding information and 6 more specified there was no health department or state allocation to poison control centers. However, nearly a third (15 of 40) skipped the item or reported they lacked information regarding such allocations. Although specific detail was lacking from 7 of the 19 reporting states, the Title V block grant was noted to support programs in 3, bioterrorism programs in 4, and state appropriations in 8; 3 states reported that more than one of these sources supported poison control centers.

Injury Prevention and Control Linkages

The State and Territorial Injury Prevention Directors Association sent a similar survey to its members in 2003 to better understand (1) the activities related to poison prevention and control that occur within their departments and (2) the relationship of the injury prevention and control programs in state public health departments to the poison control centers covering their state. Unlike the maternal and child health program, which receives a block grant for activities, there is no regular state or federal funding stream for injury prevention and control activities in states. Because of its importance in public health, every state does have a named director for these activities, regardless of the size of the program.

Of the 12 states that responded, all reported that the state's injury prevention and control activities include issues of poisoning for children, youth, and families. Little attention is directed to older adults and senior citizens. Eleven programs include education about poisonings in the materials, resources, and activities of the other injury and prevention activities in their program; two described a special focus on poisoning, such as participation in Poison Prevention Week. Two states (Georgia and Oregon) mentioned a strong link of the state program to the SAFE KIDS Coalition.

Eleven of the 12 reporting states have done analyses and/or published reports on state and local data related to poisonings. Of those with reports, most were published on an annual or biannual basis. All reports included data from deaths and hospital discharge; a few included emergency department and poison control center data. Nine stated they used the data from the poison control center, usually in the format published in an annual report. Although data from standard reports were accessible, several respondents noted that it was difficult to obtain data for specific requests and purposes from their poison control center.

One example of collaboration in sharing data for public health planning and follow-up is that between the poison control centers in Michigan and the Michigan Department of Community Health. Michigan's epidemiology unit works closely with the local poison control centers to assure reporting of occupational disease under the public health code. In addition, the Michigan poison control centers provide almost "real time" (electronically) all occupational pesticide exposure calls to the Michigan pesticide surveillance system and identify data daily for syndromic surveillance for the Emergency Preparedness Program.

All these states reported that they had an ongoing relationship/partnership with the poison control centers in their state. Seven of the 12 states reported they served on the advisory committee for the center and 6 said they participated in using data on poisonings. Five said they participated in press releases with the poison control center; for example, Massachusetts participated with the poison control center in a major event on inhalant use and young teens. Five of the states (Colorado, Indiana, Massachusetts, Michigan, and New Jersey) had a contractual relationship with the poison control centers and managed the use of state/federal funds for poison control center services. For example, in Colorado the state general funds specified for the poison control center are managed by the Emergency Medical/Trauma Services Unit. Massachusetts uses Title V Maternal and Child Health Block Grant Funds to jointly purchase services with Rhode Island; the contract is managed by the Injury Prevention and Control Division.

States responding to the Injury Prevention and Control questionnaire included four that had not provided information through the Maternal and Child Health (MCH) questionnaire or in their state MCH block grant materials. Nearly all show at least some degree of involvement with their state or regional poison control center (40 of 44); three-fourths (33 of 44) have established working relationships between their centers and state public health. Although the degree of health department involvement with the state or regional poison control center is common, the form and extent of involvement is variable and commonly linked to the extent of center financial support.

Emergency Preparedness and Response

Poison control centers can play an important role in all-hazards preparedness and response. They are considered to be a vital part of the continuum of necessary emergency services needed for all Americans facing the threat of bioterrorism and can serve as part of the nation's surveillance and first response system (http://www.hhs.gov/budget/hrsa_bioterror.html [statement by Duke]). The challenges of preparing for and responding to an act of terrorism are significant, and may include identifying, responding to, and recovering from acts of terrorism and other public health emergencies (Levy and Sidel, 2003). Cooperation among public health officials, emergency managers, first responders, and health care providers is a top priority, and the role each plays in responding to an emergency is essential to effectively respond to a crisis, minimize loss of life, and control the spread of disease and chaos. In the face of new and emerging threats, more effort is needed to strengthen the health care system's ability to detect and respond to such public health emergencies. The full spectrum of health care providers, including poison control centers, have potential roles in preparing for and responding to the possibility of bioterrorism or other public health emergencies.[1]

---

[1]In this context, the terms bioterrorism and other public health emergencies includes other forms of terrorism, such as use of chemical agents against the public.

Present-day planning for possible acts of terrorism are strengthening the public health infrastructure and, to some degree, improving poison control centers' ability to respond to natural disasters, chemical releases, disease outbreaks, and other public health emergencies. Based on funding priorities of the federal government since September 2001, regional and perhaps national emergency planning and response promise to become an important component of poison control center services, although there is a need to also consider the implications for long-term support for the centers to provide this service.

An Institute of Medicine and National Research Council (1999) report on chemical and biological terrorism indicated that in most plausible chemical terrorism scenarios, the rapid onset of toxic effects would lead to highly localized collections of victims within minutes or hours:

> A network of regional poison control centers is well established, however, and, if its personnel were educated about military chemical weapons, would be well suited for surveillance. Poison control centers are also obvious candidates to serve as regional data and resource coordinating centers in incidents involving multiple sites or large numbers of patients (p. 7).

This report also cited "a glaring need" to strengthen disease surveillance to prepare for the threat of bioterrorism and address emerging pathogens (p. 74). Better preparation of the nation's clinicians for the roles they will play in responding to a bioterrorist attack is also necessary.

The 2001 Presidential Task Force on Citizen Preparedness in the War on Terrorism recommended that poison control centers be used as a source of public information and public education regarding potential biological, chemical, and nuclear domestic terrorism (Pub. L. No. 106–174).

An Agency for Healthcare Research and Quality Evidence Review (http://www.ahrq.gov/clinic/tp/biotrp.htm) indicated that the lack of strong evidence on how to train clinicians for public health events represents an important gap in bioterrorism preparedness. The information that public health officials need for preparation for and response to a bioterrorism event can be considered in relation to the decisions they must make: the interpretation of surveillance data; the investigation of outbreaks; the institution of epidemiologic control measures; and the issuance of surveillance alerts. Communication decisions relate to the specific information that needs to be conveyed to other public health officials, clinicians, the media, and other decision makers.

The Health Resources and Services Administration (HRSA) (2004) has indicated that poison control centers can serve as part of the nation's surveillance and first response system. For example, during the anthrax incident that took place in Florida, the Florida Poison Control Center was able to provide the public with information about anthrax. Many infectious disease agents are often difficult to identify initially because the signs may be nonspecific (Ashford et al., 2003). Illnesses may be scattered geographically and occur in a number of different jurisdictions at once, depending on the source and mechanism of the initial infection. Mounting an effective, timely, and coordinated response requires health information and the involvement of a variety of health professionals, including poison control centers. In addition to preparedness for and response to biological, chemical, and nuclear threats or exposures, daily management of hazardous materials incidents and chemical contamination in coordination with public safety services has become an important activity of poison control centers (Burgess et al., 1999; Kirk et al., 1994).

State and local plans for public health and hospital emergency preparedness are beginning to acknowledge roles for poison control centers. These programs, funded by cooperative agreement grants from CDC and HRSA, help the public health system and hospitals prepare for acts of bioterrorism, outbreaks of infectious disease, and other public health threats and emergencies. The HRSA-funded program is known as the National Bioterrorism Hospital Preparedness Program (NBHPP). Its guidance to applicants specifies that poison control centers be involved in statewide bioterrorism preparedness and response planning (U.S. Department of Health and Human Services, 2003).

To better understand the involvement of poison control centers in state and local plans for public health preparedness and response, the Committee requested HRSA to conduct a review of the year 2003 applications to the NBHPP. The program's 2003 awardee applications were reviewed to learn the prevalence of mentions of poison control centers and actual statements mentioned in the plans by state and local jurisdictions. Of the applications from states, 92 percent (or all but four states) mentioned poison control centers in their applications. Additionally, all four applications from cities and half the applications from other funded agencies (e.g., U.S. territories) mentioned poison control centers. Although an approximation for the extent of involvement of poison control centers in the NBHPP, the number of mentions within the plans was also analyzed. Table 9-2 shows a considerable variation in the number of mentions of poison control centers across states and local jurisdictions.

A qualitative analysis of roles for poison control centers mentioned in the NBHPP proposals to HRSA identified the following:

- Serving as members of advisory committees;
- Reporting syndromic and diagnostic data suggestive of terrorism on a 24/7 basis;
- Dealing with chemical hazards;

TABLE 9-2 Poison Control Centers Mentions in Awardee Applications—Summary

Number of Mentions	Cities[a]	States	Other Funded Entities	Total
1–2	2	7	2	11
3–5	1	15	1	17
6–9	1	11	0	12
10+	0	13	1	14
Poison control center not mentioned	0	4	4	8
TOTALS	4	50	8	62

[a]Includes Washington, DC.

- Assisting with regional surge capacity;
- Conducting other forms of surveillance;
- Functioning as a call center or serving as a central clearinghouse for information on toxicology, antidotes and treatments, and decontamination procedures;
- Helping with plans to receive and distribute the Strategic National Stockpile;
- Allowing the poison control centers to access relevant data and communications in secure data exchanges with public health departments; and
- Providing consultation on detection and treatment of biological and chemical terrorism.

It is evident from this review of plans that some public health departments are better integrated with poison control centers than others in support of bioterrorism preparedness and improvement of capacity to effectively respond to other public health critical incidents.

Others have mentioned the possible role of centers in support of CDC's ChemPack program, in view of its characteristics, including 24-hour-a-day/7-day-a-week coverage and specially trained staff. At the national level, the American Association of Poison Control Centers' central office has obtained federal funding to analyze Toxic Exposure Surveillance System (TESS) data on a real-time basis as a surveillance tool and to enhance public health reporting. The incremental cost of sustained improvements to TESS to each of the poison control centers is unknown and has been absorbed by the individual centers.

Poison control centers also serve as a resource to their local emergency medical services (EMS) systems. Emergency department physicians, nurses, paramedics, and emergency medical technicians often call poison control centers when they have questions about treatment or prevention. If hands-on medical treatment is necessary, the centers call an ambulance, stay on the line until the ambulance arrives, and give treatment advice to the emergency care providers, as appropriate. Emergency medical services agencies typically are regulated by state departments of health. It is important for poison control centers to communicate with the EMS system regarding protocols for management of relevant exposures to avoid the potential for conflicting information and to contribute to developing systems that assure accessible and timely treatment for victims of poisonings.

Finally, poison control centers should develop cooperative arrangements with community and institutional pharmacists. These individuals are in a position to recognize and report symptoms of exposure to biological or chemical agents because they are often the first health care providers contacted by patients, particularly when persons seek advice on over-the-counter treatments for flu-like illnesses. They are well positioned to detect emerging or unusual patterns of disease and surges in sales of medications that might suggest an attack (Edge et al., 2002; MacKenzie et al., 1995). Should emergencies arise, whether in an urban or rural area, there is usually a pharmacy within 5 miles of nearly any household to serve as a point of access (The National Conference on Pharmaceutical Organizations, 2002).

Federal Agencies Involved with Poison Prevention and Control

There is no single point of accountability for poison prevention and control activities within the federal public health system. There are currently eight departments of the Cabinet (U.S. Department of Health and Human Services [DHHS], U.S. Department of Housing and Urban Development, U.S. Department of Agriculture, U.S. Department of Homeland Security, U.S. Department of Labor, U.S. Department of Justice, U.S. Department of Transportation, and U.S. Environmental Protection Agency), as well as the U.S. Consumer Product Safety Commission and the U.S. Office of National Drug Control Policy, involved with poison prevention and control activities.

The primary leadership for the public health system at the federal level resides in the DHHS. The three major agencies within DHHS are the CDC, HRSA, and Substance Abuse and Mental Health Services Administration (SAMHSA). Currently, federal money dedicated specifically for poison control activities is administered through the Maternal and Child Health Bureau in HRSA and the Center for Injury Prevention in CDC. Other parts of DHHS involved in poison prevention and control include the Agency for Toxic Substances and Disease Registry (CDC), Food and Drug Administration, Centers for Medicare and Medicaid Services, Agency for Healthcare Research and Quality, and National Institutes of Health (see Figure 9-1 for DHHS organizational structure). Federal departments, agencies, and commissions involved in some aspect of poisoning prevention and control are described in the appendix to this chapter.

In addition to dedicated funds for poison control centers and activities, there are several other major funding sources that flow from the federal level to states that can and are used to support poison prevention and control activities at state and local levels. These are the Title V Maternal and Child Health Block Grant, the Substance Abuse Block Grant, the Preventive Health Services Block Grant, Medicaid, and the State Children's Health Insurance Program. State data collection activities for vital records are supported via cooperative agreements with the Center for Vital Statistics in CDC in addition to the block grants listed. Resources for the Poison Prevention and Control System, as well as the poison control

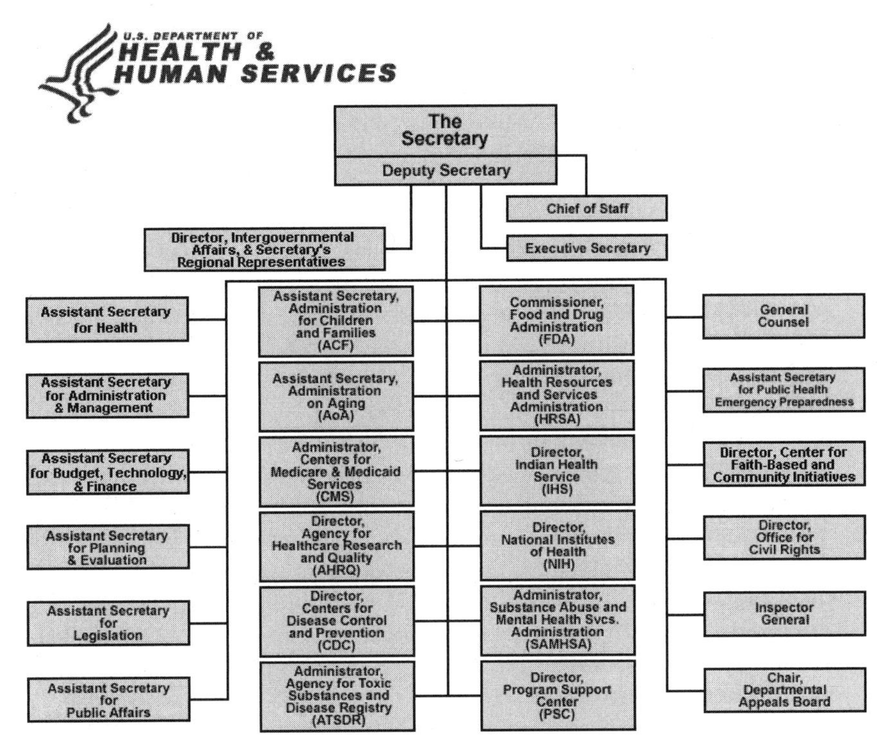

FIGURE 9-1 Department of Health and Human Services organizational chart.
SOURCE: http://www.dhhs.gov/about/orgchart.html.

centers, have become available recently through state cooperative agreements for emergency preparedness and response from CDC and HRSA.

Research investigations and surveillance activities are supported through many of the DHHS agencies as well as other agencies and departments outside of DHHS. The primary agencies are CDC, the National Institutes of Health, and the Environmental Protection Agency. In addition, regulatory, policy, and planning activities related to poisoning prevention and control occur in a number of federal agencies. Table 9-3 gives an overview of the types of poisonings addressed (e.g., alternative therapies, pesticides, occupational exposures, terrorism, drugs of abuse) by the various federal agencies.

SUMMARY AND CONCLUSIONS

Although a variety of agencies at the federal, state, and local levels have responsibility for one or more components of the Poison Prevention and Control System, there is currently no uniform, clear point of accountability at any level of government. There are uneven linkages and collaboration among the various agencies responsible for data collection and analysis, research and evaluation, policy and regulatory development, health education and other prevention activities, and clinical services and quality standards, as well as financing and payment for services across the agencies involved. Two federal agencies (CDC and HRSA) have funds earmarked for poison control centers, but these funds have not been directed to and are not sufficient to support the core activities for the proposed Poison Prevention and Control System. Furthermore, no federal agency has research funds specifically allocated for poisoning; this lack of support makes it difficult to develop a comprehensive picture of the epidemiology of poisonings or to understand the best way to deliver poison prevention and control services.

To achieve the ultimate goal of preventing poisonings, as well as to improve the outcomes for those who are poisoned, the Committee envisions the need for a clear, single point of accountability at each level of government. The responsible agencies would ensure the accomplishment of all of the core functions or essential services as they relate to poison prevention and control (Table 9-1). This does not mean that the responsible agencies would perform all of the functions within their agency. However, they would (1) take responsibility for the plan for accomplishing the activities needed to ensure the system is in place with a set of uniform standards across the country, and (2) convene and work with the other agencies, including the existing poison control center network, to implement the plan. Furthermore, the responsible agencies at the state and federal levels should work in partnership to develop a set of perfor-

TABLE 9-3 Guide to Federal Regulatory, Policy, and Planning Authorities That Focus on Aspects of Poisoning

Issue	Examples of Federal Agencies That Focus on Issue
Adult and pediatric poisoning and drug overdose	• Centers for Disease Control and Prevention (National Center for Injury Prevention) (DHHS) • Centers for Medicare and Medicaid Services (DHHS) • Substance Abuse and Mental Health Services Administration (DHHS) • Health Resources and Services Administration (DHHS)
Alternative therapies	• Food and Drug Administration (DHHS)
Biologicals and infectious disease	• Centers for Disease Control and Prevention (DHHS)
Consumer and household products	• Consumer Product Safety Commission
Drugs of abuse	• National Institute on Drug Abuse (NIH, DHHS) • Office of National Drug Control Policy • Substance Abuse and Mental Health Services Administration (DHHS)
Environmental releases, natural events, health effects	• Environmental Protection Agency • Agency for Toxic Substances and Disease Registry
Occupational exposures	• Occupational Safety and Health Administration (DOL) • Mine Safety and Health Administration (DOL)
Pesticides	• Environmental Protection Agency • Occupational Safety and Health Administration (DOL)
Prescription drug and over-the-counter side effects	• Food and Drug Administration (DHHS)
Terrorism	• Centers for Disease Control and Prevention (DHHS) • Department of Homeland Security
Veterinary	• Centers for Disease Control and Prevention (DHHS) • Department of Agriculture

mance standards for all components of the system. One possible model for the development of performance measures for a state-federal partnership is the Title V Maternal and Child Health Block Grant, which is administered by states, and the federal grants for MCH activities administered by the Maternal and Child Health Bureau in HRSA. The Secretary of the Department of Health and Human Services should designate a lead agency for poison prevention and control at the federal level and the governor of a state should designate a lead agency within the state. The most likely appointed lead within a state government is the health or public health entity.

For the Poison Prevention and Control System to be implemented and continuously improved in the most effective manner, resources are needed to carry out the mandate. Given the numerous priorities for scarce resources within public health agencies, it is most likely that the functions related to poison prevention and control will not be accomplished without the appropriate resources. Public health initiatives with a clear mandate and resources made available for both the state and federal activities are the most successful; examples are the Maternal and Child Health Block Grant that provides funds to all states with clear performance measures established. Another example is the funds from CDC to states for immunization and communicable disease reporting.

In addition to the funds required by each poison control center to implement the core activities, the Committee estimates that at least $30 million would be needed for activities to assure that all essential services of public health related to poisoning could be accomplished. This estimate includes $10 million for state-level activities and $20 million for federal-level activities. Approximately $200,000 would be allocated to each state for primary prevention activities and for a state poison prevention and control system coordinator whose responsibilities would include coordination of public education efforts and a plan for their evaluation. The grant would be given to the lead agency appointed by the governor. This estimate is based on the level of support needed by the states to coordinate and administer other activities. Title V supplemental funding provides a model for the allocation of small grants to each state's Maternal and Child Health program. Our proposal is modeled after the process used by MCHB to provide every state's Title V agency with supplemental resources ($100,000) to develop an early childhood system of care. The set of performance measures for the state grants should be determined by a federal-state partnership process and complement the performance measures for poison control centers in each state.

Rough estimates of the funds required for federal-level activities include $3 million for the development and maintenance of quality assurance and improvement mechanisms for every component of the Poison Prevention and Control System, including assessment of clinical practice; $3 million for training activities for health providers outside the poison control center who require training in toxicology, including emergency department workers such as nurses, physicians, and emergency medical technicians; and $4 million for a clearinghouse for primary prevention materials and resources (including media campaigns, material development, and dissemination, as discussed in Chapter 8). These estimates are based on similar activities funded for other content areas of public health by CDC, HRSA, and SAMHSA. Finally, a minimum of $10 million is needed for research that would cover a broad range of topics, including basic science, epidemiology, population-based studies, clinical trials, health services research, primary prevention, and program evaluation. A focus of federal-level activities should be on translation of research and evaluation studies into best practices and regulatory changes. More emphasis on the translation of findings into population-based strategies that decrease poisonings is key; one example is federal regulations (see Chapter 8). The largest reductions in unintentional poisonings to date are attributable to the safety caps on medications. Research should be both field initiated and program specific so that the gaps in science related to the various aspects of the Poison Prevention and Control System are filled.

In sum, the funds needed by state and federal agencies to assure a Poison Prevention and Control System are in addition to the $100 million estimated to be needed by the poison control centers. Resources for both centers and the federal/state infrastructure will be required to build and maintain the comprehensive system needed to ensure that poison prevention and control activities will be present in every community in the country.

Appendix 9-A

Federal Agencies Involved with Poison Prevention and Control

Descriptions of the federal departments and their agencies that have activities related to poison prevention and control are presented in this appendix. The activities of the U.S. Department of Health and Human Services and its many agencies (Centers for Disease Control and Prevention, Agency for Toxic Substances and Disease Registry, Health Resources and Services Administration, Substance Abuse and Mental Health Services Administration, Food and Drug Administration, Centers for Medicare and Medicaid Services, Agency for Healthcare Research and Quality, National Institutes of Health) are described first. Information about activities of the U.S. Department of Housing and Urban Development, U.S. Department of Agriculture, U.S. Department of Homeland Security, U.S. Department of Labor, U.S. Department of Justice, U.S. Consumer Product Safety Commission, U.S. Environmental Protection Agency, and Office of National Drug Control Policy follow.

U.S. DEPARTMENT OF HEALTH AND HUMAN SERVICES

The U.S. Department of Health and Human Services (DHHS) is the U.S. government's principal agency for protecting the health of all Americans. The Office of the Secretary provides department leadership. The department provides essential human services, especially for those who are unable to provide for themselves. DHHS is the largest grant-making agency in the federal government, offering about 60,000 per year. Within its 300 programs, those most involved with poisoning prevention and control are the Centers for Disease Control and Prevention, Health Services and Resources Administration, Substance Abuse and Mental Health Services Administration, Food and Drug Administration, Centers for Medicaid and Medicare Services, Agency for Healthcare Research and Quality, and the National Institutes of Health (http://www.hhs.gov).

Centers for Disease Control and Prevention

The Centers for Disease Control and Prevention (CDC) is recognized as the lead federal agency for protecting the health and safety of Americans at home and abroad. CDC serves as the national leader for developing and implementing disease prevention and control, environmental health, and health promotion and education activities designed to improve the health of the citizens of the United States. Working with national and world partners, CDC monitors general health, detects and investigates health problems, conducts research to enhance prevention, develops and advocates sound public health policies, implements prevention strategies, and provides leadership and training.

With respect to lead poisoning, CDC initiated the Childhood Lead Poisoning Prevention Program (CLPPP), which develops programs and policies to prevent childhood lead poisoning; educates the public and health care providers about childhood lead poisoning; provides funding to state and local health departments to screen children for elevated blood lead levels and to ensure follow-up; develops neighborhood-based efforts to prevent childhood lead poisoning; and supports research to determine the effectiveness of prevention efforts.

CDC has joined with the U.S. Department of Housing and Urban Development, the U.S. Environmental Protection Agency, and other agencies to develop a federal interagency strategy to identify and control lead paint hazards, identify and care for children with elevated blood lead levels, to survey elevated blood lead levels in children to monitor progress, and perform research to further improve childhood lead poisoning prevention methods (http://www.cdc.gov).

CDC also has 12 centers, institutes, and offices, some of which perform additional work in poison prevention and control. Information about those programs follows.

National Center for Environmental Health

CDC's National Center for Environmental Health (NCEH) provides national leadership to promote health and quality of life by preventing or controlling diseases, birth defects, disabilities, or deaths that result from interactions between individuals and their environment. NCEH conducts research in the laboratory and the field to investigate the effects of the environment on health. The center also helps domestic and international agencies and organizations prepare for and respond to natural, technological, humanitarian, and terrorism-related environmental emergencies.

The center's Emergency and Environmental Health Services (EEHS) program provides national and international leadership for coordinating, delivering, and evaluating emergency and environmental public health services. To improve public health practices, EEHS offers consultation, technical assistance, and training to state and local health departments and to federal and international agencies on environmentally related health issues. The program also responds to national and international emergencies and provides support during environmental threats.

EEHS works with other federal agencies for an integrated national approach in preventing childhood lead poisoning. The program assists in the development and evaluation of state and community childhood lead poisoning prevention programs, maintains a system for collecting and sharing data on lead poisoning, and conducts and evaluates scientific research on childhood lead poisoning.

The EEHS Emergency Preparedness and Response Branch (EPRB) coordinates CDC's activities in helping state and local health departments assure public health readiness in their emergency preparedness and response efforts. EPRB offers scientific public health guidance for emergency preparedness operations and identifies and shares best practices from academic training and field operations for all-hazards preparedness and response.

The EEHS Chemical Demilitarization Branch (CDB) ensures that the health and safety of workers and the general population are protected during the handling and destruction of the nation's chemical weapons. CDB reviews all chemical weapons elimination plans, works closely with the U.S. Department of Defense throughout their disposal process, and evaluates the capacity of the local communities to medically respond to any related emergencies.

Another NCEH program is the Division of Environmental Hazards and Health Effects (EHHE), which conducts surveillance and investigative studies to develop knowledge regarding ways to prevent or control health problems associated with exposure to air pollution, radiation, and other toxicants. EHHE also addresses natural, technological, or terrorist disasters.

The EHHE Health Studies Branch (HSB) investigates the health effects associated with exposure to environmental hazards and natural, technological, or terrorist disasters. HSB develops and evaluates strategies for preventing human exposure to such hazards and disasters, minimizing the effects of the exposure when it does occur.

The EEHE Environmental Health Tracking Branch collects, integrates, analyzes, and interprets data about environmental hazards, exposure to environmental hazards, and health effects potentially related to exposure to environmental hazards. This information is provided to federal, state, and local agencies, which can use this information to plan, implement, and evaluate public health actions to prevent and control environment-related diseases (http://www.cdc.gov/nceh).

National Center for Health Statistics

CDC's National Center for Health Statistics (NCHS) is the nation's principal health statistics agency. NCHS compiles statistical information to guide actions and policies to improve the health status of the population and important subgroups. Data are collected from birth and death records, medical records, and interview surveys, and through direct physical exams and laboratory testing. NCHS provides important surveillance information that helps identify and address critical health problems. The center has data available on injury and poisoning episodes and hospitalizations in the United States, as well as other useful reports (http://www.cdc.gov/nchs).

National Center for Injury Prevention and Control

CDC's National Center for Injury Prevention and Control (NCIPC) is the lead federal agency for injury prevention and for reducing injury, disability, death, and costs associated with injuries outside the workplace. NCIPC works closely with other federal agencies; national, state, and local organizations; state and local health departments; and research institutions to prevent and control injuries. The center uses scientific methods to prevent injuries, studying factors to decrease risk, designing and evaluating intervention strategies, and taking steps to ensure that proven strategies are implemented in communities nationwide. NCIPC also provides specific resources on poisoning, poison control, and poisoning prevention (http://www.cdc.gov/ncipc).

National Institute for Occupational Safety and Health

CDC's National Institute for Occupational Safety and Health (NIOSH) is the federal agency responsible for research and recommendations for preventing work-related injury and illness. NIOSH implements and maintains a system of surveillance for major workplace illnesses, injuries, exposures, and health and safety hazards. It promotes prevention activities through workplace evaluations, interventions, and recommendations, and provides workers, employers, the public, and the occupational safety and health community with information, training, and capacity to prevent occupational injuries and illnesses. NIOSH provides facts on topics such as chemical safety, lead, pesticide illness, injury surveillance, and "take-home toxins." In addition, NIOSH offers databases and information resources on chemical hazards and injury.

NIOSH conducts investigations of possible health hazards in the workplace through the Health Hazard Evaluation (HHE) Program. A typical HHE involves studying a workplace following a written request from employees, employee representatives, or employers to determine whether there is a health hazard caused by exposure to hazardous materials—

chemical or biological—in the workplace (http://www.cdc.gov/niosh/homepage.html).

Agency for Toxic Substances and Disease Registry

The Agency for Toxic Substances and Disease Registry (ATSDR) serves the public by providing health information to prevent harmful exposures and diseases related to toxic substances. ATSDR offers public health assessments of waste sites, health consultations regarding specific hazardous substances' health surveillance and registries, responses to emergency releases of hazardous substances, applied research, information development and dissemination, and education and training concerning hazardous substances.

ATSDR has a hazardous substance research and health effects database called HazDat. HazDat is the scientific and administrative database developed to provide access to information on the release of hazardous substances from Superfund sites or from emergency events and on the effects of hazardous substances on the health of human populations. HazDat contains information such as community health concerns, ATSDR public health threat categorizations, ATSDR recommendations, exposure routes, and physical hazards at the site/event. The agency is currently being integrated with CDC's National Center for Environmental Health (http://www.atsdr.cdc.gov).

Health Resources and Services Administration

The Health Resources and Services Administration assures the availability of quality health care to low-income, uninsured, isolated, vulnerable, and special needs populations. HRSA's goal is for Americans to have 100 percent access to health care without any disparities. HRSA has a consumer education program that provides health-related information for families to live healthier lives. This program includes contact information for poison control. Furthermore, HRSA has awarded funds to support the work of poison control centers as well as bioterrorism aid for states.

HRSA's Maternal and Child Health Bureau promotes and improves the health of mothers and children by working in partnership with states, communities, public–private partners, and families. MCHB administers seven programs, one of which is the Poison Control Centers Program (developed as a result of the Poison Control Center Enhancement and Awareness Act), jointly administered with CDC. MCHB and CDC have created a nationwide toll-free telephone number system, a nationwide media educational campaign, and a grant program to develop and improve infrastructure elements of the regional poison control centers (http://www.hrsa.gov).

Substance Abuse and Mental Health Services Administration

The Substance Abuse and Mental Health Services Administration (SAMHSA) is the nation's health care delivery system to provide substance abuse prevention, addiction treatment, and mental health services for people at risk for or experiencing substance abuse or mental illnesses. SAMHSA builds partnerships with states, communities, and private organizations to address the needs of individuals with substance abuse and mental illnesses and to identify and respond to the community risk factors that contribute to these illnesses.

SAMHSA's programs support the adoption and adaptation, as well as the evaluation, of evidence-based, high-quality diagnostic, treatment, and prevention service practices. Under its block grant program, SAMHSA encourages the states and territories to address state and local substance abuse and mental health needs by supporting implementation and maintenance of specific service programs and assesses and reports on progress, needs, and ongoing activities. SAMHSA's data collection and analysis activities—including the National Survey of Drug Use and Health and other data—gather, aggregate, assess, and report on trends related to mental health services addiction treatment and substance abuse prevention.

The agency houses three substance abuse and mental health service- and prevention-related centers—Center for Mental Health Services, Center for Substance Abuse Prevention, and Center for Substance Abuse Treatment. SAMHSA also includes the Office of Applied Studies, the data collection and analysis hub for SAMHSA, and several other staff offices (http://www.samhsa.gov).

Food and Drug Administration

The U.S. Food and Drug Administration (FDA) promotes and protects public health by helping safe and effective products reach the market in a timely manner and monitoring products for continued safety after they are in use. The FDA ensures that the nation's food is free of chemicals or other harmful substances. The FDA also monitors dietary supplements, medical products, and biologics, and protects the public from unnecessary exposure to radiation from electronic products. FDA research provides the scientific basis for its regulatory decisions, evaluates new products, develops test methods, and provides support for product monitoring. The FDA offers information on protecting children from poisons in their homes, including medicines, cleaning products, and houseplants.

Center for Biologics Evaluation and Research

The FDA's Center for Biologics Evaluation and Research (CBER) advances public health through innovative regulations that ensure the safety, effectiveness, and timely delivery to patients of biological and related products. CBER is also responsible for an adequate and safe supply of allergenic materials and antitoxins and for the safety and efficacy of biological therapeutics. CBER plays an important role in the President's Initiative on Countering Bioterrorism, including ensuring the expeditious development and licensing of products to diagnose, treat, or prevent outbreaks from exposure to the pathogens that have been identified as bioterrorist agents (http://www.fda.gov).

Centers for Medicare and Medicaid Services

The Centers for Medicare and Medicaid Services (CMS) ensures health care security for beneficiaries, improving quality and efficiency in an evolving health care system. CMS runs Medicare, the nation's largest health insurance program, which covers nearly 40 million Americans. Medicare provides care to people age 65 or older, some people with disabilities under age 65, and people with permanent kidney failure requiring dialysis or a transplant. CMS also runs Medicaid, a health insurance program for certain low-income people that is funded and administered through a state–federal partnership. There are broad federal requirements for Medicaid, but states have a wide degree of flexibility to design their program. CMS runs the State Children's Health Insurance Program, which became available on October 1, 1997, and helps states expand health care coverage to more than 5 million of the nation's uninsured children. CMS provides information on how to guard young children against poisons (http://www.cms.gov).

Agency for Healthcare Research and Quality

The Agency for Healthcare Research and Quality (AHRQ) is the health services research arm of DHHS, complementing the biomedical research mission of its sister agency, the National Institutes of Health. AHRQ specializes in research on quality improvement and patient safety, outcomes and effectiveness of care, clinical practice and technology assessment, health care organization and delivery systems, primary care and preventive systems, and health care costs and sources of payment. AHRQ has been studying and improving links between the clinical care delivery system and the public health infrastructure to improve the nation's capacity to respond to bioterrorism (http://www.ahrq.gov).

National Institutes of Health

The National Institutes of Health (NIH) is one of the world's foremost medical research centers. NIH's goals are to foster innovative research to advance the nation's capacity to protect and improve health significantly; develop, maintain, and renew resources that will ensure the nation's capability to prevent disease; expand the knowledge base in medical and associated sciences to enhance the nation's economic well-being to ensure a high public investment in research; and exemplify and promote the highest level of scientific integrity, public accountability, and social responsibility in the conduct of science (http://www.nih.gov).

National Institute on Alcohol Abuse and Alcoholism

The National Institute on Alcohol Abuse and Alcoholism (NIAAA) supports and conducts biomedical and behavioral research on the causes, consequences, treatment, and prevention of alcoholism and alcohol-related problems. NIAAA aims its research at determining the causes of alcoholism, discovering how alcohol damages the organs of the body, and developing prevention and treatment strategies in the nation's health care system (http://www.niaaa.nih.gov).

National Institute on Drug Abuse

The National Institute on Drug Abuse (NIDA) is a national leader for research on drug abuse and addiction. NIDA supports a comprehensive research program that focuses on the biological, social, behavioral, and neuroscientific bases of drug abuse as well as its causes, prevention, and treatment. NIDA also supports research and research training on specific biomedical and behavioral effects of drugs of abuse on the body and brain; the causes and consequences of drug abuse, including morbidity and mortality in selected populations; the relationship of drug abuse to the acquisition, transmission, and clinical course of HIV/AIDS, tuberculosis, and other diseases; and the development of effective prevention and intervention strategies (http://www.nida.nih.gov).

National Institute of Environmental Health Sciences

The National Institute of Environmental Health Sciences (NIEHS) aims to reduce environment-related illnesses by understanding each component in their development and how they are interrelated. With the National Toxicology Program headquartered at NIEHS, research is conducted to help eliminate, reduce, or control many hazards, such as lead, mercury, asbestos, many industrial chemicals, food dyes, and agricultural chemicals. NIEHS also funds basic and applied research on health effects of human exposure to potentially toxic or harmful environmental agents (http://www.niehs.nih.gov).

National Institute of Child Health and Human Development

The National Institute of Child Health and Human Development conducts and supports research to advance knowledge of pregnancy, fetal development, and birth for developing strategies that prevent maternal, infant, and childhood mortality and morbidity; identify and promote the prerequisites of optimal physical, mental, and behavioral growth and development through infancy, childhood, and adolescence; and contribute to the prevention and amelioration of mental retardation and developmental disabilities (http://www.nichd.nih.gov).

U.S. DEPARTMENT OF HOUSING AND URBAN DEVELOPMENT

The U.S. Department of Housing and Urban Development (HUD) works to increase homeownership, support community development, and increase access to affordable housing free from discrimination. The Office of Healthy Homes and Lead Hazard Control (HHLHC), a HUD program, brings together health and housing professionals in a concerted effort to eliminate lead-based paint hazards in U.S. privately owned and low-income housing. HHLHC develops lead-based paint regulations, guidelines, and policies; provides technical assistance; conducts demonstrations, studies, and standards development; and maintains a community outreach program focused on disseminating program information.

HUD's Healthy Homes Initiative (HHI), run by HHLHC, protects children and their families from housing-related health and safety hazards, such as lead hazard control. HHI is a nationwide effort, and is assisted by a panel of nationally recognized experts from the private sector and federal, state, and local governments. Eligible HHI activities may include evaluating the effectiveness of hazard interventions, developing and delivering public education programs, and developing low-cost methods for hazard assessment and intervention (http://www.hud.gov).

U.S. DEPARTMENT OF AGRICULTURE

The U.S. Department of Agriculture (USDA) provides leadership on food, agriculture, natural resources, and related issues based on sound public policy, the best available science, and efficient management. USDA is responsible for the safety of meat, poultry, and egg products. It leads research in topics from human nutrition to new crop technologies that allows farmers to grow more food and fiber using less water and pesticides, brings safe drinking water to rural America, leads the federal anti-hunger effort, helps ensure open markets for U.S. agricultural products, and provides food aid to needy people overseas, as well as other tasks that help U.S. farmers, ranchers, and lands (http://www.usda.gov).

Food Safety and Inspection Service

The Food Safety and Inspection Service (FSIS) protects consumers by ensuring that meat, poultry, and egg products are safe, wholesome, and accurately labeled. FSIS regulates meat, poultry, and eggs sold in interstate commerce and reinspects imported products to ensure that they meet U.S. safety standards. FSIS sets requirements for labels and certain slaughter and processing activities. FSIS also tests for microbiological, chemical, and other types of contamination and conducts epidemiological investigations in cooperation with CDC based on reports of foodborne health hazards and disease outbreaks (http://www.fsis.usda.gov).

Office of Public Health and Science

The Office of Public Health and Science (OPHS) provides scientific analysis, data, and recommendations on all matters involving public health and science that are of concern to the FSIS. OPHS assures scientifically sound food safety programs and policies to reduce or eliminate foodborne illness. OPHS experts monitor and analyze production processes; identify and evaluate potential foodborne hazards; determine estimates of risk to human health; respond to recognized, emerging, or potential threats to the food supply; investigate the origin of hazards; coordinate the recall of products when necessary; and provide emergency preparedness for foodborne problems (http://www.fsis.usda.gov/OPHS/ophshome.htm).

Agricultural Marketing Service

The Agricultural Marketing Service (AMS) administers programs that facilitate the fair, efficient marketing of U.S. agricultural products, including food, fiber, and specialty crops. One AMS program, the Science and

Technology Program, collects and analyzes data about pesticide residue levels in agricultural commodities. It administers the Pesticide Record-keeping Program, which requires all certified private applicators of federal restricted-use pesticides to maintain records of all applications. The records will be put into a database to help analyze agricultural pesticide use (http://www.ams.usda.gov).

U.S. DEPARTMENT OF HOMELAND SECURITY

The U.S. Department of Homeland Security (DHS) develops and coordinates the implementation of a comprehensive national strategy to secure the United States from terrorist threats or attacks. Working with executive departments and agencies, state and local governments, and private entities, DHS ensures an adequate strategy for detecting, preparing for, protecting against, responding to, and recovering from terrorist threats or attacks within the United States. DHS coordinates the development of monitoring protocols and equipment use for detecting the release of biological, chemical, and radiological hazards; prevention of unauthorized access to, development of, and unlawful importation of chemical, biological, radiological, nuclear, or other related materials; and containment and removal of biological, chemical, radiological, or other hazardous materials in the event of a terrorist threat or attack involving such hazards (http://www.dhs.gov/dhspublic).

U.S. DEPARTMENT OF LABOR

The U.S. Department of Labor (DOL) fosters and promotes the welfare of job seekers, wage earners, and retirees of the United States by improving working conditions, advancing employment opportunities, protecting retirement and health care benefits, helping employers find workers, and tracking changes in employment, prices, and other national economic measurements. DOL administers a variety of federal labor laws, including those that guarantee workers' rights to a safe and healthful working environment (http://www.dol.gov).

Occupational Safety and Health Administration

The Occupational Safety and Health Administration (OSHA) programs are designed to save lives, prevent injuries, and protect the health of U.S. workers, in partnership with more than 100 million working men and women and their 6.5 million employers. OSHA protects workers with its strong enforcement program and prevents on-the-job injuries and illness through outreach, education, and compliance assistance.

OSHA has several cooperative programs, such as the Alliance Program, which allows trade or professional organizations, businesses, labor organizations, educational institutions, and government agencies that share an interest in workplace safety and health to collaborate with the administration to prevent injuries and illnesses in the workplace. The Strategic Partnership Program targets strategic areas and includes partnerships that target specific hazards in specific geographic areas. Voluntary Protection Program worksites have achieved exemplary occupational safety and health (http://www.osha.gov).

Mine Safety and Health Administration

The Mine Safety and Health Administration (MSHA) enforces compliance with mandatory safety and health standards to end fatal accidents, reduce the frequency and severity of nonfatal accidents, minimize health hazards, and promote improved safety and health conditions in U.S. mines.

MSHA's programs include Coal Mine Safety and Health, which is responsible for enforcing the Mine Act at all coal mines. The Act addresses activities such as site inspections, investigations of fatal and serious accidents and complaints of hazardous conditions reported by miners, and development of improved safety and health standards. Metal and Nonmetal Safety and Health enforces the Mine Act at all metal and nonmetal mining operations in the United States. The Directorate of Educational Policy and Development implements MSHA's education and training programs, which are designed to promote safety and health in the U.S. mining industry. Finally, the Directorate of Technical Support provides expertise to assist MSHA, the states, and the mining industry in the resolution of safety and health issues. Technical Support conducts field investigations, studies, and analyses. Equipment and materials used in mines are also evaluated and approved by the directorate. Emergency response capabilities in mines, including onsite analysis and decision-making assistance for crisis management, are also evaluated (http://www.msha.gov).

Bureau of Labor Statistics

The Bureau of Labor Statistics is the principal fact-finding agency for the federal government for economics and statistics, producing impartial, timely, and accurate data about the social and economic conditions of the United States, its workers, workplaces, and the workers' families (http://www.bls.gov).

U.S. DEPARTMENT OF JUSTICE

The U.S. Department of Justice (DOJ) enforces the law and defends the interests of the United States according to the law; provides federal leadership in preventing and controlling crime; seeks just punishment for those guilty of unlawful behavior; administers and enforces the nation's immigration laws fairly and effectively; promotes fair and impartial administration of justice for all U.S. citizens; and protects the United States from the threat of terrorism (http://www.usdoj.gov).

National Drug Intelligence Center

The National Drug Intelligence Center (NDIC) is both a component of DOJ and a member of the intelligence community. It is the center for strategic counterdrug intelligence. NDIC assists national policy makers and law enforcement with strategic domestic drug intelligence; supports the intelligence community counterdrug efforts; and produces national, regional, and state drug threat assessments. Among the many products produced by NDIC, *The National Drug Threat Assessment* is an annual report that provides information such as the current primary drug threat to the nation, fluctuations in consumption levels, and the effects of particular drugs on abusers and society as a whole (http://usdoj.gov/ndic).

Drug Enforcement Administration

The Drug Enforcement Administration enforces controlled substance laws and regulations of the United States by investigating and preparing for the prosecution of major violators of controlled substance laws at interstate and international levels; investigating and preparing for the prosecution of criminals and drug gangs that perpetrate violence and terrorism; managing a national drug intelligence program in cooperation with federal, state, local, and foreign officials; and other activities (http://www.dea.gov).

Environment and Natural Resources Division

DOJ's Environment and Natural Resources Division (ENRD) enforces federal civil and criminal environmental laws and defends environmental challenges to government programs and activities, representing the United States in matters concerning the stewardship of the nation's natural resources and public lands. The Environmental Crimes Section is responsible for prosecuting individuals and corporations for violating laws that protect the environment, such as those that compel clean-up of hazardous waste sites (http://www.usdoj.gov/enrd).

U.S. CONSUMER PRODUCT SAFETY COMMISSION

The U.S. Consumer Product Safety Commission (CPSC) protects the public from unreasonable risks of serious injury or death from certain types of consumer products under the agency's jurisdiction, such as toys, cribs, power tools, cigarette lighters, and household chemicals. CPSC is committed to protecting consumers and families from products that pose an electrical, chemical, or mechanical hazard, or can injure children. CPSC has a hotline to report a dangerous product or a product-related injury and a Poison Lookout Checklist (http://www.cpsc.gov).

U.S. ENVIRONMENTAL PROTECTION AGENCY

The U.S. Environmental Protection Agency (EPA) works to protect human health and to safeguard the natural environment. EPA is one of the nation's leaders in environmental science, research, education, and assessment efforts, addressing emerging environmental issues and advancing the science and technology of risk assessment and management. Many of the substances regulated by EPA are poisonous to both humans and the environment. Some of the most common cases of human poisoning are from pesticides, lead, and mercury. EPA provides information to help prevent poisoning from these substances and other poisons in the home. EPA also has information on topics such as chemical and radiation accidents, accident preparedness and prevention, emergency preparedness and response, a radiological emergency response team, and the Toxics Release Inventory (TRI) (http://www.epa.gov).

OFFICE OF NATIONAL DRUG CONTROL POLICY

The Office of National Drug Control Policy (ONDCP) establishes the policies, priorities, and objectives for the U.S. drug control program. The director of ONDCP produces the National Drug Control Strategy to reduce illicit drug use and the manufacturing and trafficking of drugs, drug-related crime and violence, and drug-related health consequences. For fiscal year 2004, the National Drug Control Strategy proposes three core priorities: (1) stopping drug use before it starts, (2) healing America's drug users, and (3) disrupting the illicit drug market (http://www.whitehousedrugpolicy.gov).

Part III

Conclusions and Recommendations

10

Conclusions and Recommendations

The Institute of Medicine's Committee on Poison Prevention and Control was charged by the Health Resources and Services Administration (HRSA) to consider a "systematic approach to understanding, stabilizing, and providing long-term support for poison prevention and control services in the United States" by reviewing the past and current approaches to the provision of these services in terms of:

1. The scope of services provided, including consumer telephone consultation, technical assistance and/or hospital consultation for the care of patients with life-threatening poisonings, and education of the public and professionals;
2. The coordination of poison control centers with other public health, emergency medical, and other emergency services;
3. The strengths and weaknesses of various organizational structures for poison control centers and services, including a consideration of personnel needs;
4. Approaches to providing the financial resources for poison prevention and control services;
5. Methods for assuring consistent, high-quality services, including the certification of centers and methods of evaluation; and
6. Current and future data systems and surveillance needs.

This broad charge led the Committee to take a systems approach, viewing poison control centers within the public health and medical care systems, and reconsidering the organizational structure of poison control centers to serve the needs of the nation. Addressing the charge also demanded that the Committee define clearly what is meant by a "poisoning." Recognizing the controversies in the field and the fact that there is no universally agreed-upon definition, we adopted an operational definition, using the categories that are used by agencies and organizations that currently monitor the problem in the population. The Committee's operational definition of poisoning subsumes "damaging physiological effects of ingestion, inhalation, or other exposure to a range of pharmaceuticals, illicit drugs, and chemicals, including pesticides, heavy metals, gases/vapors, and common household substances, such as bleach and ammonia" (Centers for Disease Control and Prevention, 2004, p. 233).

The Committee concluded, based on its research and discussions, that the current network of poison control centers does not constitute the complete "system" of poison prevention and control services needed by the nation in the 21st century. Such a system must provide the best prevention and patient care services for the diverse population of Americans who are exposed to hazardous substances and protect the nation from the threats associated with biological and chemical terrorist events and other public health emergencies. The Committee therefore based its report on a proposed Poison Prevention and Control System that included a network of poison control centers as a vital, but not exclusive, element. The Committee also concluded that in order to fulfill their pivotal role in the overall system, poison control centers must be more stable financially and better integrated and coordinated for performance of their public health roles.

SCOPE OF CORE POISON PREVENTION AND CONTROL ACTIVITIES

Poison control centers are the fundamental building blocks of the proposed Poison Prevention and Control System. A regional distribution of such centers will satisfy the need to distribute medical toxicological leadership across the United States to address the diversity of poison exposures and to provide firsthand consultation to hospitals and physicians. The interaction among regionally based centers will promote innovation and the sharing of best practices. Finally, a regionalized system should provide enough redundancy in skills and resources to meet surge needs and potential equipment failures. Therefore, the Committee carefully examined the activities, functions, performance, and organizational structures of current poison control centers. Based on the information and analyses provided in Chapters 5 through 9, a core set of activities was defined that constitutes the essential functions of the network of poison control centers within the larger system envisioned by the Committee.

Although these activities are already being carried out, it is essential to identify them as a set of core activities so that they become the basis for consistent funding under the aegis of the proposed expanded federal legislation. These activities are considered by the Committee to be core because (1) they represent critical components of current and future poison control efforts; (2) the structure of poison control centers and expertise of their staffs make them uniquely capable of performing these activities (i.e., there are no other organizations in the public health and health care arena that can perform these activities at the same level of excellence and cost); and (3) they provide an infrastructure to which other related activities can readily be added as required. The notion of core activities does not imply that poison control centers should confine their activities solely to these areas. The addition of other activities should be based on local capabilities and opportunities for funding. Examples include understanding clinical toxicology research or providing training for health care students who are not specifically focused on careers in medical or clinical toxicology.

Recommendations

1. All poison control centers should perform a defined set of core activities supported by federal funding that is tied to the provision of these activities. The core activities include (1) manage telephone-based poison exposure and information calls; (2) prepare and respond to all-hazards emergency needs (especially biological or chemical terrorism or other mass exposure events); (3) capture, analyze, and report exposure data; (4) train poison control center staff, including specialists in poison information and poison information providers; (5) carry out continuous quality improvement; and (6) integrate their services into the public health system. In addition, a subset of poison control centers should train medical toxicologists; this is considered a core activity for only a subset of poison control centers because their involvement is necessary for the certification of this specialty. A subset of poison control centers should also assist in the training of pharmacists through clinical toxicology fellowships that prepare them for poison control center management positions.

2. Poison control centers should collaborate with state and local health departments to develop, disseminate, and evaluate public and professional education activities. Poison control centers alone cannot fulfill the need for public and professional education related to poisoning prevention and treatment and all-hazards response. Public health agencies already have the authorities, networks, and administrative mechanisms to carry out broad educational efforts, as they do for the prevention of other injuries and for other public health campaigns.

COORDINATION OF POISON CONTROL CENTERS WITH OTHER PUBLIC HEALTH ENTITIES

At the heart of the Committee's proposal for a Poison Prevention and Control System is the integration of the current network of poison control centers into the broader public health system. As discussed in Chapter 9, the accountability for the establishment and maintenance of a population-based poison prevention and control system is currently diffuse, involving multiple levels of government. Although there are several programs or components within public health agencies that are relevant to poison prevention and control, some of which currently interact with the poison control centers, there is no clarity concerning the roles of each entity in the integrated system. This has resulted in inefficient interactions among federal, state, and local public health agencies and poison control centers that have limited their potential contributions to prevention of poisoning and promotion of health.

To achieve the ultimate goal of preventing poisonings, as well as to improve the outcomes for those who are poisoned, the Committee envisions the need for a clear, single point of accountability at each level of government. The responsible agencies would assure the accomplishment of all the public health core functions or essential services as they relate to poison prevention and control. This does not mean that the responsible agencies would perform all the functions within their respective agencies. However, they would (1) take responsibility for developing the plan to accomplish the activities needed to assure that the system is in place, with a set of uniform standards across the country; (2) convene and work with the other agencies, including the existing poison control center network, to implement the plan; and (3) work in partnership to develop a set of performance standards for all components of the system. One possible model for the development of performance measures for a state-federal partnership is the Title V Maternal and Child Health (MCH) Block Grant, which is administered by states, and the federal grants for MCH activities, which are administered by the Maternal and Child Health Bureau in HRSA. This partnership has been in place for 5 years and has successfully developed and implemented performance criteria and data reporting mechanisms.

Recommendations

3. The U.S. Department of Health and Human Services (DHHS) and the states should establish a Poison Prevention and Control System that integrates poison control centers with public health agencies, establishes performance measures, and holds all parties accountable for

protecting the public. At the federal level, the Secretary of Health and Human Services should designate the lead agency for this purpose; at the state level, the governor of each state should formally designate the appropriate lead (e.g., injury prevention directors from the public health entity).

a. The Secretary of DHHS should assure integration of the existing regional network of poison control centers with the public health system.

b. The Secretary of DHHS should create a single national repository of legislation, model prevention and education programs, website designs, and best practices material. Technical assistance should be provided for website design, content, navigation, and maintenance, maximizing the individual centers' identity and contributions. Materials should be evaluated for quality and impact on intended audiences. For maximum effectiveness, their content should reflect the range of cultures and languages in the United States.

c. The governor should assure that relevant all-hazards emergency preparedness and response activities are integrated with the Poison Prevention and Control System.

4. The Centers for Disease Control and Prevention, working with HRSA and the states, should continue to build an effective infrastructure for all-hazards emergency preparedness, including bioterrorism and chemical terrorism. A specific activity of this effort is to evaluate, through an objective structured review, the use of the Toxic Exposure Surveillance System as a source of case detection to all-hazards surveillance.

STRENGTHS AND WEAKNESSES OF POISON CONTROL CENTER ORGANIZATIONAL STRUCTURES

Early in its information gathering, the Committee weighed the options of conducting an in-depth analysis of all poison control centers or relying on existing survey data available from the American Association of Poison Control Centers (AAPCC) supplemented by case studies of a sample of centers varying with regard to size, cost, efficiency, and penetrance (number of human exposure calls per 1,000 population). The Committee's assessment was that the existing data should be adequate to address the questions raised by HRSA about the organization and financing of the centers. However, as the analysis progressed, it became clear that the information available to the Committee was not sufficient to fully address this aspect of the charge. No data on service quality and outcomes had been systematically collected by the centers. Data on local variations in salaries and rent were not readily available. As a result, the Committee's analysis presents preliminary findings that are useful in defining the information needed for a full-scale, definitive study of organizational efficiency and effectiveness.

As noted in Chapter 6, a number of published studies provide cost-effectiveness and cost-benefit analyses of various aspects of the poison control center system and some take account of the potential reduction in morbidity and mortality as benefits. In many of these studies the lack of data presents a challenge. Nonetheless, taken as a whole, this literature makes a convincing case that, at least in terms of treatment management guidance for the public, poison control centers save the health care system economic resources and save members of the public time, lost wages, and anxiety. The Committee found no studies that compare cost-effectiveness of service delivery models among poison control centers.

Also noted in Chapter 6, the Committee found a wide range of service delivery models, organizational structures, and financing arrangements among poison control centers that successfully deliver core services. Although an earlier study conducted on six poison control centers suggested possible economies of scale for service areas of 2 million people or more, the Committee found little conclusive evidence from its own analysis that economies of scale operate with respect to size of population served and poison control center costs. Costs were best predicted by variables related to staffing patterns and wage rates rather than hardware expenses, population served, or funding source. More complete data are needed to further explore this important concern.

The Committee's qualitative analysis of 10 poison control centers indicated that the more efficient centers had lower staff turnover rates with fewer concerns about salaries and were more likely to (1) participate in partnerships or joint ventures in the community, (2) have written strategic plans specific to the poison control center, and (3) be organizationally affiliated with a private institution. Furthermore, the more efficient centers were less likely to cite problems related to complex reporting and accountability and problems of balancing core poison control functions with other activities such as research and bioterrorism response and preparedness. These results provide some indications of desirable (e.g., written strategic plans and participation in joint ventures) and undesirable organizational characteristics. It is important to note that these analyses were based solely on population served, cost per human exposure call, and penetrance.

The existing data are insufficient for the development of either contractual specifications or performance measures for a new Poison Prevention and Control System. The Committee suggests new data-gathering efforts to obtain original financial and performance data from existing poison control centers. These data are needed to guide future public funding of core activities.

Recommendation

5. HRSA should commission a systematic management review focusing on organizational determinants of cost, quality, and staffing of poison control centers as the foundation for the future funding of this program. This analysis should include the following elements:

a. The development of new indicators of quality and impact of poison control center services.

b. The implications of different organizational structures and funding accountabilities on service quality and impact.

c. The role of center size and governance in poison control center service quality and impact.

d. The impact of regional differences on poison control center operational cost.

e. How staffing patterns, recruitment, and retention of poison control center staff affect cost, quality, and impact of poison control centers.

f. An economic evaluation of poison control centers to determine whether economies of scale exist among them.

FINANCIAL SUPPORT FOR THE POISON PREVENTION AND CONTROL SYSTEM

Poison Control Centers

As noted in Chapter 6, poison control centers are currently funded by a patchwork of sources (including federal, state, institutional, and private) that are subject to budget cuts and changing priorities every year. Across the states there are 29 separate funding sources: 6 percent of total poison control center funding comes from federal and state Medicaid programs, 3 percent from federal block grants, and 8 percent from other federal programs, for a total of 17 percent from federally associated programs. Approximately 44 percent of total funding comes from states, with many different approaches to state funding, ranging from line-item appropriation to state-funded universities to telephone surcharges. Hospitals represent 15 percent of total funding (either as host institutions or network members), another 3 percent of funding comes from a wide range of donations and grant sources, and 20 percent comes from myriad other sources.

Because of the lack of consistent, reliable funding sources, poison control centers report that significant time is spent in raising revenues and that there has been substantial instability in funding. As financial pressures on state governments and health systems have risen, the willingness of traditional funders to continue to provide revenues has diminished, leaving many centers facing great uncertainty, budget pressures, and cutbacks.

Initial efforts to stabilize the delivery of poison center services to the public and health care professionals were provided by the Poison Control Center Enhancement and Awareness Act of 2002; however, the funds appropriated through this legislation have not been sufficient. In 2001, AAPCC reported $104 million in total funding for poison control centers. In a separate analysis the Committee estimated a similar amount by multiplying the cost per human exposure call by call volume (see Chapter 6). The Committee concludes that the most effective approach to stabilization is through federal funding of approximately $100 million to support the core activities. This funding could reduce or replace the support for core activities provided by many of the current funding sources; however, it would not reduce the need for state and local funding to support non-core services.

Recommendation

6. Congress should amend the current Poison Control Center Enhancement and Awareness Act to provide sufficient funding to support the proposed Poison Prevention and Control System with its national network of regional poison control centers. Support for the core activities at the *current* level of service is estimated to require more than $100 million annually. Extension of services to include the growing all-hazards emergency needs (especially biological or chemical terrorism) and enhancements to current surveillance and data collection activities will require additional support and should be supplemented as appropriate to such mandates. The funding could be channeled either through a direct federal grant or a federal-state matching process. Performance measures for poison control center services must be specified and monitored by the funding agencies involved. Separate funding will be required to support activities performed at the federal and state levels.

State and Local Infrastructure

For the Poison Prevention and Control System to be implemented and continuously improved in the most effective manner, resources must be made available to carry out the mandate. Public health initiatives with a clear mandate and resources available to both federal and state agencies are the most successful. In addition to the funds required by each poison control center to implement the core activities, the Committee estimates an amount roughly on the magnitude of $30 million to assure that all the essential services of public health related to poisoning could be accomplished. This estimate includes approximately $10 million in the form of grants to each state to support a poison prevention coordinator's office whose responsibilities would include coordination of public education efforts and a plan for their evaluation, and $20 million for federal-level activities, including (1) development and maintenance of quality assurance and improvement mechanisms for every component of the Poison Prevention and Control System; (2) training activities for health providers outside the poison control centers who require training in toxicology, such as emergency department workers and emergency medical technicians; (3) a clearinghouse for primary prevention materials and resources; and (4) research and the translation of research and evaluation studies into best practices and regulatory changes. Federal estimates are based on similar public health programs funded by the Centers for Disease Control and Prevention (CDC) and the Health Resources and Services Administration.

Recommendation

7. Congress should amend existing public health legislation to fund a state and local infrastructure to support an integrated Poison Prevention and Control System. The Committee at this time is not able to provide a precise estimate of the required level of support for such a federal and state program. The Committee recommends that the Secretary of Health and Human Services should develop a budget proposal to support the costs of training, research, data archiving and reporting, quality assurance, and public education (including state-level coordination of prevention education and the creation of a central repository of best model programs). This amount is in addition to the $100 million needed to support poison control core services.

ASSURE HIGH-QUALITY POISON CONTROL CENTER SERVICE

Certification of poison control centers is currently the responsibility of AAPCC, and the centers are required to join this organization to become certified. A more accepted model for certification of health care professionals or programs is for it to be the responsibility of an independent agency, rather than an organization in which the applicants are paying members. (For example, medical toxicologists are certified by a board that is a member of the American Board of Medical Specialties rather than by a toxicology organization.) With the continued development of poison control centers and their increased integration into the public health system, alternative certification processes will offer advantages over the current system, including greater independence of the process from the participants, wider input from the health care community, and wider recognition of the skills and contributions of poison control centers and their personnel.

Recommendation

8. A fully external, independent body should be responsible for certification of poison control centers and specialists in poison information. This body should be separate from the professional organizations representing them.

NATIONAL DATA SYSTEM AND SURVEILLANCE NEEDS

A Uniform Definition of Poisoning

Among the most important functions of the Poison Prevention and Control System will be the collection and provision of poison exposure and surveillance data to the nation's health authorities. The Committee's analysis focused on existing data resources, including national surveys, which have been designed at least in part for epidemiological tracking purposes, or can be readily exploited for such purposes. Although electronic medical records systems may hold promise for augmenting existing data and surveillance resources in the future, they were not included because of issues of sensitivity, specificity, data access, data coding, scope of use, and data requirements as they pertain to surveillance for poisonings.

The Committee found many barriers to the effective operation of a comprehensive data and surveillance system and to the provision and utilization of the information by agencies at the federal, state, and local levels (details of this analysis are presented in Chapter 7). The steps to ameliorate this situation are complex, but there is a pressing need for change. The Committee recommends that these be addressed at the same time that the legislative, financing, and organizational reforms are being implemented.

Recommendation

9. The Secretary of Health and Human Services should instruct key agencies to convene an expert panel to develop a definition of poisoning that can be used in surveillance activities (including the Toxic Exposure Surveillance System) and ongoing data collection studies. Furthermore:

a. The Secretary should ask the World Health Organization to review and reform the *International Classification of Diseases* codes for poisoning, thereby addressing the discrepancies and complexities identified in the current classification.

b. The Secretary should require agencies that sponsor existing surveillance and data collection instruments to use a common definition of poisoning that allows comparability across data collection efforts.

c. The National Center for Health Statistics (NCHS) should review the methodology of its existing surveys to maximize the value of their survey data for poison prevention and control.

d. Other agencies collecting health-related data at the federal level outside NCHS, and at the state level, should enhance their surveys or surveillance data systems to better gather and interpret data related to poisoning injury and risk factors.

Privacy Barriers to Data Collection

New patient protections provided by the Health Insurance Portability and Accountability Act and state privacy regulations have placed substantial limitations on sharing health care data. This situation is exacerbated by the fact that there are many misconceptions among health care professionals regarding the conditions under which such data are available.

Recommendation

10. DHHS should undertake a targeted education effort to improve health provider awareness of poisoning data collection as it relates to the Health Insurance Portability and Accountability Act (HIPAA) and state privacy regulations to mitigate their unintended chilling effect on poison control center consultation, including follow-up. DHHS should review and resolve the negative impact of HIPAA and state privacy regulations on poison control center functions, including toxicology consultations and outcomes evaluation.

Availability of TESS Data

The Toxic Exposure Surveillance System is a proprietary data and surveillance system owned by AAPCC. Using funding from the CDC, AAPCC recently developed the capability to provide real-time surveillance through TESS based on input from the poison control centers. The Committee recognizes that this system was established and has been significantly strengthened through the initiative of AAPCC. However, there is now enough evidence to suggest that a private system cannot meet the national need for timely data in this area. Despite federal funding, the computer code for TESS is owned by a private company, further complicating its use and distribution.

Recommendation

11. The Director of the Centers for Disease Control and Prevention should ensure that exposure surveillance data generated by the poison control centers and currently reported in the Toxic Exposure Surveillance System are available to all appropriate local, state, and federal public health units and to the poison control centers on a "real-time" basis at no additional cost to these users. These data should also be publicly accessible with oversight mechanisms and privacy guarantees and at a cost consistent with other major public use systems such as those currently managed by the National Center for Health Statistics.

Research Needs

The Committee made an attempt, within the constraints of the available literature and data systems, to document the magnitude of the poisoning problem and its cost, in terms of health care outcomes, to the nation. The results of this analysis are provided in Chapter 3. We concluded that despite limitations in the data, poisoning is a far greater problem than has been generally recognized and it deserves a higher level of scrutiny and support. The Committee has provided rough estimates that at best need to be refined to become the basis of policy. Therefore, as a first step, the Committee recommends a baseline assessment of the magnitude and cost of poisoning. Furthermore, the Committee found a dearth of research on poisoning and poison control center operations and encourages funding of research in this area.

Recommendation

12. Federally funded research should be provided for (1) studies on the epidemiology of poisoning, (2) the prevention and treatment of poisoning and drug overdose, (3) health services access and delivery, (4) strategies to improve regulations and facilitate researchers' input into regulatory procedures, and (5) the cost efficiency of the new Poison Prevention and Control System on population-based outcomes for general and specific poisonings.

a. CDC should take the lead in marshalling the relevant data pertaining to the epidemiology of poisoning. It should produce a comprehensive report estimating the national incidence of poisoning morbidity and mortality, exploiting its existing data sources. Within the centers, the National

Center for Injury Prevention and Control (NCIPC) could lead this effort, coordinating data needs with NCHS. Data sources should include TESS, the National Health Interview Survey, the National Electronic Injury Surveillance System, the Drug Abuse Warning Network, MedWatch, and others.

b. The Agency for Healthcare Research and Quality (AHRQ) and CDC should be directed to undertake a rigorous economic analysis of the overall direct and indirect health care costs of poisoning and drug overdose.

c. The Secretary of Health and Human Services should encourage funding by appropriate agencies, such as CDC and the Consumer Product Safety Commission, to ensure the needed flow of information from toxicology researchers in poison control centers on prevention problems and strategies to regulators and to encourage the study and development of new regulatory strategies and initiatives to reduce poisonings.

d. Researchers should be funded through grants from appropriate institutes such as the National Institutes of Health, the National Library of Medicine, AHRQ, and CDC/NCIPC, to study prevention and treatment of poisonings and drug overdose, health service access and delivery, and the cost efficiency and clinical impact of the Poison Prevention and Control System.

References

Accreditation Council for Graduate Medical Education. (2003). Reports—Programs by specialty. Available: <http://www.acgme.org/adspublic/reports/program_specialty_list.asp>.

Agran, P.F., Anderson, C., Winn, D., Trent, R., Walton-Haynes, L., and Thayer, S. (2003). Rates of pediatric injuries by 3-month intervals for children 0 to 3 years of age. *Pediatrics, 111*, 683–692.

Ajzen, I. (1985). From intentions to actions: A theory of planned behavior. In J. Kuhl and J. Bechmann (Eds.), *Action control: From cognition to behavior* (pp. 11–39). New York: Springer-Verlag.

Ajzen, I. (1991). The theory of planned behavior. *Organizational Behavior and Human Decision Processes, 50*, 179–211.

Ajzen, I., and Madden, T.J. (1986). Prediction of goal-directed behavior: Attitudes, intentions, and perceived behavioral control. *Journal of Experimental Social Psychology, 22*, 453–474.

American Association of Poison Control Centers. (1998). Criteria for certification of poison centers and poison center systems.

American Association of Poison Control Centers. (2001). Instructions for the American Association of Poison Control Centers Toxic Exposure Surveillance System (TESS), Effective November 1, 2001, Required Definitions and Fields for TESS 2002. Unpublished.

American Association of Poison Control Centers. (2002a). *2002 annual report of the American Association of Poison Control Centers Toxic Exposure Surveillance System.* Washington, DC: Author.

American Association of Poison Control Centers. (2002b). *The Poison Line. 20*, 12.

American Association of Poison Control Centers. (2003a). *Membership directory.* Revised as of September 14, 1998. Washington, DC: Author.

American Association of Poison Control Centers. (2003b). *Public Education Committee membership directory, year 2003–2004.* Washington, DC: Author.

American Association of Poison Control Centers. (2003c). *Poison prevention education materials resource guide.* Public Education Committee. Washington, DC: Author.

American Public Health Association. (1998). *How states are collecting and using cause of injury data.* Injury Control and Emergency Health Services Section. San Francisco: Trauma Foundation/San Francisco Injury Center.

Anderson, I.B., Mullen, W.H., Meeker, J.E., Khojasteh-Bakht, S.C., Oishii, S., Nelson, S.D., and Blanc, P.D. (1996). Pennyroyal toxicity: Measurement of toxic metabolites in two cases and review of the literature. *Annals of Internal Medicine, 124,* 726–734.

Anonymous. (1982). Unintentional and intentional injuries—United States. *Morbidity and Mortality Weekly Report, 31*(18):240, 245–240, 248.

Arena, J.M. (1959). Safety closure caps. *Journal of the American Medical Association, 169,* 1187–1188.

Arena, J.M. (1983). The pediatrician's role in the poison control movement and poison prevention. *American Journal of Diseases of Children, 137*(l), 870–873.

Ashford, D.A., Kaiser, R.M., Bales, M.E., Shutt, K., Patrawalla, A., McShan, A., et al. (2003). Planning against biological terrorism: Lessons from outbreak investigations. *Emerging Infectious Diseases, 9*(5) [serial online]. Available: <http://www.cdc.gov/ncidod/EID/vol9no5/02-0388.htm>.

Bain, K. (1954). Death due to accidental poisoning in young children. *Journal of Pediatrics, 44,* 616–623.

Bandura, A. (1977). Self-efficacy: Toward a unifying theory of behavioral change. *Psychological Review, 84,* 191–215.

Bandura, A. (1986). *Social foundations of thought and action: A social cognitive theory.* Englewood Cliffs, NJ: Prentice-Hall.

Bandura, A. (1991). Self-efficacy mechanism in physiological activation and health-promoting behavior. In J. Madden (Ed.), *Neurobiology of learning, emotion and affect* (pp. 229–269). New York: Raven.

Bandura, A. (1994). Social cognitive theory and exercise of control over HIV infection. In R.J. DiClemente and J.L. Peterson (Eds.), *Preventing AIDS: Theories and methods of behavioral interventions* (pp. 25–29). New York: Plenum Press.

Barton, E.D., Tanner, P., Turchen, S.G., et al. (1995). Ciguatera fish poisoning a southern California epidemic. *Western Journal of Medicine, 163,* 31–35.

Bates, D.W., Evans, R.S., Murff, H., Stetson, P.D., Pizziferri, L., and Hripcsak, G. (2003). Detecting adverse events using information technology. *Journal of the American Medical Informatics Associations, 19*(2), 115–128.

Becker, M.H. 1974. The health belief model and personal health behavior. *Health Education Monographs, 2,* 324–508.

Bennett, C.L., Weinberg, P.D, Rozenberg-Ben-Dor, K., et al. (1998). Thrombotic thrombocytic purpura associated with ticlopidine. *Annals of Internal Medicine, 128,* 541–544.

Berkowitz, Z., Barnhart, H.X., and Kaye, W.E. (2003). Factors associated with severity of injury resulting from acute releases of hazardous substances in the manufacturing industry. *Journal of Occupational and Environmental Medicine, 45,* 734–742.

Blanc, P.D., and Olson, K.R. (1986). Occupationally related illness reported to a regional poison control center. *American Journal of Public Health, 76*(11), 1303–1307.

Blanc, P.D., Rempel, D., Maizlish, N., Hiatt, O., and Olson, K.R. (1989). Occupational illness: Case detection by poison control surveillance. *Annals of Internal Medicine, 111,* 238–244.

Blanc, P.D., Maizlish, N., Hiatt, P., Olson, K.R., and Rempel D. (1990). Occupational illness and the poison control center: Referral patterns and service needs. *Western Journal of Medicine, 152,* 181–184.

Blanc, P.D., Galbo, M., Hiatt, P., and Olson, K.R. (1991). Morbidity following acute irritant inhalation in a population-based study. *Journal of the American Medical Association, 266,* 664–669.

Blanc, P.D., Galbo, M., Hiatt, P., Olson, K.R., and Balmes, J.R. (1993a). Symptoms, lung function and airway responsiveness following irritant inhalation. *Chest, 103*, 1699–1705.

Blanc, P.D., Jones, M.R., and Olson, K.R. (1993b). Surveillance of poisoning and drug overdose through hospital discharge coding, poison control center reporting, and the Drug Abuse Warning Network. *American Journal of Emergency Medicine, 11*, 14–19.

Blanc, P.D., Saxena, M., and Olson, K.R. (1993c). Drug detection and trauma cause: A case control study of fatal injuries. *Clinical Toxicology, 32*, 137–145.

Blanc, P.D., Kearney, T.E., and Olson, K.R. (1995). Underreporting of fatal cases to a regional poison control center. *Western Journal of Medicine, 162*(6), 505–509.

Bonnie, R.J., Fulco, C.E., and Liverman, C.T., eds. (1999). *Reducing the burden of injury advancing prevention and treatment.* Committtee on Injury Prevention and Control. Institute of Medicine. Washington, DC: National Academy Press.

Bosse, G.M. (1994). Nebulized sodium bicarbonate in the treatment of chlorine gas inhalation. *Journal of Toxicology, Clinical Toxicology, 32*, 233–241.

Botticelli, J.T., and Pierpaoli, P.G. (1992). Louis Gdalman, pioneer in hospital pharmacy poison information services. *American Journal of Hospital Pharmacy, 49*(6), 1445–1450.

Brogan H., and Lobell, D. (1999). Decreasing accidental poisonings through effective education. *Journal of Toxicology, Clinical Toxicology, 37*(5), 584–585.

Brubacher, J.R., Ravikumar, P.R., Bania, T., Heller, M.B., and Hofman, R.S. (1996). Treatment of toad venom poisoning with digoxin-specific Fab fragments. *Chest, 110*, 1282–1288.

Burda, A.M., and Burda, N.M. (1997). The nation's first poison control center: Taking a stand against accidental childhood poisoning in Chicago. *Veterinary and Human Toxicology, 39*(2), 115–119.

Burgess, J.L., Pappas, G.P., and Robertson, W.O. (1997). Hazardous materials incidents: The Washington Poison Center experience and approach to exposure assessment. *Journal of Occupational and Environmental Medicine, 39*(8), 760–766.

Burgess, J.L., Kirk, M., Borron, S.W., and Cisek, J. (1999). Emergency department hazardous materials protocol for contaminated patients. *Annals of Emergency Medicine, 34*(2), 205–212.

Calvert, G.M., Sanderson, W.T., Barnett, M., Blondell, J.M., and Mehler, L.N. (2001). Surveillance of pestidice-related illness and injury in humans. In *Handbook of Pesticide Toxicology, Second Edition: Volume I, Principles.* New York: Academic Press.

Centers for Disease Control and Prevention. (1998). *Preventing emerging infectious diseases: A strategy for the 21st century.* Atlanta: Author.

Centers for Disease Control and Prevention. (2001a). National estimates of nonfatal injuries treated in hospital emergency departments—United States, 2000. *Morbidity and Mortality Weekly Review, 50*, 340–346.

Centers for Disease Control and Prevention. (2001b). *Updated guidelines for evaluating public health surveillance systems.* Atlanta: Author.

Centers for Disease Control and Prevention. (2001c). *Public health's infrastructure: A status report.* Prepared for the Appropriations Committee of the United States Senate. Atlanta: Author.

Centers for Disease Control and Prevention. (2003). Nonfatal dog bite-related injuries treated in hospital emergency departments—United States, 2001. *Morbidity and Mortality Weekly Report, 52*, 605–610.

Centers for Disease Control and Prevention. (2004). Unintentional and undetermined poisoning deaths—11 states, 1990–2001. *Morbidity and Mortality Weekly Report, 53*, 233–238.

Chyka, P.A. (2000). How many deaths occur annually from adverse drug reactions in the United States? *American Journal of Medicine, 109*, 122–130.

Chyka, P.A., and Somes, G.W. (2001). Poison control centers and state-specific poisoning mortality rates. *Medical Care, 39*(7):654–660.

Cisternas, M. (2003). Epidemiology of poisoning estimates from u.s. public use data sources: An analysis. Paper prepared for the Committee on Poison Prevention and Control, Institute of Medicine.

Cobb, N., and Etzel, R. (1991). Unintentional carbon monoxide-related deaths in the United States, 1979 through 1988. *Journal of the American Medical Association, 266,* 659–663.

Cone, E.J., Fant, R.V., Rohay, J.M., et al. (2003). Oxycodone involvement in drug abuse deaths: A DAWN-based classification scheme applied to an oxycodone postmortem database containing over 1,000 cases. *Journal of Analytical Toxicology, 27,* 57–67.

Czaja, P.A., Skoutakis, V.A., Wood, G.C., and Autian, J. (1979). Clinical toxicology consultation by pharmacists. *American Journal of Hospital Pharmacy, 36*(8), 1987–1089.

Dart, R.C. (2003). 2002 Rocky Mountain Poison and Drug Center Annual Report. Presentation to the Committee on Poison Prevention and Control, Institute of Medicine.

Davidson, P.J., McLean, R.L., Kral, A.H., Gleghorn, A.A., Edline, B.R., and Moss, A.R. (2003). Fatal heroin-related overdose in San Francisco, 1997–2000: A case for targeted intervention. *Journal of Urban Health, 80,* 261–271.

Davis, C.O., Cobaugh, D.J., Leahey, N.F., and Wax, P.M. (1999). Toxicology training of paramedic students in the United States. *American Journal of Emergency Medicine, 17*(2), 138–140.

Donovan, R.J. (1995). Steps in planning and developing health communication campaigns: A comment on CDC's framework for health communication. *Public Health Reports, 110*(2), 215–218.

Durch, J.S., Bailey, L.A., and Stoto, M.A., eds. (1997). *Improving health in the community: A role for performance monitoring.* Committee on Using Performance Monitoring to Improve Community Health. Institute of Medicine. Washington, DC: National Academy Press.

Dyer, J.E., Roth, B., and Hyma, B.A. (2001). Gamma-hydroxybutyrate withdrawal syndrome. *Annals of Emergency Medicine, 37,* 147–153.

Edge, V.L., Lim, G.H., Aramini, J.J., et al. (2002). Development of an Alternative Surveillance Alert Program (ASAP): Syndromic surveillance of gastrointestinal illness using pharmacy over-the-counter sales. National Syndromic Surveillance Conference, New York Academy of Medicine, Mew York City Department of Health and Mental Hygiene and the Centers for Disease Control and Prevention, September 23–24, New York City.

Fawcett, S.B. (1995). Using empowerment theory in collaborative partnerships for community health and development. *American Journal of Community Psychology,* 23(5), 677–698.

Fingerhut, L.A., and Cox, C.S. (1998). Poisoning mortality, 1985–1995. *Public Health Reports, 113,* 221–235.

Fishbein, M., Middlestadt, S.E., and Hitchcock, P.J. (1991). Using information to change sexually transmitted disease-related behaviors: An analysis based on the theory of reasoned action. In J.N. Wasserheit, S.O. Aral, and K.K. Holmes (Eds.), *Research issues in human behavior and sexually transmitted diseases in the AIDS era* (pp. 243–257). Washington, DC: American Society for Microbiology.

Fisher, L. (1981). The next five years: The goals and objectives of poison control systems in New York State. *Veterinary and Human Toxicology, 23*(2), 103–107.

Fisher, L. (1986). New York State Regional Poison Control Centers Injury Control Network legislation. *Veterinary and Human Toxicology, 28*(6), 545–546.

Fisher, L., VanBuren, J., Nitzkin, J., Lawrence, R., and Swackhamer, R. (1981). Highlight Results of the Monroe County Poison Prevention Demonstration Project: An Empirical Model for Preventing Childhood Accidental Poisoning. Paper prepared for presentation at the Annual Joint Meeting of the American association of Poison Control Centers, American Academy of Clinical Toxicology, and Canadian Academy of Clinical and Analytical Toxicology, Minneapolis, MN, August 5.

Fisher, L., VanBuren, J., Lawrence, R.A., Nitzkin, J.L., Oppenheimer, B., Sinacore, J., Matteson, J., and Ennis, A. (1986). Genesee Region Poison Prevention Project: Phase II. *Veterinary and Human Toxicology, 28*(2), 123–126.

Fleming, L.E., Gomez-Marin, O., Zheng, D., Ma, F., and Lee, D. (2003). National Health Interview Survey mortality among U.S. farmers and pesticide applicators. *American Journal of Industrial Medicine, 43*, 227–233.

Food and Drug Administration. (1957). *The clinical toxicology of commercial products.* Washington, DC: Author.

Food and Drug Administration. (1973). *Evaluation of the poison control system.* Rockville, MD: Author.

Funk, A., Schier, J., Belson, M., Patel, M., Rubin, C., Watson, W., et al. (2003). Toxicosurveillance: Utilization of the Toxic Exposure Surveillance System for detection of potential chemical terrorism events (abstract). In *Proceedings of the Second Annual New York City Syndromic Surveillance Meeting*, October 23 and 24, 2003.

Geller, R.G., and Lopez, P.G. (1999). Poison center planning for mass gatherings: The Georgia Poison Center Experience with the 1996 Centennial Olympic Games. *Clinical Toxicology, 37*(3), 315–319.

German, R.R. (2001). Updated guidelines for evaluating public health surveillance systems—Recommendations from the Guidelines Working Group. *Morbidity and Mortality Weekly Report, 50*(RR13), 1–35.

Goldfrank, L.R., Flomenbaum, N., Lewis, N.A., Howland, M.A., Hoffman, R.S., and Nelson, L.S., eds. (2002). *Goldfrank's Toxicologic emergencies, seventh edition.* New York: McGraw-Hill.

Gotsch, K.E., and Thomas, J.D. (2002). National data sources for the surveillance of poisonings (Abstract). *Journal of Toxicology, Clinical Toxicology, 40*, 641.

Green, L., and Kreuter, M. (1991). *Health promotion planning, second edition.* Mountain View: Mayfield Publishing Co.

Griffin, M., Barrera-Garcia, V., Thompson, G., and Watson, B. (2001). Language and Barriers to Poison Center Utilization. Slide presentation.

Harchelroad, F., Clark, R.F., Dean, B., and Krenzelok, E.P. (1990). Treated vs reported toxic exposure: Discrepancies between a poison control center and a member hospital. *Veterinary and Human Toxicology, 32*, 156–159.

Hardin, J.S., Wessinger, G.R., Proksch, J.W., and Laurenzana, E. (2002). A single dose of monoclonal anti-phencyclidine IgG offers long-term reductions in phencyclidine behavioral effects in reates. *Journal of Pharmacology and Experimental Therapeutics, 302*, 119–126.

Harrison, D.L., Draugalis, J.R., Slack, M.K., and Tong, T.G. (1995). The production model as a basis for conducting economic evaluations of regional poison control centers. *Journal of Toxicology, Clinical Toxicology, 33*(3), 233–237.

Harrison, M.D.L., Draugalis, J.R., Slack, M.K., and Langley, P.C. (1996). Cost-effectiveness of regional poison control centers. *Archives of Internal Medicine, 156*(Dec 9/23), 2601–2608.

Health Resources and Services Administration. (2004). Review of poison control centers in the 2003 NBHPP awardee applications. HRSA Special Programs Bureau, Division of Healthcare Emergency Preparedness. Washington, DC: U.S. Department of Health and Human Services.

Heard, S.E. (2003). The California Poison Control System: Update for the Institute of Medicine. Presentation to the Committee on Poison Prevention and Control, Institute of Medicine.

Henneberger, P.K., Metayer, C., Layne, L.A., and Althouse, R. (2002). Nonfatal work-related inhalations: Surveillance data from hospital emergency departments, 1995–1996. *American Journal of Industrial Medicine, 38*, 140–148.

Hoffman, B.B. (2001). Catecholamines, sympathomimetic drugs, and adrenegic receptor antagonists. In J.G. Hardman and L.E. Limbird (Eds.), *Goodman and Gilman's: The pharmacological basis of therapeutics, tenth edition.* New York: McGraw-Hill.

Hoffman, R.S. (2002). Poison information centers and poison epidemiology. In L.R. Goldfrank, N.E. Flomenbaum, N.A. Lewis, M.A. Howland, R.S. Hoffman, and L.S. Nelson (Eds.), *Goldfrank's toxicologic emergencies, seventh edition* (pp. 1747–1752). New York: McGraw-Hill.

Home Safety Council. (2002). *The state of home safety in America: The facts about unintentional injury in the home.* Wilkesboro, NC: Author.

Honigman, B., Lee, J., Rothschild, J., Light, P., Pulling, R.M., Yu, T., and Bates, D.W. (2001). Using computerized data to identify adverse drug events in outpatients. *Journal of the American Medical Informatics Association, 8*(3), 254–266.

Hoppe-Roberts, J.M., Lloyd, L.M., and Chyka, P.A. (2000). Poisoning mortality in the United States: Comparison of national mortality statistics and poison control center reports. *Annals of Emergency Medicine, 35*(5), 440–448.

Horton, D.K., Berkovitz, Z., and Kaye, W.E. (2002). The public health consequences from acute chlorine releases, 1993–2000. *Journal of Occupational and Environmental Medicine, 44*, 906–913.

Hoyt, B.T., Rasmussen, R., Griffin, S., and Smilkstein, M.J. (1999). Poison center data accuracy: A comparison of rural hospital chart data with the TESS data base. *Academy of Emergency Medicine, 6*, 851–855.

Hurt, R.D. (1985). The poison squad. *Timeline, 2*(1), 64–70.

Institute of Medicine. (1988). *The future of public health.* Committee for the Study of the Future of Public Health. Division of Health Care Services. Washington, DC: National Academy Press.

Institute of Medicine. (2002a). *The future of the public's health in the 21st century.* Committee on Assuring the Health of the Public in the 21st Century. Board on Health Promotion and Disease Prevention. Washington, DC: National Academy Press.

Institute of Medicine. (2002b). *Speaking of health: Assessing health communication strategies for diverse populations.* Committee on Communication for Behavior Change in the 21st Century: Improving the Health of Diverse Populations. Board on Neuroscience and Behavioral Health. Washington, DC: National Academy Press.

Institute of Medicine and National Research Council. (1999). *Chemical and biological terrorism: Research and development to improve civilian medical response.* Committee on R&D Needs for Improving Civilian Medical Response to Chemical and Biological Terrorism Incidents. Washington, DC: National Academy Press.

Journal of Toxicology, Clinical Toxicology. (2001). Abstracts of the 2001 North American Congress of Clinical Toxicology Annual Meeting. *39*(5), 473–568.

Journal of Toxicology, Clinical Toxicology. (2002). Abstracts of the 2002 North American Congress of Clinical Toxicology Annual Meeting. *40*(5), 599–698.

Journal of Toxicology, Clinical Toxicology. (2003). Abstracts of the 2003 North American Congress of Clinical Toxicology Annual Meeting. *41*(5), 641–752.

Kelly, N.R., Kirkland, R.T., Holmes, S.E., Ellis, M.D., Delclos, G., and Kozinetz, C.A. (1997). Assessing parental utilization of the poison center: An emergency center-based survey. *Clinical Pediatrics, 36*(8), 467–473.

Kelly, N.R., Huffman, L.C., Mendoza, F.S., and Robinson, T.N. (2003). Effects of a videotape to increase use of poison control centers by low-income and Spanish-speaking families: A randomized, controlled trial. *Pediatrics, 111*(1), 21–26.

King, W.D. (1991). Pediatric injury surveillance: Use of a hospital discharge database. *Southern Medical Journal, 84*, 342–348.

King, W.D., and Palmisano, P.A. (1991). Poison control centers: Can their value be measured? *Southern Medical Journal, 84*(6), 722–726.

Kirk, M.A., Cisek, J., and Rose, S.R. (1994). Emergency department response to hazardous materials incidents. *Concepts and Controversies in Toxicology, 12*(2), 461–481.

Klein-Schwartz, W., and Smith, G.S. (1997). Agricultural and horticultural chemical poisonings: Mortality and morbidity in the United States. *Annals of Emergency Medicine, 29*, 232–238.

Kozel, N.J. (1990). Epidemiology of drug abuse in the United States: A summary of methods and findings. *Bull PAHO, 24*, 53–62.

KRC Research. (2003). Omnibus Tracking Survey (slide presentation) to the American Association of Poison Control Centers. April.

KRC Research and Consulting. (2001). Slide presentation to the American Association of Poison Control Centers on a study of nine focus groups, comparing areas that currently use the Mr. Yuk logo with those that do not. March 13.

Krenzelok, E.P. (1998). Editorial commentary: Do poison centers save money . . . ? What are the data? *Clinical Toxicology, 36*(6), 545–547.

Krenzelok, E.P., and Dean, B.S. (1988). Hazardous Substance Center: A poison center's workers right to know program. *Veterinary and Human Toxicology, 20*(1), 18–20.

Krenzelok, E.P., and Mvros, R. (2003). Initial impact of toll-free access on poison center call volume. *Veterinary and Human Toxicology, 45*(6), 325–327.

Kroner, B.A., Scott, R.B., Waring, E.R., and Zanaga, J.R. (1993). Poisoning in the elderly: Characterization of exposures reported to a poison control center. *Journal of the American Geriatric Society, 41*, 842–846.

Krueger, A.-M. (2003). Low Income and Culturally Diverse Populations: Special Programs. Presentation to the Committee on Poison Prevention and Control, Institute of Medicine.

Krummen, K., Tsipis, G.B., Siegel, E., and Bottei, E. (1999). Accuracy of drug abuse call patterns in predicting prescription drug abuse (abstract). *Journal of Toxicology, Clinical Toxicology, 37*(5), 643–644.

Landen, M.G., Castle, S. Notle, K.B., et al. (2003). Methodological issues in the surveillance of poisoning, illicit drug overdose, and heroin overdose deaths in New Mexico. *American Journal of Epidemiology, 157*, 273–278.

Leikin, J.B., and Krenzelok, E.P. (2001). In M.D. Ford, K.A. Delaney, L.J. Ling, and T. Erickson (Eds.), *Clinical toxicology* (pp. 111–114). New York: W.B. Saunders Company.

Levy, B.S., and Sidel, V.W. (2003). *Terrorism and public health.* New York: Oxford.

Lewis, C. (2002). The "Poison Squad" and the advent of food and drug regulation. *FDA Consumer, 36*(6), 12–15.

Liller, K.D., Craig, J., Crane, N., and McDermott, R.J. (1998). Evaluation of a poison prevention lesson for kindergarten and third grade students. *Injury Prevention, 4*, 218–221.

Linakis, J.G., and Frederick, K.A. (1993). Poisoning deaths not reported to the regional poison control center. *Annals of Emergency Medicine, 22*, 1822–1828.

Litovitz, T. (1998). The TESS database—Use in product safety assessment. *Drug Safety, 18*(1), 9–19.

Litovitz, T.L., Klein-Schwartz, W., Rodgers, G.C., et al. (2002). 2001 annual report of the American Association of Poison Control Centers Toxic Exposure Surveillance System. 2002. *The American Journal of Emergency Medicine, 20*(5), 391–452.

Lovejoy, F.H., Jr., Robertson, W.O., and Woolf, A.D. (1994). Poison centers, poison prevention, and the pediatrician (Part 1 of 2). *Pediatrics, 94*(2), 220–224.

MacKenzie, W.R., Schell, W.L., Blair, K.A., Addiss, D.G., Peterson, D.E., Hoxie, N.J. et al. (1995). Massive outbreak of waterborne *Cryptosporidium* infection in Milwaukett, Wisconsin: Recurrence of illness and risk of secondary transmission. *Clinical Infectious Diseases, 21*, 57–62.

Manoguerra, A.S. (1976). The poison control center—Its role. *American Journal of Pharmacology and Education, 40*(4), 382–382.

Marder, S., Winkler, T., Tadaki, K., Bobbink, S., and Robertson, W.O. (2001). Decoding "drug imprints" at the millennium: A proposal to increase accuracy and reduce costs. *Veterinary and Human Toxicology, 43*(1), 46–47.

Martin, J.M., and Arena, J.M. (1939). Lye poisoning and stricture of the esophagus: A report of 50 cases. *Southern Medical Journal, 32*, 286–290.

McCaig, L.F., and Burt, C.W. (1999). Poisoning-related visits to emergency departments in the United States, 1993–1996. *Clinical Toxicology, 37*(7), 817–826.

McGuigan, M.A. (1997). Quality management for poison centers. *Journal of Toxicology, Clinical Toxicology, 35*(3), 283–293.

McIntrie, M.S., Angle, C.R., Ekins, B.R., Mofensen, H., Rauber, A., and Scherz, R. (1984). Trends in childhood poisoning: A collaborative study 1970, 1975, 1980. *Journal of Toxicology, Clinical Toxicology, 21*, 321–331.

McKnight, R.H., Levine, E.J., and Rodgers, G.C. (1994). Detection of green tobacco sickness by a regional poison center. *Veterinary and Human Toxicology, 36*, 505–510.

Miller, T.R., and Lestina, D.C. (1997). Costs of poisoning in the united states and savings from poison control centers: A benefit-cost analysis. *Annals of Emergency Medicine, 29*(2), 239–245.

Mofenson, H.C. (1975). The American Association of Poison Control Centers (founded 1958). *Clinical Toxicology, 8*(1), 77–79.

Morbidity and Mortality Weekly. (1996). Scopolamine poisoning among heroin users—New York City, Newark, Philadelphia, and Baltimore, 1995 and 1996. *45*(22), 457–480.

Morton, W.S. (1998). Hawaii Poison Center: Forty years of saving lives and health costs. *Hawaii Medical Journal, 57*, 440–442.

Mrvos, R., Dean, B.S., and Krenzelok, E.P. (1988). A poison center's emergency response plan. *Veterinary and Human Toxicology, 30*(2), 138–140.

Nathan, A.R., Olson, K.R., Everson, G.W., Kearney, T.E., and Blanc, P.D. (1992). Effects of a major earthquake on calls to regional poison control centers. *Western Journal of Medicine, 156*(3), 278–280.

The National Conference of Pharmaceutical Organizations. (2002). A 21st century system of terrorism defense. June White Paper.

North American Congress of Clinical Toxicology. (2003). Abstracts from the 2002 North American Congress of Clinical Toxicology Annual Meeting. *Journal of Toxicology, Clinical Toxicology, 41*, 641–752.

Oderda, G.M., and Klein-Schwartz, W. (1985). Public awareness survey: The Maryland Poison Center and Mr. Yuk, 1981 and 1975. *Public Health Reports, 100*(3), 278–282.

Olson, K.R., Phillips, K.A., Kearney, T.E., et al. (1999). Cost effectiveness analysis of a regional poison control center. CDC Contract: U50/CCU910980. San Francisco, CA: California Poison Control System.

Olson, K.R., et al. (2003). *Poisoning and drug overdose, fourth edition.* New York: McGraw-Hill.

Orr, M.F., Kaye, W.E., Zeitz, P., Powers, M.E., and Rosenthal, L. (2001). Public health risks of railroad hazardous emergency events. *Journal of Occupational and Environmental Medicine, 43*, 94–100.

Osborn, H. (2004). Ethanol. In L.R. Goldfrank, et al., *Toxicological emergencies, fifth edition* (pp. 813–824). Norwalk, CT: Appleton and Lange.

Osterloh, J.D., and Kelly, T.J. (1999). Study of the effect of lactational bone loss on blood lead concentrations in humans. *Environmental Health Perspective, 107*, 187–194.

Palmer, M.E., Haller, C., McKinney, P.E., et al. (2003). Adverse events associated with dietary supplements: An observational study. *The Lancet, 361.*

Palmisano, P.A. (1981). Targeted intervention in the control of accidental drug overdoses by children. *Public Health Reports, 96*(2), 150–156.

Patel, M.M., Tsutaoka, B.T., Banerji, S., Blanc, P.D., and Olson, K.R. (2002). ED utilization of computed tomography in a poisoned population. *American Journal of Emergency Medicine, 20,* 212–217.

Pentel, P.R., Scarlet, W., Ross, C.A., Landon, J., Sidki, A., and Keyler, D.E. (1995). Reduction of desipramine cardiotoxicity and prolongation of survival in rats with the use of polyclonal drug-specific antibody Fab fragments. *Annals of Emergency Medicine, 26,* 334–341.

Phillips, K.A., Homan, R.K., Luft, H.S., et al. (1997). Willingness to pay for poison control centers. *Journal of Health Economics, 16,* 343–357.

Phillips, K.A., Homan, R.K., Hiatt, P.H., et al. (1998). The costs and outcomes of restricting public access to poison control centers: Results from a natural experiment. *Medical Care, 36,* 271–280.

Poison Control Center Advisory Work Group. (1996). *Final Report: The Poison Control Center Advisory Work Group.* Submitted to the National Center for Injury Prevention and Control (CDC) and the Maternal and Child Health Bureau (HRSA). Atlanta: Centers for Disease Control and Prevention.

Polivka, B.J., Elliott, M.B., and Wolowich, W.R. (2002). Comparison of poison exposure data: NHIS and TESS data. *Clinical Toxicology, 40*(7), 839–845.

Pond, S.M., Olson, K.R., Woo, O.F., et al. (1986). Amatoxin poisoning in northern California, 1982–1983. *Western Journal of Medicine, 145,* 204–209.

Powell, E.C., and Tanz, R.R. (2002). Adjusting our view of injury risk: The burden of nonfatal injuries in infancy. *Pediatrics, 110,* 792–796.

Reddy, U.P., Yee, S., Evanoff, J., et al. (1999). Enhancing poison prevention by pre-emptive family education: A randomized prospective study. *Pediatrics (Supplement to Pediatrics, Part 3 of 3), 104*(3), 701–702 (Abstract 13).

Rice, D.P., MacKenzie, E.J., Jones, A.S., et al. (1989). *Cost of injury in the United States: A report to Congress.* San Francisco: Institute for Health and Aging, University of California; Injury Prevention Center, Johns Hopkins University.

Roberts, C.D. (1996). Data quality of the Drug Abuse Warning Network. *American Journal of Drug and Alcohol Abuse, 22,* 389–401.

Rodgers, G.B. (1996). The safety effects of child-resistant packaging for oral prescription drugs. *Journal of the American Medical Association, 275*(21), 1661–1665.

Rodgers, G.B. (2002). The effectiveness of child-resistant packaging for aspirin. *Archives of Pediatric and Adolescent Medicine, 156* (September).

Rodriguez, J.G., and Sattin, R.W. (1987). Epidemiology of poisonings leading to hospitalization in the United States, 1979–1983. *American Journal of Preventive Medicine, 3,* 164–170.

Rosenstock, I.M., Strecher, V.J., and Becker, M.H. (1994). The health belief model and HIV risk behavior change. In R.J. DiClemente and J.L. Peterson (Eds.), *Preventing AIDS: Theories and methods of behavioral interventions* (pp. 5–24). New York: Plenum Press.

Roth, B., Woo, O., and Blanc, P.D. (1999). Early metabolic acidosis and coma following massive acetaminophen ingestion. *Annals of Emergency Medicine, 33,* 452–456.

Rouse, B.A. (1996). Epidemiology of illicit and abused drugs in the general population, emergency department drug-related episodes, and arrestees. *Clinical Chemistry, 42,* 1330–1336.

Rumack, B.H. (1975). POISINDEX: An emergency poison management system. *Drug Information Journal, 9*(2–3), 171–180.

Rumack, B.H., Ford, P., Sbarbaro, J., Bryson, P., and Winokur, M. (1978). Regionalization of poison centers—A rational role model. *Clinical Toxicology, 12*(3), 267–275.

Russell, S.L., and P.A. Czajka. (1984). Comparison of poison control statutes in the United States. *American Journal of Hospital Pharmacy, 41,* 481–484.

Saunders, S.K., Kempainen, R., and Blanc, P.D. (1996). Outcomes of ocular exposures reported to a regional poison control center. *Journal of Toxicology, Cut and Ocular Toxicology, 15,* 249–259.

Scherz, R.G., and Robertson, W.O. (1978). The history of poison control centers in the United States. *Clinical Toxicology, 12*(3), 291–296.

Seifert, S.A., Von Essen, S., Jacobitz, K., Crouch, R., and Lintner, C.P. (2003). Do poison centers diagnose organic dust toxic syndrome? *Journal of Toxicology, Clinical Toxicology, 41,* 115–117.

Shepherd, G., and Klein-Schwartz, W. (1998). Accidental and suicidal adolescent deaths in the United States, 1979–1994. *Archives of Pediatric and Adolescent Medicine,* 1181–1185.

Smith, G.S., Langlois, J.A., and Buechner, J.S. (1991). Methodological issues in using hospital data to determine the incidence of hospitalized injuries. *American Journal of Epidemiology, 134,* 1146–1158.

Smith, S.M., Colwell, L.S., and Sniezek, J.E. (1985). An evaluation of external cause-of-injury codes using hospital records fróm the Indian Health Service, 1985. *American Journal of Public Health, 80,* 279–281.

Soslow, A.R., and Wolf, A.D. (1992). Reliability of data sources for poisoning deaths in Massachusetts. *American Journal of Emergency Medicine, 10,* 124–127.

Spiller, H.A., and Bosse, G.M. (2003). Prospective study of morbidity associated with snakebite envenomation. *Journal of Toxicology, Clinical Toxicology, 41,* 125–130.

Spiller, H.A., and Krenzelok, E.P. (1997). Epidemiology of inhalant abuse reported to two regional poison centers. *Journal of Toxicology, Clinical Toxicology, 35,* 167–174.

Suarez, L., Nichols, D.C., Pulley, L., Brady, C.A., and McAlister, A. (1993). Local health departments implement a theory-based model to increase breast and cervical cancer screening. *Public Health Reports, 108*(4), 477–482.

Sumner, D., and Langley, R. (2000). Pediatric pesticide poisoning in the Carolinas: An evaluation of the trends and proposal to reduce the incidence. *Veterinary and Human Toxicology, 42,* 101–103.

Symonds, J., and Robertson, W. (1967). Drug identification: Use of coded imprint. *Journal of the American Medical Association, 199*(9), 664–665.

Taylor, V. (1994). Medical community involvement in a breast cancer screening promotional project. *Public Health Reports, 109*(4), 491–500.

Teitelbaum, D. (1968). New directions in poison control. *Clinical Toxicology, 1*(1), 3–13.

Thacker, S.B., and Berkelman, R.L. (1988). Public health surveillance in the United States. *Epidemiological Review, 10,* 164–190.

Thompson, D.F., Trammel, H.L., Robertson, N.J., and Reigart, J.R. (1983). Evaluation of regional and nonregional poison centers. *The New England Journal of Medicine, 398*(4), 191–194.

Thurman, P.A. (2003). Detection of drug-related adverse events in hospitals. *Expert Opinion on Drug Safety, 2*(5), 447–449.

Trestrail, J.H., III, (2003). What is a "certified" regional poison center? Presentation to the Committee on Poison Prevention and Control, Insitute of Medicine.

Troutman, W.G., and Wanke, L.A. (1983). Advantages and disadvantages of combining poison control and drug information centers. *American Journal of Hospital Pharmacy, 40*(7), 1219–1222.

United States Pharmacopoeia. (2002). Summary of information submitted to MEDMARX in the year 2002. Available: <http://store.usp.org> and <http://www.usp.org/medmarx/overview.html>.

U.S. Department of Health and Human Services. (2002). *Mortality from the Drug Abuse Warning Network, 2001.* Substance Abuse and Mental Health Services Administration Office of Applied Studies. Available: <http://www.samhsa.gov>.

U.S. Department of Health and Human Services. (2003). *Emergency department trends from the Drug Abuse Warning Network, final estimates 1995–2002.* Substance Abuse and Mental Health Services Administration Office of Applied Studies. Available: <http://www.samhsa.gov>.

Valent, F., McGwin, G., Bovenzi, M., and Barbone, F. (2002). Fatal work-related inhalation of harmful inhalation of harmful substances in the United States. *Chest, 121*, 969–975.

Veltri, J.C., and Litovitz, T.L. (1984). 1983 annual report of the American Association of Poison Control Centers National Data Collection System. *American Journal of American Medicine, 2*, 420–443.

Vernberg, K., Culver-Dickinson, P., and Spyker, D.A. (1984). The deterrent effect of poison-warning stickers. *American Journal of Diseases of Children, 138*(11), 1018–1120.

Walton, W. (1982). An evaluation of the Poison Prevention Packaging Act. *Pediatrics, 69*(3), 363–370.

Wan, C., Cardus, L., McGreevy, B., Lewis, V., Johnson, J., and Robertson, W.O. (1993). Content audit of POISINDEX. *Veterinary and Human Toxicology, 35*(2), 168–169.

Wanke, L.A., Burton, B., and Putnam, T.S. (1988). Financial support for poison control centers: A unique partnership with a chain drug store corporation. *Veterinary and Human Toxicology, 30*(5), 168–169.

Watson, B., and Villarrreal, R. (2002). Increased Aware[ness] of Poison Center Services in the Hispanic Community (CFDA#93.253–HRSA h4B MC 00052-01). Slide Presentation.

Watson, W., Litovitz, T., Rodgers, G., Klein-Schwartz, W., Youniss, J., Rose, S., Borys, D., and May, M. (2003). 2002 annual report of the American Association of Poison Control Centers Toxic Exposure Surveillance System. *American Journal of Emergency Medicine, 21*(5), 353–421.

Wax, P.M., and Donovan, J.W. (2000). Fellowship training in medical toxicology: Characteristics, perceptions, and career impact. *Journal of Toxicology, Clinical Toxicology, 38*(6), 637–642.

Weisskopf, M.G., Drew, J.M., Hanrahan, L.P., Anderson, H.A., and Haugh, G.S. (2003). Hazardous ammonia releases: Public health consequences and risk factors for evacuation and injury, United States, 1993–1998. *Journal of Occupational and Environmental Medicine, 45*, 197–204.

Wheatley, G.M. (1953). A formula for child safety. *Ohio State Medical Journal, 49*, 609–613.

Wolf, A., Lewander, W., Filipoone, G., and Lovejoy, F. (1987). Prevention of childhood poisoning: Efficacy of an educational program carried out in an emergency clinic. *Pediatrics, 80*(3), 359–363.

Woolf, A., and Shaw, J. (1998). Childhood injuries from artificial nail primer cosmetic products. *Archives of Pediatric and Adolescent Medicine, 152*, 41–46.

Woolf, A., Wieler, J., and Greenes, D. (1997). Costs of poison-related hospitalizations at an urban teaching hospital for children. *Archives of Pediatrics and Adolescent Medicine, 151*, 719–723.

World Health Organization. (1989). *International statistical classification of diseases. Ninth revision (3 volumes). Clinical Modifications.* Commission on Professional Hospital Activities. Ann Arbor, MI: Author.

World Health Organization. (1992–1994). *International statistical classification of diseases and related health problems tenth revision (3 volumes).* Geneva: Author.

Youniss, J., Litovitz, T., and Villanueva, P. (2000). Characterization of U.S. poison centers. *Veterinary and Human Toxicology, 42*(1), 43–53.

Zuvekas, A., Scarpulla Nolan, L., Azzouzi, A., Tumaylle, C., and Ellis, J. (1997). An analysis of potential economies of scale in poison control centers. Final Report for health policy research. Washington, DC: Center for Policy Research, The George Washington University Medical Center.

Appendix A

Contributors

Bruce Anderson
Maryland Poison Center

Leo Artalejo
West Texas Regional Poison Center

Suzanne Barone
Consumer Product Safety Commission

Marty Belson
Centers for Disease Control and Prevention (NCIPC)

John Benitez
Fingerlakes Regional Poison and Drug Information Center

Angel Bivens
Maryland Poison Center

Jerome Blondell
Environmental Protection Agency

Greg Bogdan
Rocky Mountain Poison and Drug Center

Lisa Booze
Maryland Poison Center

Doug Borys
American Association of Poison Control Centers

Delon Brennen
Johns Hopkins University

Al Bronstein
Rocky Mountain Poison and Drug Center

Dan Budnitz
Centers for Disease Control and Prevention (NCIPC)

Keith Burkhart
American College of Emergency Physicians

Jessica Cates
Massachusetts Department of Public Health

Peter Chyka
American Board of Applied Toxicology

Miriam Cisternas
MGC Data Services

Laurel Copeland
University of Michigan

Richard Dart
Rocky Mountain Poison and Drug Center

Suzanne Doyon
Maryland Poison Center

Lois Fingerhut
National Center for Health Statistics

Tracy Finlayson
University of Michigan

Holly Hackman
Massachusetts Department of Public Health

Christina Hantsch
Illinois Poison Center

Stacy Harper
Centers for Disease Control and Prevention (NCIPC)

Stuart Heard
California Poison Control System

Mark Johnson
Alaska Department of Health and Social Sciences

Edwin Kilbourne
Centers for Disease Control and Prevention (ASTD)

Wendy Klein-Schwartz
Maryland Poison Center

Ann-Marie Krueger
Banner Poison Control Center

Ruth Lawrence
Fingerlakes Regional Poison and Drug Information Center

Toby Litovitz
American Association of Poison Control Centers

Robin Malinowski
Illinois Poison Center

Steven Marcus
New Jersey Poison Control Center

Jude McNally
Arizona Poison and Drug Center

Maria Mercurio-Zappala
New York City Poison Control Center

Maureen Metzger
University of Michigan

Rick Niemeier
Centers for Disease Control and Prevention (NIOSH)

Mary Powers
Wisconsin Poison Center

William Robertson
Washington Poison Control Center

Maria Rudis
Society of Critical Care Medicine

Emilio Saenz
West Texas Regional Poison Center

Donna Seger
American Academy of Clinical Toxicology

Greene Shepherd
North Texas Poison Center

Monique Sheppard
Pacific Institute for Research and Evaluation

Cathy Smith
Wisconsin Poison Center

Soheil Soliman
University of Michigan

Rosanne Soloway
American Association of Poison Control Centers

Ernest Stremski
Wisconsin Poison Center

Philip Talboy
Centers for Disease Control and Prevention (NCIPC)

Michael Thompson
Mississippi Poison Control Center

John Trestrail
DeVos Children's Hospital Regional Poison Center

Robert Waddell II
National Association of Emergency Technicians

Mike Wahl
Illinois Poison Center

Evelyn Waring
Virginia Poison Center

William Watson
American Association of Poison Control Centers

Paul Wax
American College of Medical Toxicology

Sara Welch
Georgia Poison Center

Mary Willy
Office of Drug Safety, Food and Drug Administration

Kathryn Wruk
Rocky Mountain Poison and Drug Center

SPONSORS

Byron Bailey, Health Resources and Services Administration

Carol Delany, Health Resources and Services Administration

David Heppel, Health Resources and Services Administration

Richard J. Smith III, Health Resources and Services Administration

Peter Van Dyck, Maternal and Child Health Bureau

Appendix B

Committee and Staff Biographies

Bernard Guyer, M.D., M.P.H. ***(Chair)*****,** is Zanvyl Kreiger Professor of Children's Health and former chair of the Department of Population and Family Health Science in the Bloomberg School of Public Health at Johns Hopkins University. He is a member of the Institute of Medicine, and served as chair of the Committee on Immunization Finance Policies and Practices. Dr. Guyer's research interests are in maternal and child health, including work on prevention of childhood injuries, childhood immunization and primary care systems, strategies for preventing infant mortality, and maternal and child health program organization, finance, and implementation. His research examines the span of children's health conditions, including etiological factors, prevention strategies, systems of health care, and the social and policy contexts for child health and well-being.

Jeffrey Alexander, Ph.D., is the Richard Carl Jelinek Professor of Health Management and Policy in the School of Public Health, University of Michigan. He also holds positions as professor of organizational behavior and human resources management, School of Business; faculty associate, Survey Research Center, Institute for Social Research; and research scientist, Veterans Affairs Health Services Research and Development Center. He received his Ph.D. in Sociology (organization theory) from Stanford University in 1980, after earning a Master's degree in Health Services Administration from Stanford in 1976. His teaching and research interests focus on organizational change in the health care sector, multi-institutional systems, governance and physician participation in institutional management, and policy making. His research has focused extensively on interorganizational arrangements in the health care sector and includes studies of integrated health care systems, public–private partnerships, physician–organization arrangements, and multihospital systems. His recent publications have appeared in *Health Services Research, The Milbank Quarterly, Medical Care Research and Review, Administrative Science Quarterly*, and *Journal of Health and Social Behavior*.

Paul D. Blanc, M.D., M.S.P.H., is a professor of medicine and Endowed Chair in Occupational and Environmental Medicine in the Division of Occupational and Environmental Medicine at the University of California–San Francisco. He is a member of the Toxic Air Contaminant Scientific Review Panel of the California Air Resources Board and associate medical director of the California Poison Control System, San Francisco Division. Dr. Blanc is board certified in occupational medicine and internal medicine and holds a certificate of added qualifications in medical toxicology. His current research interests are in the epidemiology of occupational lung disease and occupational toxicology, especially in terms of pulmonary responses. He has authored several reports based on poison control surveillance.

Dennis Emerson, R.N., B.S., is a nurse in the emergency department of Saint Luke's Regional Medical Center in Boise, ID. He has served as an emergency department nurse for 23 years and was a nurse at the Idaho Poison Control Center for 8 years. Mr. Emerson is an active member of the Emergency Nurses Association and former member of the Pediatric Resource Group. He is an instructor for three courses: Emergency Nursing Pediatric Course, Trauma Nurse Core Course, and Advanced Cardiac Life Support. He is a Certified Poison Information Specialist and an instructor in hospital operations training for hazardous material exposure.

Jerris R. Hedges, M.D., is professor and chair, Department of Emergency Medicine, Oregon Health and Science University, Portland. He is internationally recognized for his contributions to the development of emergency medicine as a scientific discipline. His book, *Clinical Procedures in Emergency Medicine*, helped define the discipline and set the standard for evidence-based practice texts addressing clinical procedures. Dr. Hedges' research broadly encompasses the field, and his work on transcutaneous cardiac pacing in the early 1980s helped introduce that modality into daily emergency practice. His evaluations of Oregon trauma care have helped clarify the impact of systems of trauma care on clinical outcomes.

Dr. Hedges is the current president of the Association of Academic Chairs of Emergency Medicine. He is a member of the Institute of Medicine.

Mark Scott Kamlet, Ph.D., is provost at Carnegie Mellon University and H. John Heinz III Professor of Economics and Public Policy. His primary research interests include health policy, federal budget policy, and statistical methodology (econometrics and decision analysis). Professor Kamlet served on the expert panel that published the book *Cost-Effectiveness in Health and Medicine: A Report of the Expert Panel on Cost-Effectiveness in Health and Medicine*. This panel was charged with developing national guidelines for conducting cost-effectiveness analysis for health policy. He has also served on National Institutes of Health panels to propose national guidelines for population-based genetic screening for cystic fibrosis and for newborn screening for metabolic disorders.

Angela Mickalide, Ph.D., C.H.E.S., is program director of the National SAFE KIDS Campaign, Washington, DC, a nonprofit organization dedicated to preventing unintentional injuries among children ages 14 and under. In this capacity, she ensures the scientific rigor of the Campaign's injury prevention programs addressing traffic injuries, fire and burns, drowning, falls, airway obstruction, and poisoning for more than 600 SAFE KIDS coalitions and chapters in all 50 states and the District of Columbia. Dr. Mickalide is an adjunct associate professor of prevention and community health at the George Washington University School of Public Health and Health Services. Dr. Mickalide has published 50 articles, book chapters, and research reports and has delivered more than 200 presentations concerning injury prevention, health education, and clinical preventive services. She serves on the editorial boards of *Injury Prevention, American Journal of Preventive Medicine,* and *Health Promotion Practice* and served previously on the editorial boards of *Patient Education and Counseling* and *Health Education and Behavior.*

Paul Pentel, M.D., is a professor in the Departments of Medicine and Pharmacology at the University of Minnesota Medical School and chief of the Division of Clinical Pharmacology at Hennepin County Medical Center. He serves as president of the Minneapolis Medical Research Foundation, and is a past president of the American College of Medical Toxicology. Dr. Pentel's research interests include the mechanisms and treatment of antidepressant and stimulant drug toxicity, drug dependence, and the pharmacokinetic determinants of drug dependence. His current focus is nicotine and tobacco pharmacology, and the development of treatment medications for tobacco dependence. He is certified in internal medicine and medical toxicology.

Barry H. Rumack, M.D., served as director of the Rocky Mountain Poison and Drug Center in Denver from 1974 through 1991 and is now director emeritus. He has been professor of pediatrics at the University of Colorado School of Medicine and is now Clinical Professor. He founded Poisindex®, a toxicology database, and developed other databases as part of Micromedex, which is now part of the Thomson Corporation. He served as president of the American Association of Poison Control Centers from 1982 to 1984 and was chairman of the American Board of Medical Toxicology from 1988 to 1990. He has been a member of the Colorado State Board of Health and has served on advisory committees to the Consumer Product Safety Commission, the Food and Drug Administration, the Governor of Colorado for the Rocky Mountain Arsenal and Lowry Landfill, and others. He also serves as an arbitrator for the American Arbitration Association. His specialty is pediatrics and general medical toxicology, with specific interests in acetaminophen and mushrooms.

David P. Schor, M.D., M.P.H., F.A.A.P., is chief of the Division of Family and Community Health Services in the Ohio Department of Health. Prior to taking this position, he held positions as health promotion and education director, medical advisor, and maternal and child health director in the Nebraska Department of Health and Human Services, in which he implemented program expansions in tobacco control, cardiovascular disease and injury prevention, and asthma programs; developed methods and reports for child death review and asthma surveillance; and staffed or led consensus and task force panels for diabetes treatment, human genetics technologies, and state infant mortality. Dr. Schor is certified by the University of Oklahoma School of Public Health as an instructor in epidemiology and biostatistics and has served as a consultant to the federal Maternal and Child Health Bureau in the area of grant reviews and site visits.

Daniel Adrian Spyker, M.D., Ph.D., is the director of clinical pharmacology at Genentech, Inc., South San Francisco, CA. He was an academic internist at the University of Virginia until he joined the Food and Drug Administration (FDA) in 1990. At the FDA, Dr. Spyker served as medical officer in the Center for Drug Evaluation and Research Pilot Drug Evaluation staff from 1990 to 1993, after which he joined the Center for Diseases and Radiological Health as deputy director in the Division of Cardiovascular, Respiratory, and Neurological Devices. He is an active and enthusiastic participant in the design and reporting of clinical trials and in drug and device labeling, in the design of information systems relating to toxicology and pharmacokinetics, and in the application of biostatistical models and statistical analysis. He has a Ph.D. in electrical engineering;

his medical specialties are internal medicine, medical toxicology, and clinical pharmacology.

Andy Stergachis, Ph.D., R.Ph., is professor of epidemiology and affiliate professor of pharmacy, Northwest Center for Public Health Practice, School of Health and Community Medicine, University of Washington, where his focus is on public health emergency preparedness and response. Previously, he was professor and chair of the Department of Pharmacy, School of Pharmacy, University of Washington. He is a former member of the Board of the Washington Poison Center. His research and practice have been at the intersection of public health and proper medication use and he has done extensive work in postmarketing drug evaluations. Dr. Stergachis serves as a member of the National Committee for Quality Assurance's Asthma Measurement Advisory Panel and the Agency for Healthcare Research and Quality Health Systems Research Study Section. He served as a member of the Institute of Medicine Committee on Interactions of Drugs, Biologics, and Chemicals in U.S. Military Forces. He is the former chief pharmacist for drugstore.com. He was awarded the 2002 Pinnacle Award by the American Pharmaceutical Association Foundation for his career contributions toward improving quality of care through the medication use process.

David J. Tollerud, M.D., M.P.H., is professor and chair of the Department of Environmental and Occupational Health Sciences at the School of Public Health and Information Sciences, University of Louisville, Kentucky. Dr. Tollerud has medical training in internal medicine, pulmonary and critical care medicine, and occupational medicine, and research expertise in environmental and occupational health, epidemiology, immunology, and injury prevention. He currently serves on the Institute of Medicine's Board on Health Promotion and Disease Prevention and chairs the National Research Council's Committee on the Superfund Site Assessment and Remediation in the Coeur d'Alene River Basin. He has served on numerous Institute of Medicine committees and chaired a series of committees that evaluated the health effects of Agent Orange in Vietnam veterans.

Deborah Klein Walker, Ed.D., recently joined Abt Associates, Inc., as a principal associate in the Health Services, Research and Evaluation practice area. For the prior 15 years, Dr. Walker was at the Massachusetts Department of Public Health, where she most recently was the associate commissioner for programs and prevention and was responsible for programs in maternal and child health, health promotion, and disease prevention; primary care and community health programs; minority health;

and data integration and information systems. She is currently an adjunct professor at the Boston University School of Public Health and an adjunct lecturer at the Harvard School of Public Health. Dr. Walker is a past president of the Association of Maternal and Child Health Programs and a former board member of the American Public Health Association. Dr. Walker's research and policy interests include child and family policy, program implementation and evaluation, public health practice, disability policy, community health systems, health outcomes, and data systems.

Liaison

Mary Jane England, M.D., is president of Regis College. She trained in psychiatry at Boston's University Hospital and at San Francisco's Mt. Zion Hospital, completing her child and adolescent residency at Boston University and Boston City Child Guidance Clinic. She has served as director of child psychiatry at Brighton's St. Elizabeth Hospital of Boston, then as director of clinical psychiatry at the Brighton-Allston Mental Health Clinic, and from 1974 to 1976 as director of planning and manpower for Children's Services in the Massachusetts Department of Mental Health. In 1976 she assumed the position of associate commissioner, Massachusetts Department of Mental Health, and was a consultant to and chairperson of the Human Resources Policy Committee at the National Institute of Mental Health. As commissioner of the Massachusetts Department of Social Services, she designed and implemented a new state agency replacing the Department of Public Welfare to improve the quality of services and ensure citizen involvement as she improved social services and child welfare policy. In 1983 she was lured to Harvard's Kennedy School of Government as associate dean and director of the Lucious N. Littauer Master in Public Administration Program. From 1995 to 1996, Dr. England served as the president of the American Psychiatric Association.

Staff

Anne S. Mavor (*Program Director*) is the study director for the Committee on Poison Prevention and Control. Her previous National Research Council and Institute of Medicine work has included studies on health care messages for diverse populations, workplace activities and their relationship to musculoskeletal disorders, occupational analysis and the enhancement of human performance, the changing nature of work, and the implications of youth values, aptitudes, and opportunities for military recruiting and retention. She is the staff director for the Committee on

Human Factors and has extensive experience in cognitive psychology and information system design. She has an M.S. in experimental psychology from Purdue University.

Susan McCutchen (*Research Associate*) has been on staff at The National Academies for over 20 years and worked in several Academy divisions and with many different boards, committees, and panels in those units. The studies in which she has participated have covered a broad range of subjects and focused on a variety of issues related to international affairs, technology transfer, aeronautics, natural disasters, education, needle exchange, the polygraph, and human factors. She has assisted in the production of a large number of Academy publications. A French major, with minors in English, Italian, and Spanish, she has a B.A. degree from Ohio's Miami University, and an M.A. degree from Kent State University.

Index

A

Access to services, 131, 137, 139, 194, 202, 205, 271, 282, 292
  see also Internet; Public education/ outreach; Telephone services; Utilization of health care services
  cost factors, 140, 141
  emergency service personnel, 114, 126, 282
  ethnic minorities, 43
  historical perspectives, 84
  mass poisonings, 114
  nonpoisoning information needs, 117
  pharmacists, 230
  public education and, 202, 205, 208, 214
  research on, 10, 18, 19, 118, 270, 274, 275, 316, 317
  toxicologists, 87
Accountability
  see also Quality control
  government organizational assurance structure, 10, 12, 282, 284-286
  poison control centers, 12, 13, 100, 107, 310, 311
  Poison Prevention and Control System (proposed), 273, 282, 284-286, 308-309
Accreditation Council for Graduate Medical Education, 122
Adolescents, 3, 29, 223, 224
Advocacy, 10, 11, 25, 31, 39, 203, 204, 205, 218, 271, 274, 289
Age factors, 29
  see also Children
  adolescents, 3, 29, 223, 224
  elderly persons, 3, 29, 43, 51-53, 55, 58, 59, 60, 64, 99
    public education materials, 223
  incidence data, 48, 51-60 (passim), 63, 64, 68, 69
  public education, 209
Agency for Healthcare Research and Quality, 19, 180, 283, 294-295, 317
Agency for Toxic Substances and Disease Registry (ATSDR), 39, 90, 115, 126, 189, 195, 283, 292
  Hazardous Substances Emergency Events Surveillance System, 181, 198
Alcohol use and abuse, 30, 99
  costs of, 29
  deaths due to poisoning, 3, 28, 68
  definition of poisoning, 4, 45-46, 79
  incidence data, 68, 74
Allergic responses, 4, 45, 46, 223, 294
American Academy of Clinical Toxicology, 83

American Association of Poison Control Centers (AAPCC), 6, 17, 25, 32, 49, 115, 118-119, 123, 134, 309
see also Toxic Exposure Surveillance System
certification of centers, 2, 15-16, 26, 84, 127-129, 130, 160
cost-effectiveness/cost-benefit analysis, 137, 142-170
definition of poisoning, 34-35
ethnic diversity issues, 208
funding for centers, 312
historical perspectives, 83, 90, 91, 92, 97-98, 100
incidence data, 48, 90, 91-92, 98, 184-190
public education efforts, 209, 210-212, 219, 234
quality assurance, 127-128, 130
telephone services, 113
American Board of Applied Toxicology, 127
American Board of Medical Specialists, 84, 128, 130
American Board of Medical Toxicology, 83, 128
American College of Emergency Physicians, 83
American Medical Association, 80-81, 86, 230
American Nurses Association, 230
American Pharmacists Association, 80-81, 230
Animals, poisoning of, 110, 119, 144, 165, 186
Animals, venomous, *see* Insect/snake bites
Anthrax, 25, 46, 99, 114, 132, 133
Antidotes, 120
poison control centers, 112, 114, 116, 134, 205, 213, 281
stockpiles, 132, 134
Arizona, 214
Arsenic, 117, 134-135, 177
Aspirin, 99, 204
Association of Maternal and Child Health Programs, 33, 276
A Su Salud, 208, 215

B

Behavioral Risk Factor Examination Survey, 182, 199
Biological and chemical terrorism, 3, 5, 7-8
see also Mass poisonings
anthrax, 25, 46, 99, 114, 132, 133
Centers for Disease Control and Prevention (CDC), 85, 114-115
committee methodology, 2, 26, 33
committee recommendations, 9, 305, 309
Department of Homeland Security, 134, 282, 285, 288, 298
funding for response services, 14, 85
Haddon Matrix, 203
historical perspectives, 85, 99
poison control centers, 9-10, 25, 35, 37, 85, 99, 100, 107, 114-115, 116, 127, 132-133, 278-282
Poison Prevention and Control System (proposed), 12, 26, 35, 37, 38, 309
Block grants, 14, 15, 92-93, 147, 148, 311
Maternal and Child Health Block Grants, 11-12, 98, 276, 283, 285-286, 308
Bureau of Labor Statistics, 299-300
Census of Fatal Occupational Injuries, 181, 198

C

California, 7, 31, 36, 113, 143, 214, 215
Canada, 212
Case series studies, 119, 191
see also Medical examiners/coroners
Census of Fatal Occupational Injuries, 181, 198
Center for Biologics Evaluation and Research, 294
Centers for Disease Control and Prevention (CDC), 12
see also National Center for Health Statistics
Behavioral Risk Factor Examination Survey, 182, 199
biological and chemical terrorism, 85, 114-115
epidemiology, 18, 48, 176, 316-317
human costs of poisoning, 19
National Center for Environmental Health, 189, 289-290
National Center for Injury Prevention and Control, 18, 196, 291, 316-317

National Health and Nutrition Examination Survey, 182, 199
National Institute for Occupational Safety and Health, 33-34, 198, 228, 285, 291-292
National Vital Statistics System, 50, 66-68, 179, 192-193
poison control centers and, 25, 35, 85, 97, 114-115, 122, 134
Poison Prevention and Control System (proposed), 12, 15, 39, 282, 287, 315-317 (passim)
public education, 207, 237-238
role in poison prevention, overview, 288-292
surveillance methodology, 176
Toxic Exposure Surveillance System (TESS), 6, 17, 18, 189, 315-316, 317
Centers for Medicare and Medicaid Services, 294
Certification of professionals, 15, 123, 127-128, 130
emergency medical personnel, 84
toxicologists, 15, 24(n.1), 38, 118-119, 122, 125
Chemical terrorism, *see* Biological and chemical terrorism
Children, 3, 4, 5, 27, 29
Association of Maternal and Child Health Programs, 33, 276
Head Start program, 211
historical perspectives, 53, 81, 84, 87, 96, 99
incidence data, 51-53, 55, 58, 59, 60, 64
death rates, 3, 50, 66-68, 74, 85, 86
lead poisoning, 120, 227, 228, 230, 276, 289, 290, 296
legislation to protect, 23, 96
packaging, child-resistant caps, 84-85, 99, 106, 204, 205, 210, 218, 287
poison control centers, 4, 5, 84, 87, 96, 99, 106
public education, 209, 210-212, 216, 219, 234-268 (passim), 276, 277, 317
The Clinical Toxicology of Commercial Products, 83, 87, 92, 95
Coding of data, 50-51, 61, 63, 73, 74-75, 78-79, 99, 178-184, 186, 187-188
see also E-codes; *International Classification of Diseases*
morbidity and mortality statistics, 50, 66-68
Confidentiality, *see* Privacy and confidentiality
Consumer Product Safety Commission, 23, 35, 39, 48, 219, 229, 301
Core activities
see also Telephone services
poison control centers, 8-15 (passim), 31, 37-38, 106-131, 169, 173, 306-307, 308, 310
funding, 11-12, 14-15, 39, 307, 310
Poison Prevention and Control System (proposed), 270-275, 284, 306-307, 308
population-based, 10, 11
quality assessment/assurance, 270-272
telephone services discussed as, 14, 107-113, 116, 123, 125-127, 130, 307
toxiciology training, 15, 24(n.1), 38, 118-119, 122, 125
Coroners, *see* Medical examiners/coroners
Cost and cost-benefit factors
access to services, 140, 141
American Association of Poison Control Centers, determination of, 137, 142-170
interagency coordination, 157, 159
poison control centers, 9, 13-14, 92, 96, 136-171 (passim), 191, 310, 311
public education, 158, 163, 164, 166, 206
quality assurance, 158
staff, 136, 143-158 (passim), 161-163, 170-171, 310, 311
telephone services, 143, 144, 152-155, 160
Poison Prevention and Control System (proposed), 13, 18, 19, 311-313
poisoning, economic burden, 3, 19, 27, 29, 96, 137-142, 310, 317
methodology for determining morbidity, 27, 28, 34
safety products, low-cost distribution, 204-205
survey data, general, 180
Cultural factors
see also Race/ethnicity
languages, 43-44, 113, 164, 208, 210, 214-215
poison control centers, 11, 43-44, 113, 208-209

D

Data needs, 18-19, 78-79, 176-200
  see also Definitional issues; Epidemiology; National Center for Health Statistics; Research methodology; Surveillance
  Agency for Toxic Substances and Disease Registry, 39
  coding of data, 50-51, 61, 63, 73, 74-75, 78-79, 99, 178-184, 186, 187-188
  committee charge/methodology, 2, 5, 26
  committee recommendations, 18-19, 305, 314-317
  cost-effectiveness studies, 142, 153-155, 170, 310
  Drugdex, 92
  historical perspectives, 90-92, 95, 98
  Micromedex, 91, 92, 113
  organizational factors, general, 44, 309-310
  poison control centers, 10, 11, 12-13, 25, 37, 90-92, 95, 98-99, 106-108, 117-118
    quality control, 129, 309-310
  Poisindex, 91, 92, 113
  Poison Prevention and Control System (proposed), 13, 16-17, 37, 39, 305
  privacy issues, 17, 18, 112, 169, 192, 214-215, 315
  Toxic Exposure Surveillance System, 6, 17-19 (passim), 178, 184-190, 199-200
Death rates, 18
  see also Suicide and attempts
  alcohol poisoning, 3, 28, 68
  annual, from poisoning, 3, 28, 48
  children, 3, 50, 66-68, 74, 85, 86
  data coding, 50-51
  incidence data, general, 64, 66, 69-72, 74, 88, 94, 181
  medical examiners/coroners, 179, 193-194, 197, 199, 200, 273
  mortality vital statistics, 50, 66-68, 74, 179, 181, 192-193
Definitional issues
  see also Core activities
  data coding, 50-51, 61, 63, 73, 74-75, 78-79, 99, 178-184, 186, 187-188
  epidemiology, 2, 43
  Haddon Matrix, 115, 201-204
  poisoning, 2, 3, 4, 16-17, 27, 28, 34-35, 43, 45-48, 50-51, 73, 79, 306, 314-315
  prevention, primary and secondary, 205
  surveillance, 3, 4, 176-177
  toxicology, 2, 3, 4, 24(n.1), 27, 45-46
Demographic factors, general, 43, 48
  see also Age factors; Cultural factors; Geographic factors; Race/ethnicity
  committee charge, 2, 26
  gender factors, 48, 51, 52, 55-60 (passim), 63, 64, 68
  incidence data, 48-68 (passim), 76
  poison control centers, 7, 209
Department of Agriculture, 297-298
Department of Health and Human Services (DHHS), 19
  see also Centers for Disease Control and Prevention; Food and Drug Administration; Health Resources and Services Administration; National Center for Health Statistics
  Agency for Healthcare Research and Quality, 19, 180, 283, 294-295, 317
  Centers for Medicare and Medicaid Services, 294
  definition and classification issues, 16-17, 314-315
  legislation, national repository, 12, 309
  Maternal and Child Health Bureau, 1, 2, 25, 26, 285
  National Institute of Child Health and Human Development, 296
  National Institute of Environmental Health Sciences, 296
  National Institute on Drug Abuse, 295
  National Institutes of Health, 120, 283, 295-296
  Poison Prevention and Control System (proposed), 12, 282-283, 285-286, 308-309, 313
  privacy issues, 17, 315
  role in poison prevention, overview, 12, 282-283, 285-286, 288-292
  Substance Abuse and Mental Health Services Administration (SAMHSA), 181, 182, 196, 229, 282, 287, 293

Department of Homeland Security, 134, 282, 285, 288, 298
Department of Housing and Urban Development, 282, 288, 289, 296
Department of Justice, 282, 288, 300
Department of Labor, 198, 282, 288, 298
  Bureau of Labor Statistics, 299-300
    Census of Fatal Occupational Injuries, 181, 198
  Mine Safety and Health Administration, 285, 299
  Occupational Safety and Health Administration, 198, 285, 298-299
Diagnosis
  see also Incidence and prevalence; Symptoms
  definitional issues, 3, 27, 46, 50-61 (passim)
  definitional issues, classification codes, 50-51, 61, 63, 73, 74-75, 78-79, 99, 178-184, 186, 187-188
  poison control centers, core activities, 10, 109
Dosage factors, 3
  see also Drugs
  data sources, 184-185
  definitional issues, 4, 23, 43
  Haddon Matrix, 202
  legislation, 23
Drug Abuse Warning Network, 19, 181, 196-197, 200
Drug Enforcement Administration, 300
Drugdex, 92
Drugs
  see also Alcohol use and abuse; Packaging and labeling; Substance abuse
  definition of overdose, 27, 43, 45, 46, 51, 78-79
  household, 2, 81, 84, 87
  incidence of overdose, 50, 51, 61, 62, 64, 67
  overdoses during hospitalization, 184
  research on overdose, 18, 19, 184, 197-198, 317
  suicides and attempts, 3, 5, 27, 29-30, 51, 61, 68, 71
  telephone services, 110, 164
  unintentional overdose, 3, 30

E

E-codes, 50-63 (passim), 70, 76, 78
Economic factors
  see also Cost and cost-benefit factors; Funding
  poison control centers, 3, 6, 7, 136-171
    geographic factors, 142, 143, 152-153, 162
  prevention incentives, 204-205
Education, *see* Professional education; Public education/outreach
Elderly persons, 3, 29, 43, 51-53, 55, 58, 59, 60, 64, 99
  public education materials, 223
Emergency and Environmental Health Services, 289-290
Emergency medical services (EMS)
  see also Biological and chemical terrorism
  certification of specialists, 84
  HAZMAT response, 112, 114, 126, 130, 133
  historical perspectives, 83-84, 92, 96
  incidence data, 53, 54, 59-61, 65-66, 72, 73, 74, 75, 78, 179, 180
    National Electronic Injury Surveillance Survey (NEISS), 4, 17-18, 48, 49, 65-66
  organizational factors, 6, 39, 309
  poison control centers and, 2, 6, 24, 30, 38, 60, 107, 108, 112, 114, 116, 278-282
    access by EMS personnel, 114, 126, 282
  Poison Prevention and Control System (proposed), 38, 39
  state emergency department data, 179
Emergency Medical Services Systems Act, 24, 84
Environment and Natural Resources Division (Department of Justice), 300
Environmental Protection Agency, 98, 219, 226, 227, 273, 282, 284, 285, 288, 289, 301
Epidemiology, 1, 24, 25, 27-30, 48-68, 176, 177
  see also Death rates; Incidence and prevalence; Surveillance
  annual poisoning cases, 3, 28

Centers for Disease Control and Prevention (CDC), 18, 48, 176, 316-317
committee study methodology, 32
definitional issues, 2, 43
historical perspectives, 81
public education materials, 221
research recommendations, 18-19
Essential Services of Public Health, 10, 272
Ethnic groups, *see* Race/ethnicity
External cause of injury codes, *see* E-codes

F

Federal government, 6
see also Funding; Legislation; *specific departments and agencies*
agencies involved in poison prevention, overview, 282-284, 288-301
committee study methodology, 32-33
definitions and classification systems, 3, 4
poison control centers and, 1-2, 11-12, 35-36, 39, 92-93, 95-96
Poison Prevention and Control System (proposed), 12, 15, 35-36, 39
public education efforts, 219-231
Food and Drug Administration (FDA), 35, 39, 81, 82, 283, 293-294
MedWatch, 181, 197
National Clearinghouse for Poison Control Centers (NCHPCC), 83, 87, 91, 92, 95-96, 97, 184
poison control centers, 84, 87-90, 93, 94-97, 122
historical perspectives, 81, 82, 84, 86
Food and Safety Inspection Service, 297
Food poisoning, 297-298
definitional issues, 46, 51, 53
historical perspectives, 80-81
incidence data, 66, 191
sentinel events, 118
Funding
see also Research recommendations
American Association of Poison Control Centers (AAPCC), 312
biological and chemical terrorism response, 14, 85
block grants, 14, 15, 92-93, 147, 148, 311
core activities, 11-12, 14-15, 39, 307, 310
committee study charge/methodology, 2, 26, 32
federal agencies responsible, overview, 282
Maternal and Child Health Block Grants, 11-12, 98, 276, 283, 285-286, 308
Medicaid, 14, 147, 148, 214, 285, 288, 294, 311
Medicare, 93, 285, 288, 294
poison control centers, 3, 6, 24-25, 26, 32, 36, 83, 130, 147-152, 159-160, 165, 283-286
committee conclusions/ recommendations, 7, 9, 11-12, 13, 14-15, 307, 310, 311-313
core activities, 11-12, 14-15, 39, 307, 310
historical perspectives, 85, 90, 92-93, 95, 97-98, 100, 103-105
hospital funding of, 36, 93
private sector, 148-149
public education, 11, 93
quality assurance, 129, 130, 159
regionalization, 37
state government, 24, 36, 92-93, 98, 136-137, 147, 148, 284, 311, 313
Poison Prevention and Control System (proposed), 8, 12, 14-15, 36, 39, 283-287 (passim), 305, 307
regionalization, 37
technical assistance, 12
public education, general, 11, 93, 211, 286, 313
research, general, 18-19, 90, 316-317
Toxic Exposure Surveillance System, 17-18, 90

G

Gender factors, incidence data, 48, 51, 52, 55-60 (passim), 63, 64, 68
Geographic factors
see also Population-based factors
incidence data, regional, 52, 53, 55, 56, 58, 59, 64, 68, 71
national telephone service, 25, 52, 55, 92, 97, 103, 113, 210, 211
poison control centers
case studies, 132-135

economic factors, 142, 143, 152-153, 162
general, 93, 95
regionalization, 7-8, 10, 36-37, 88-89, 93, 94, 142, 143, 306, 312
telephone services, 127

H

Haddon Matrix, 115, 201-204
Hazardous Substances Emergency Events Surveillance System, 181, 198
Hazardous Substances Labeling Act, 96
HAZMAT response, 112, 114, 126, 130, 133
Health care professionals
see also Nurses; Pharmacists; Physicians; Poison control center staffing; Professional education
certification, 15, 123, 127-128, 130
emergency medical personnel, 84
toxicologists, 15, 24, 83, 122
medical examiners/coroners, 179, 193-194, 197, 199, 200, 273
poison control centers, relationships with outside, 107
poison information providers/specialists, 123, 124-125, 127, 130
telephone services, 110-112
public education materials, 221-231 (passim)
toxicologists, 15, 24(n.1), 83, 87, 122
Health departments
local, 35, 114, 274-282, 289, 307
state, 35, 93, 133, 274-282, 289, 307
Health Insurance Portability and Accountability Act, 17, 112, 169, 191, 192, 315
Health Resources and Services Administration (HRSA)
biological and chemical terrorism, 85, 114-115
committee study charge, 1, 2, 26, 305
Maternal and Child Health Block Grants, 11-12, 98, 276, 283, 285-286, 308
poison control centers, 11-12, 15, 25, 35, 39, 97
Poison Prevention and Control System (proposed), 12, 13, 15, 39, 282, 285, 287, 309
role in poison control, overview, 292-293
Healthcare Cost and Utilization Program National Inpatient Sample, 180
Healthy People 2000, 5, 25, 276
Healthy People 2010, 9-10, 25, 30, 269, 273
Historical perspectives, 23-26, 30, 80-105
access to services, 84
American Association of Poison Control Centers (AAPCC), 83, 90, 91, 92, 97-98, 100
children, 53, 81, 84, 87, 96, 99
committee study charge/methodology, 1, 2
epidemiology, 81
data needs, 90-92, 95, 98
definition of poisoning, 3
emergency medical services, 83-84, 92, 96
food adulteration, 80-81
Food and Drug Administration (FDA), 81, 82, 84, 86
incidence of poisonings, 3, 30, 43-44, 48-75 (passim), 79, 270
legislation, 23, 24, 30, 31, 82
pharmacists, 80-81
poison control centers, 2, 5-6, 24-25, 30, 81-84, 86-96, 159
Food and Drug Administration (FDA), 81, 82, 84, 86
funding, 85, 90, 92-93, 95, 97-98, 100, 103-105
organizational factors, 84, 87-90, 93-100
state government role, 86, 87, 88-89, 91-93, 98
telephone services, 81, 83, 88, 92, 94, 97, 103
regulatory issues, 80-81, 84-85, 87, 94, 95, 96, 98
surveillance, 90
terrorism, 85, 99
toxic exposure surveillance system, 90, 91, 92
Home Safety Council, 206
Hospitals and hospitalization, 8, 14, 306
see also Emergency medical services
bioterrorism preparedness, 33
committee study charge/methodology, 26, 28, 33

incidence data, 3, 28, 49, 53, 54, 58, 66, 72
National Hospital Ambulatory Care Survey, 4, 49, 51, 54, 57-63, 71, 72, 73-74, 75, 78, 180, 195
National Hospital Discharge Survey, 4, 49, 63-65, 72, 74, 78, 179, 195
state/local hospital discharge data systems, 194
overdoses during hospitalization, 184
poison control center funding, 36, 93
Household products, 2, 3, 81, 84, 87, 88, 206
see also Packaging and labeling

I

Illinois, 86, 87, 91
Incidence and prevalence, 3, 5, 6, 18, 27-30, 44, 48-79, 94, 95
see also Death rates
alcohol poisoning, 3, 28, 68, 74
American Association of Poison Control Centers (AAPCC), 48, 90, 91-92, 98, 184-190
children, 51-53, 55, 58, 59, 60, 64
death rates, 3, 50, 66-68, 74, 85, 86
data coding, 50-51, 61, 63, 73, 74-75, 78-79, 99, 178-184, 186, 187-188
data needs, 176-199 (passim)
definitional issues, 47-48
drug overdoses, 50, 51, 61, 62, 64, 67
emergency medical services, 53, 54, 59-61, 65-66, 72, 73, 74, 75, 78, 179, 180
National Electronic Injury Surveillance Survey (NEISS), 4, 17-18, 48, 49, 65-66
public education, evaluation of impacts, 218
food poisoning, 66, 191
historical perspectives, 3, 30, 43-44, 48-75 (passim), 79, 270
hospital data, 3, 28, 49, 53, 54, 58, 66, 72
National Hospital Ambulatory Care Survey, 4, 49, 51, 54, 57-63, 71, 72, 73-74, 75, 78, 180, 195
National Hospital Discharge Survey, 4, 49, 63-65, 72, 74, 78, 179, 195
state/local hospital discharge data systems, 194
International Classification of Diseases (ICD), use in incidence surveys, 50-51, 55, 57, 58, 59, 60, 62, 63, 76, 77, 192-193
physicians, data from, 53, 54, 55, 65, 72, 77, 78
poison control center call volume, 110, 143, 144, 164, 187, 309
racial/ethnic breakdowns, 48, 52-60 (passim), 63, 68, 70
referral data, 59, 60, 63-66 (passim), 72, 74
regional breakdowns, 52, 53, 55, 56, 58, 59, 64, 68, 71
research methodology, 53-79 (passim)
substance abuse, 62, 67
suicide attempts, 51, 61, 68, 71
symptoms, 46, 55, 56, 58, 59, 61
telephone services, data on, 52, 53, 55, 69, 70-71, 72, 73-74, 94, 178, 184-190
race/ethnicity data, 208-209
Toxic Exposure Surveillance System (TESS), 66, 68-74, 76
Injury prevention and control field
committee study methodology, 1, 26
definition of poisoning, 3, 45, 50
Haddon Matrix, 115, 201-204
medical directors, 12, 33
poison control center organization, 23, 122
Poison Prevention and Control System (proposed), 40
public education materials, 224
state public health departments, linkage, 277
Insect/snake bites, 61, 62, 65, 66, 74, 131, 191
Interagency coordination, 3, 26, 30-31, 35-36, 39, 100, 114-115, 132-135, 165-166, 172-173, 273, 276-278, 284, 289
committee recommendations, 6, 10-11, 12, 17, 18-19, 305, 307, 308-309
cost effectiveness, 157, 159
public education, 205-206, 211, 219-232
International Classification of Diseases (ICD), 3, 16-17, 27, 34, 47, 78-79, 314-315
incidence surveys, 50-51, 55, 57, 58, 59, 60, 62, 63, 76, 77, 192-193
surveys, various, 179-181, 195
vital statistics mortality data, 66-67

Internet
American Association of Poison Control Centers (AAPC), public education site, 211-212, 216
educational materials published by non-poison center sources, 221-231
federal agency websites, 288-301 (passim)
National Poison Prevention Week, 219
poison control centers, 112, 113, 118, 120, 215-216, 234-268
surveillance surveys, various, 12, 183
technical assistance for site development, 12, 309
Toxic Exposure Surveillance System (TESS), 49-50, 183

L

Labeling, *see* Packaging and labeling
Languages, non-English, 43-44, 113, 127, 164, 208, 210, 214-215
Lead poisoning, 120, 227, 228, 230, 276, 289, 290, 296
Legislation
see also Advocacy; Funding; Regulatory issues
Emergency Medical Services Systems Act, 24, 84
Food and Drug Act, 81, 82
Hazardous Substances Labeling Act, 96
Health Insurance Portability and Accountability Act (HIPAA), 17, 112, 169, 191, 192, 315
historical perspectives, 23, 24, 30, 31, 82
Maternal and Child Health Block Grants, 11-12
national repository, committee recommendation, 12, 309
Poison Control Center Enhancement and Awareness Act, 6, 14-15, 25, 97-98, 101-105, 312
Poison Prevention and Control System (proposed), 39, 312, 313
Poison Prevention Packaging Act, 23, 94, 96, 204
Local government
health departments, 35, 114, 274-282, 289, 307
hospital discharge data systems, 194
poison control centers and, 1-2, 5, 9, 114, 148, 307, 312-313
Poison Prevention and Control System (proposed), 273-282

M

Managing directors, 31, 32, 33, 84, 123, 127, 133, 159, 224
Mass media, 95, 103, 107, 116, 207, 210, 286, 307
National Poison Prevention Week, 95, 103, 209, 219
Mass poisonings
see also Biological and chemical terrorism
access to poison services, 114
HAZMAT response, 112, 114, 126, 130, 133
legislation on, 103
poison control centers, 107, 112, 114, 127, 191
Poison Prevention and Control System (proposed), 37
Maternal and Child Health Block Grants, 11-12, 98, 276, 283, 285-286, 308
Maternal and Child Health Bureau, 1, 2, 25, 26, 285
Medicaid, 14, 147, 148, 214, 285, 288, 294, 311
Medicare, 93, 285, 288, 294
Medical directors
background of, 84
injury prevention directors, 12, 33
poison control centers, 84, 145
committee methodology, 31, 32, 33
Medical examiners/coroners, 179, 193-194, 197, 199, 200, 273
Medication poisoning, *see* Drugs
MEDMARX, 197-198
MedWatch, 181, 197
Men, *see* Gender factors
Methodology, *see* Epidemiology; Quality control; Research methodology; Surveillance
Micromedex, 91, 92, 113
Mine Safety and Health Administration, 285, 299
Minnesota Poison Control System, 189-190
Minority groups, *see* Race/ethnicity

Morbidity
alcohol poisoning, 36
committee study methodology, 27, 28, 34, 43
cost of poisoning, methodology, 27, 28, 34
definitional issues, 28
national reduction goals, 5
Mortality, *see* Death rates; Suicide and attempts
Mortality Vital Statistics, 50

N

National Ambulatory Medical Care Survey (NAMCS), 4, 48-49, 51, 54, 55-57, 71, 73, 77, 78, 180, 195
National Bioterrorism Hospital Preparedness Programs, 33
National Center for Environmental Health, 189, 289-290
National Center for Health Statistics (NCHS), 4, 17, 18, 32, 39, 47-48, 290-291, 315, 317
see also National Ambulatory Medical Care Survey; National Hospital Ambulatory Medical Care Survey; National Hospital Discharge Survey; National Vital Statistics System
National Center for Injury Prevention and Control, 18, 196, 291, 316-317
National Clearinghouse for Poison Control Centers (NCHPCC), 83, 87, 91, 92, 95-96, 97, 184
National Drug Intelligence Center, 300
National Electronic Injury Surveillance Survey (NEISS), 4, 17-18, 48, 49, 65-66, 72, 181, 196, 317
National Health and Nutrition Examination Survey, 182, 199
National Health Interview Survey (NHIS), 4, 17-18, 48, 51-55, 68-74, 76, 182, 192, 317
National Hospital Ambulatory Medical Care Survey (NHAMCS), 4, 49, 51, 54, 57-63, 71, 72, 73-74, 75, 78, 180, 195
National Hospital Discharge Survey (NHDS), 4, 49, 63-65, 72, 74, 78, 179, 195
National Household Survey on Drug Abuse, 182
National Institute for Occupational Safety and Health, 33-34, 198, 228, 285, 291-292
National Institute of Child Health and Human Development, 296
National Institute of Environmental Health Sciences, 296
National Institute on Drug Abuse, 295
National Institutes of Health, 120, 283, 295-296
National Poison Prevention Week, 95, 103, 209, 219
National Vital Statistics System, 50, 66-68, 179, 192-193
Northern New England Poison Center, 134-135
Nurses
American Nurses Association, 230
poison control centers
calls to, 112, 132
training, 6, 84
public education materials, 221, 222

O

Occupational Safety and Health Administration, 198, 285, 298-299
Office of National Drug Control Policy, 282, 285, 288, 301
Ohio, 215
Organizational factors
see also Accountability; Cost and cost-benefit factors; Funding; Quality control
committee charge, 2, 26
committee methodology, 5, 31-33, 34-35
data needs, 44, 309-310
emergency medical services, 6, 39, 309
health departments
local, 35, 114, 274-282, 289, 307
state, 35, 93, 133, 274-282, 289, 307
interagency coordination, 3, 26, 30-31, 35-36, 39, 100, 114-115, 132-135, 165-166, 172-173, 273, 276-278, 284, 289
poison control centers, 1-3, 5-7, 8-10, 12-14, 23, 24-25, 26, 30-31, 32, 114, 155-175, 309-311

see also Poison control centers staffing
affiliation (private/public), 13, 93, 100, 108, 118-119, 124, 125, 136, 157-158, 160-161, 173, 311
core activities, 8-15 (passim), 31, 37-38, 106-131, 169, 173, 306-307, 308, 310
historical perspectives, 84, 87-90, 93-100
injury prevention and control field, general, 23, 122
partnerships, 10, 11, 13, 277-278, 308
planning, 35-36, 93, 98-99, 115, 123, 133, 147, 171, 175, 177
Poison Prevention and Control System (proposed), 7-8, 35-40, 306
regionalization, 7-8, 10, 14, 36-37
systems approach, 35-36, 38, 305-306
telephone services, 107-109
Poison Prevention and Control System (proposed), 5, 7, 8, 12, 13, 35-36, 38-40, 268-301, 305-306, 308-309
Overdose, *see* Drugs

P

Packaging and labeling, 23, 30, 35, 80-81, 84-85, 94, 95, 96, 99, 196, 203, 210, 218
child-resistant caps, 84-85, 99, 106, 204, 205, 210, 218, 287
Haddon Matrix, 202-203, 204
imprint codes, 84, 89
Poison Prevention Packaging Act, 23, 94, 96, 204
Pharmacists
accessibility, 230
American Pharmacists Association, 80-81, 222
elderly persons, 99
historical perspectives, 80-81
poison control centers, 115, 212
training, 6, 9, 99
public education materials, 221
Physicians
American College of Emergency Physicians, 83
American Medical Association, 80-81, 86, 230
biochemical terrorism, poison control center consultations, 7-8, 37
incidence data, 53, 54, 55, 65, 72, 77, 78
poison control centers, 53, 112, 116, 212, 306
public education materials, 221
telephone consultations, 53, 112
toxicology training and certification of, 15, 24, 83, 122
Poisindex, 91, 92, 113
Poison Control Center Advisory Working Group study (1996), 136
Poison Control Center Enhancement and Awareness Act, 6, 14-15, 25, 97-98, 101-105, 312
Poison control centers staffing, 6, 7, 8, 9, 13, 14, 24, 37, 168, 172, 311
costs, 136, 143-158 (passim), 161-163, 170-171, 310, 311
managing directors, 31, 32, 33, 84, 123, 127, 133, 159, 224
medical directors, 31, 32, 33, 84, 145
pharmacists, 115, 212
training, 6, 9, 99
physicians, 53, 112, 116, 212, 306
professional education, 37, 123-126, 167, 307
public education, 205-206, 209, 212-213
recruitment and retention, 14, 159, 168, 169, 170, 171, 172, 310, 311
specialists in poison information/ poison information providers, 123, 124-125, 127, 130, 143-145, 153-155
training of staff, 37, 123-126, 167, 307
volunteers, 124, 161, 172, 208
Poison Prevention and Control System (proposed), 5, 7-8, 12, 269-301, 305-317
accountability issues, 273, 282, 284-286, 308-309
biological and chemical terrorism, 12, 26, 35, 37, 38, 309
Centers for Disease Control and Prevention (CDC), 12, 15, 39, 282, 287, 315-317 (passim)
concept of, 34-40, 305-306
core activities, 270-275, 284, 306-307, 308; *see also* Telephone services
cost factors, 13, 18, 19, 311-313
data needs, 13, 16-17, 37, 39, 305

Department of Health and Human Services (DHHS), 12, 282-283, 285-286, 308-309, 313
emergency medical services, 38, 39
federal government, general, , 12, 15, 35-36, 39; *see also specific departments and agencies*
funding, 8, 12, 14-15, 36, 39, 283-287 (passim), 305, 307
regionalization, 37
technical assistance, 12
local government role, 273-282
mass poisonings, 37
organizational factors, 5, 7, 8, 12, 13, 35-36, 38-40, 268-301, 305-306, 308-309
population-base factors, 18, 287
professional education, 15, 37
public education, 12, 15, 37, 39, 274
quality control, 8, 13, 37, 270-272, 274-275, 284-285, 305, 308-309, 313-314
regionalization of centers, 7-8, 10, 36-37, 306, 312
state government role, 12, 273-282, 285-286, 309
surveillance, 14-15, 37, 305
Toxic Exposure Surveillance System (TESS), 35, 309
Poison Prevention Council, 219
Poison Prevention Packaging Act, 23, 94, 96, 204
Poison Prevention Week, *see* National Poison Prevention Week
Poisoning, 2, 3, 4, 16-17, 27, 28, 34-35, 43, 45-48, 50-51, 73, 79, 306, 314-315
Department of Health and Human Services, committee recommendations, 16-17, 314-315
diagnosis, 3, 27, 46, 50-61 (passim)
dosage, 4, 23, 43
overdose, 27, 43, 45, 46, 51, 78-79
Toxic Exposure Surveillance System (TESS), 34-35, 45-46, 79
Policy, *see* Advocacy; Legislation; Regulatory issues
Population-based factors
see also Demographic factors; Geographic factors
committee study charge, 2, 26
poison control centers, 35, 164, 169-170, 171
core activities, 10, 11
population at risk, 29-30
regionalization, 7-8, 10, 36-37
Poison Prevention and Control System (proposed), 18, 287
PRECEDE/PROCEED model, 206-207
Prevalence, *see* Incidence and prevalence
Privacy and confidentiality, 17, 18, 112, 169, 192, 214-215, 315
Private sector
see also Toxic Exposure Surveillance System
Drugdex, 92
epidemiological data, 38, 90
Poisindex, 91, 92, 113
poison control centers and, 1-7, 13, 14, 17, 35, 117, 122, 160, 164
affiliation (private/public), 13, 93, 100, 108, 118-119, 124, 125, 136, 157-158, 160-161, 171, 173-174, 311
funding, 148-149
Professional education, 26
certification of professionals, 15, 123, 127-128, 130
cultural competence, 11, 43-44, 113, 208-209
educational materials, 221-231
poison control centers, 6, 9, 10, 11, 15, 24, 25, 81, 83, 87, 107, 115-116, 117, 118-119, 121-131 (passim), 133, 158, 162-163
staff of, 37, 123-126, 167, 307
telephone services, 110-112
Poison Prevention and Control System (proposed), 15, 37
privacy regulations, 17, 315
toxicologists, 15, 24(n.1), 38, 118-119, 122, 125
Program Evaluation Logic Model, 207-208, 214
Public education/outreach, 201-268
see also Advocacy; Mass media; Packaging and labeling; Telephone services
access to services, enhancement, 202, 205, 208, 214
American Association of Poison Control Centers (AAPCC), 209, 210-212, 219, 234
Centers for Disease Control and Prevention (CDC), 207, 237-238

children, 209, 210-212, 216, 219, 234-268 (passim), 276, 277, 317
federal government, general, 219-231
funding for, general, 11, 93, 211, 286, 313
Haddon Matrix, 115, 201-204
language factors, 43-44, 113, 164, 208, 210, 214-215
materials by organizations other than poison centers, 219-231
pharmacists, 221
poison control centers, 9, 10, 11, 37, 89, 93, 95, 107, 116-117, 120-121, 123-124, 204, 205-268, 307
best practices, 213-215
biological/chemical terrorism, 115
children, 209, 216, 217, 234-268 (passim)
cost and cost-effectiveness, 158, 163, 164, 166, 206
educational materials available from, 233-268
ethnic diversity and, 43-44, 113, 208, 210, 214-215
quality control, 206-209, 217-219
staff types and roles, 205-206, 209, 212-213
state government role, 213-215, 234-268
Poison Prevention and Control System (proposed), 12, 15, 37, 39, 274
quality control, 206-209, 217-219, 231-232
racial/ethnic factors, 43-44, 113, 208-209, 214, 215
theoretical models, 206-208

Q

Quality control
see also Accountability; Core activities; Cost and cost-benefit factors
American Association of Poison Control Centers (AAPCC), 127-128, 130
assessment as core public health function, 270-271
assurance as core public health function, 271-272
best practices, 11, 15, 37, 213-215, 232, 275, 287, 313
certification of professionals, 15, 123, 127-128, 130
emergency medical personnel, 84
toxicologists, 15, 24, 83, 122
committee study charge, 2, 26
data coding, 50-51, 61, 63, 73, 74-75, 78-79, 99, 178-184, 186, 188
data on, 12-13, 129, 309-310
data quality, 176, 187-188
national tool free telephone service, 211
poison control centers, 37, 95, 107, 127-130, 167, 171, 174, 206-208, 313-314
best practices, 11, 15, 37, 213-215
certification of, 2, 15-16, 26, 84, 127-129, 130, 160
committee recommendations, 8, 10, 11, 13, 15-16, 309, 310
cost factors, 158
public education, 206-209, 217-219
telephone services, 211, 217
Poison Prevention and Control System (proposed), 8, 13, 37, 270-272, 274-275, 284-285, 305, 308-309, 313-314
public education, 206-209, 217-219, 231-232
theoretical models
Haddon Matrix, 115, 201-204
program development and evaluation, 206-208, 214

R

Race/ethnicity, 43
see also Cultural factors
access to services, 43
incidence data, 48, 52-60 (passim), 63, 68, 70
language factors, 43-44, 113, 164, 210, 214-215
A Su Salud, 208, 215
public education and, 43-44, 113, 208-209, 214, 215
Reducing the Burden of Injury: Advancing Prevention and Treatment, 25
Referrals
incidence data, 59, 60, 63-66 (passim), 72, 74
poison control centers, 166-167
telephone services, 110, 112, 116, 131, 137, 157

self-referrals to emergency departments, 138-139
Regional factors, *see* Geographic factors
Regulatory issues, 18
see also Legislation; Packaging and labeling
advocacy, 10, 11, 25, 31, 39, 203, 204, 205, 218, 271, 274, 289
data coding, 315
data sources on regulation fulfillment, 186
historical perspectives, 80-81, 84-85, 87, 94, 95, 96, 98
privacy, 17, 18, 112, 169, 192, 214-215, 315
telephone services, 112
Reporting, 6
definition of poisoning, 46
National Clearinghouse for Poison Control Centers (NCHPCC), 97
poison control centers, 117-118, 310
sentinel events, 117-118, 188-190
Research methodology
see also Data needs; Definitional issues; Epidemiology; Quality control; Surveillance; Theoretical models
committee charge/methodology, 1, 2, 5-6, 26, 30-33, 34, 172-175, 305-306, 309
committee reviewers, ix-x
incidence data, 53-79 (passim)
Poison Control Center Advisory Working Group study (1996), 136-137
poison control centers
cost-benefit/cost-effectiveness, 142, 153-157, 170
used by, 118-120
Research recommendations, 18-19, 39, 287, 309-311, 314-317
access to services, 10, 18, 19, 118, 270, 274, 275, 316, 317
Rocky Mountain Poison and Drug Center, 31, 89, 116, 132-133

S

Safety equipment, 84, 203, 204-205, 218
child-resistant caps, 84-85, 99, 106, 204, 205, 210, 218, 287
Sentinel events, 106, 115, 117-118, 188-190
Sex differences, *see* Gender factors
Snake bites, *see* Insect/snake bites
Special Supplemental Nutrition Program for Women, Infants, and Children, 209
State and Territorial Injury Prevention Directors Association, 33
State government
biological and chemical terrorism, 85
block grants, 14, 15, 92-93, 147, 148, 311
classification systems, 3
committee study methodology, 33
emergency department data, 179
emergency preparedness/response, 12
health departments, 35, 93, 133, 274-282, 289, 307
hospital discharge data systems, 179, 194
Maternal and Child Health Block Grants, 11-12, 98, 276, 283, 285-286, 308
poison control centers and, 1-2, 35-36, 93, 100, 114, 116, 132-135, 161, 276-278
committee conclusions/ recommendations, 7, 9, 307, 312-313
funding, general, 24, 36, 92-93, 98, 136-137, 147, 148, 284, 311, 313
historical perspectives, 86, 87, 88-89, 91-93, 98
Maternal and Child Health Block Grants, 11-12, 98, 276
public education, 213-215, 234-268
regionalization, 7-8, 10, 36-37, 88-89, 93, 94, 142, 143, 306, 312
websites, 234-268
Poison Prevention and Control System (proposed), 12, 273-282, 285-286, 309
surveillance efforts, 17, 179, 189-190
Toxic Exposure Surveillance System and, 6, 309, 316
Substance abuse, 5, 28, 29, 99
see also Alcohol use and abuse
data coding, 51, 62, 181
Drug Abuse Warning Network, 19, 181, 196-197
Haddon Matrix, 202
incidence data, 62, 67
MEDMARX, 197-198

National Household Survey on Drug Abuse, 182
Substance Abuse and Mental Health Services Administration (SAMHSA), 181, 182, 196, 229, 282, 287, 293
Suicide and attempts, 3, 5, 27, 29-30
incidence data, 51, 61, 68, 71
Surveillance, 176-200
see also Centers for Disease Control and Prevention; Data needs; Incidence and prevalence; Reporting; Toxic Exposure Surveillance System
Agency for Toxic Substances and Disease Registry, 39, 90, 115, 126, 189, 195, 283, 292
Hazardous Substances Emergency Events Surveillance System, 181, 198
committee recommendations, 9, 16-17, 314-317
committee study charge, 2, 26
data coding, 50-51, 61, 63, 73, 74-75, 78-79, 99, 178-184, 186, 187-188
definitions and classification systems, 3, 4, 176-177
historical perspectives, 90
overview and distribution of poisonings in United States, 43-79
poison control centers, 9, 10, 11, 37, 90, 115, 190-192
Toxic Exposure Surveillance System (TESS), 111, 113, 115, 117, 134, 137, 178, 184-191, 211
Poison Prevention and Control System (proposed), 14-15, 37, 305
sentinel events, 106, 115, 117-118, 188-190
state government, 17, 179, 189-190
Symptoms
see also Diagnosis
incidence data, 46, 55, 56, 58, 59, 61

T

Technical assistance, 12, 225
federal assistance to health departments, 289
Internet, 12, 309
patient care, 26, 305
Telephone services, 5, 6, 29, 37, 106, 107-113, 116, 130, 210-211, 212
American Association of Poison Control Centers (AAPCC), 113
call volume, 110, 143, 144, 164, 187, 309
committee recommendations, 9
committee study charge, 2, 26
core activity of poison centers, 14, 107-113, 116, 123, 125-127, 130, 307
cost factors, 143, 144, 152-155, 160
definition of poisoning, 4, 34-35
drug overdoses, 51, 61, 68, 71
evaluation of, 211, 218
historical perspectives, 81, 83, 88, 92, 94, 97, 103
incidence data, 52, 53, 55, 69, 70-71, 72, 73-74, 94, 178, 184-190
race/ethnicity data, 208-209
national number, 25, 52, 55, 92, 97, 103, 113, 210, 211
physician consultations, 53, 112
quality assurance, 127-129, 167,
sentinel event reporting, 117-118, 188-190
specialists in poison information/poison information providers, 123, 124-125, 127, 130, 143-145, 153-155
Toxic Exposure Surveillance System (TESS), 111, 113, 115, 117, 134, 137, 178, 184-190, 211
Terrorism, *see* Biological and chemical terrorism
Texas, 116, 133, 214
Theoretical models
Haddon Matrix, 115, 201-204
program development and evaluation, 206-208
Toxic Exposure Surveillance System (TESS), 4, 6, 28, 29, 48, 49-50, 113, 115, 117, 127, 177
biological/chemical terrorism, 114
Centers for Disease Control and Prevention (CDC), 6, 17, 18, 189, 315-316, 317
committee/conclusions recommendations, 12, 17-18, 25, 315-316
data needs/methodology described, 6, 17-19 (passim), 178, 184-190, 199-200
definition of poisoning, 34-35, 45-46, 79

emergency preparedness, 115
funding, 17-18, 90
historical perspectives, 90, 91, 92
incidence data, 66, 68-74, 76
Internet, 66, 68-74, 76
Poison Prevention and Control System (proposed), 35, 309
telephone services, 111, 113, 115, 117, 134, 137, 178, 184-190, 211
Toxicology
see also Dosage factors
certification of professionals, 15, 24, 83, 122
committee study methodology, 1, 26
definitional issues, 2, 3, 4, 24(n.1), 27, 45-46
poison control centers, 15, 38, 87, 107-108, 118-119, 122
telephone services, 107-108
training of professionals, 15, 24(n.1), 38, 118-119, 122, 125
Training, *see* Professional education

U

United States Pharmacopoeia, 197-198
Utilization of health care services, 53, 121, 129, 169, 180, 195, 196, 198, 209
see also Access to services; Incidence and prevalence; Public education/outreach; Telephone services

V

Venom, *see* Insect/snake bites

W

Women, *see* Gender factors
World Health Organization, 16-17
see also International Classification of Diseases
World Wide Web, *see* Internet